Chevrolet Cavalier & Pontiac Sunfire Automotive Repair Manual

by Rob Maddox, Mike Stubblefield and John H Haynes

Member of the Guild of Motoring Writers

Models covered:

All Chevrolet Cavalier and Pontiac Sunfire models
1995 through 2000

(9D6 - 38016)

ABCDE
FGHIJ
K

Haynes Publishing Group
Sparkford Nr Yeovil
Somerset BA22 7JJ England

Haynes North America, Inc
861 Lawrence Drive
Newbury Park
California 91320 USA

About this manual

Its purpose

The purpose of this manual is to help you get the best value from your vehicle. It can do so in several ways. It can help you decide what work must be done, even if you choose to have it done by a dealer service department or a repair shop; it provides information and procedures for routine maintenance and servicing; and it offers diagnostic and repair procedures to follow when trouble occurs.

We hope you use the manual to tackle the work yourself. For many simpler jobs, doing it yourself may be quicker than arranging an appointment to get the vehicle into a shop and making the trips to leave it and pick it up. More importantly, a lot of money can be saved by avoiding the expense the shop must pass on to you to cover its labor and overhead costs. An added benefit is the sense of satisfaction and accomplishment that you feel after doing the job yourself.

Using the manual

The manual is divided into Chapters. Each Chapter is divided into numbered Sections, which are headed in bold type between horizontal lines. Each Section consists of consecutively numbered paragraphs.

At the beginning of each numbered Section you will be referred to any illustrations which apply to the procedures in that Section. The reference numbers used in illustration captions pinpoint the pertinent Section and the Step within that Section. That is, illustration 3.2 means the illustration refers to Section 3 and Step (or paragraph) 2 within that Section.

Procedures, once described in the text, are not normally repeated. When it's necessary to refer to another Chapter, the reference will be given as Chapter and Section number. Cross references given without use of the word "Chapter" apply to Sections and/or paragraphs in the same Chapter. For example, "see Section 8" means in the same Chapter.

References to the left or right side of the vehicle assume you are sitting in the driver's seat, facing forward.

Even though we have prepared this manual with extreme care, neither the publisher nor the author can accept responsibility for any errors in, or omissions from, the information given.

NOTE

A **Note** provides information necessary to properly complete a procedure or information which will make the procedure easier to understand.

CAUTION

A **Caution** provides a special procedure or special steps which must be taken while completing the procedure where the Caution is found. Not heeding a Caution can result in damage to the assembly being worked on.

WARNING

A **Warning** provides a special procedure or special steps which must be taken while completing the procedure where the Warning is found. Not heeding a Warning can result in personal injury.

Acknowledgements

Wiring diagrams originated exclusively for Haynes North America, Inc. by Valley Forge Technical Information Services.

© **Haynes North America, Inc. 1998, 2000**

With permission from J.H. Haynes & Co. Ltd.

A book in the Haynes Automotive Repair Manual Series

Printed in the U.S.A.

ISBN 1 56392 403 X

Library of Congress Catalog Card Number 00-109632

Contents

Haynes mechanic, author and photographer with 1997 Chevrolet Cavalier

Introduction to the Chevrolet Cavalier and Pontiac Sunfire

The Chevrolet Cavalier and Pontiac Sunfire is available in four-door sedan and two-door coupe body styles.

These models are available with two engine options. The transversely mounted 2.2-liter four-cylinder, pushrod engine or the 2.3L (1995) or 2.4L (1996 through 1998) engine with Dual Overhead-Camshafts (DOHC). Both models are equipped with a sequential multi port electronic fuel injection system.

The engine transmits power to the front wheels through either a five-speed manual transaxle or a three or four-speed automatic transaxle via a pair of driveaxles.

The Cavalier and Sunfire features a steel uni-body and independent suspension with MacPherson strut/coil spring suspension at the front and shock absorbers and coil springs at the rear. The rack and pinion steering unit is mounted behind the engine with power-assist available as optional equipment.

Standard models are equipped with power assisted front disc and rear drum brakes. All models are equipped with an Anti-lock Brake System (ABS).

Buying parts

Replacement parts are available from many sources, which generally fall into one of two categories - authorized dealer parts departments and independent retail auto parts stores. Our advice concerning these parts is as follows:

Retail auto parts stores: Good auto parts stores will stock frequently needed components which wear out relatively fast, such as clutch components, exhaust systems, brake parts, tune-up parts, etc. These stores often supply new or reconditioned parts on an exchange basis, which can save a considerable amount of money. Discount auto parts stores are often very good places to buy materials and parts needed for general vehicle maintenance such as oil, grease, filters, spark plugs, belts, touch-up paint, bulbs, etc. They also usually sell tools and general accessories, have convenient hours, charge lower prices and can often be found not far from home.

Authorized dealer parts department: This is the best source for parts which are unique to the vehicle and not generally available elsewhere (such as major engine parts, transmission parts, trim pieces, etc.).

Warranty information: If the vehicle is still covered under warranty, be sure that any replacement parts purchased - regardless of the source - do not invalidate the warranty!

To be sure of obtaining the correct parts, have engine and chassis numbers available and, if possible, take the old parts along for positive identification.

Vehicle identification numbers

Modifications are a continuing and unpublicized process in vehicle manufacturing. Since spare parts manuals and lists are compiled on a numerical basis, the individual vehicle numbers are essential to correctly identify the component required.

Vehicle Identification Number (VIN)

This very important identification number is located on a plate attached to the dashboard inside the windshield on the driver's side of the vehicle (see illustration). The VIN also appears on the Vehicle Certificate of Title and Registration. It contains information such as where and when the vehicle was manufactured, the model year and the body style.

VIN engine and model year codes

Two particularly important pieces of information found in the VIN are the engine code and the model year code. Counting from the left, the engine code letter designation is the 8th digit and the model year code designation is the 10th digit.

On the models covered by this manual the engine codes are:

4	2.2L OHV	T	2.4L OHC
D	2.3L OHC		

On the models covered by this manual the model year codes are:

S	1995	W	1998
T	1996	X	1999
V	1997	Y	2000

Vehicle Safety Certification label

The Vehicle Safety Certification label is attached to the edge of the driver's side door (see illustration). The label contains the name of the manufacturer, the month and year of production, the Gross Vehicle Weight Rating (GVWR), the Gross Axle Weight Rating (GAWR) and the certification statement.

Tire placard

The tire placard (label) is attached to the driver's side door post (see illustration). The label contains the tire label code, the tire sizes (front, rear and spare) tire pressures, tire speed rating, maximum vehicle capacity weight, total seating occupancy and the VIN.

Engine identification numbers

The engine identification numbers can be found stamped on a pad on either the left rear of the cylinder block, behind the starter (2.2L OHV engine) or on the back of the timing cover (2.3L and 2.4L OHC engines).

2.2L engine codes are stamped on the rear section of the engine block near the number 4 cylinder spark plug (see illustration).

The Vehicle Identification Number (VIN) is stamped into a metal plate fastened to the dashboard on the driver's side - it's visible through the windshield

The Vehicle Safety Certification label is affixed to the drivers side door end or post

2.3L and 1996 and 1997 2.4L engine codes are affixed on the label behind the timing cover (partial laser codes) and also stamped onto the engine block near the oil filter housing and below the starter assembly.

1998 and later 2.4L engine codes are located on the end surface of the cam cover while the VIN derivative is stamped on the bottom of the engine block in front of the transaxle extension.

Transaxle identification numbers

The transaxle identification information can be found on a bar code label located on the front of the transaxle (see illustration).

Vehicle Emissions Control Information (VECI) label

The emissions control information label is found under the hood, normally on the radiator support or the bottom side of the hood. This label contains information on the emissions control equipment installed on the vehicle, as well as tune-up specifications (see Chapter 6).

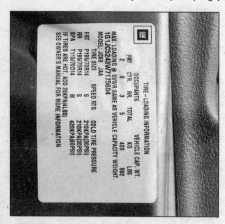

The tire placard (label) is affixed to the drivers side door end or post

Typical Engine Identification Number location

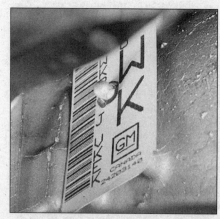

Location of the transaxle bar code label

Maintenance techniques, tools and working facilities

Maintenance techniques

There are a number of techniques involved in maintenance and repair that will be referred to throughout this manual. Application of these techniques will enable the home mechanic to be more efficient, better organized and capable of performing the various tasks properly, which will ensure that the repair job is thorough and complete.

Fasteners

Fasteners are nuts, bolts, studs and screws used to hold two or more parts together. There are a few things to keep in mind when working with fasteners. Almost all of them use a locking device of some type, either a lockwasher, locknut, locking tab or thread adhesive. All threaded fasteners should be clean and straight, with undamaged threads and undamaged corners on the hex head where the wrench fits. Develop the habit of replacing all damaged nuts and bolts with new ones. Special locknuts with nylon or fiber inserts can only be used once. If they are removed, they lose their locking ability and must be replaced with new ones.

Rusted nuts and bolts should be treated with a penetrating fluid to ease removal and prevent breakage. Some mechanics use turpentine in a spout-type oil can, which works quite well. After applying the rust penetrant, let it work for a few minutes before trying to loosen the nut or bolt. Badly rusted fasteners may have to be chiseled or sawed off or removed with a special nut breaker, available at tool stores.

If a bolt or stud breaks off in an assembly, it can be drilled and removed with a special tool commonly available for this purpose. Most automotive machine shops can perform this task, as well as other repair procedures, such as the repair of threaded holes that have been stripped out.

Flat washers and lockwashers, when removed from an assembly, should always be replaced exactly as removed. Replace any damaged washers with new ones. Never use a lockwasher on any soft metal surface (such as aluminum), thin sheet metal or plastic.

Fastener sizes

For a number of reasons, automobile manufacturers are making wider and wider use of metric fasteners. Therefore, it is important to be able to tell the difference between standard (sometimes called U.S. or SAE) and metric hardware, since they cannot be interchanged.

All bolts, whether standard or metric, are sized according to diameter, thread pitch and length. For example, a standard 1/2 - 13 x 1 bolt is 1/2 inch in diameter, has 13 threads per inch and is 1 inch long. An M12 - 1.75 x 25 metric bolt is 12 mm in diameter, has a thread pitch of 1.75 mm (the distance between threads) and is 25 mm long. The two bolts are nearly identical, and easily confused, but they are not interchangeable.

In addition to the differences in diameter, thread pitch and length, metric and standard bolts can also be distinguished by examining the bolt heads. To begin with, the distance across the flats on a standard bolt head is measured in inches, while the same dimension on a metric bolt is sized in millimeters (the same is true for nuts). As a result, a standard wrench should not be used on a metric bolt and a metric wrench should not be used on a standard bolt. Also, most standard bolts have slashes radiating out from the center of the head to denote the grade or strength of the bolt, which is an indication of the amount of torque that can be applied to it. The greater the number of slashes, the greater the strength of the bolt. Grades 0 through 5 are commonly used on automobiles. Metric bolts have a property class (grade) number, rather than a slash, molded into their heads to indicate bolt strength. In this case, the higher the number, the stronger the bolt. Property class numbers 8.8, 9.8 and 10.9 are commonly used on automobiles.

Strength markings can also be used to distinguish standard hex nuts from metric hex nuts. Many standard nuts have dots stamped into one side, while metric nuts are marked with a number. The greater the number of dots, or the higher the number, the greater the strength of the nut.

Metric studs are also marked on their ends according to property class (grade). Larger studs are numbered (the same as metric bolts), while smaller studs carry a geometric code to denote grade.

It should be noted that many fasteners, especially Grades 0 through 2, have no dis-

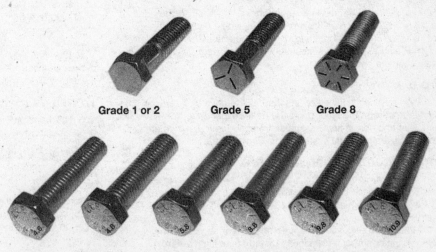

Grade 1 or 2   Grade 5   Grade 8

Bolt strength marking (standard/SAE/USS; bottom - metric)

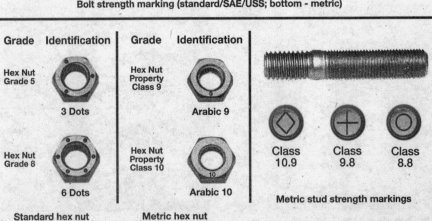

Grade	Identification	Grade	Identification
Hex Nut Grade 5	3 Dots	Hex Nut Property Class 9	Arabic 9
Hex Nut Grade 8	6 Dots	Hex Nut Property Class 10	Arabic 10

Standard hex nut strength markings

Metric hex nut strength markings

Class 10.9   Class 9.8   Class 8.8

Metric stud strength markings

00-1 HAYNES

tinguishing marks on them. When such is the case, the only way to determine whether it is standard or metric is to measure the thread pitch or compare it to a known fastener of the same size.

Standard fasteners are often referred to as SAE, as opposed to metric. However, it should be noted that SAE technically refers to a non-metric fine thread fastener only. Coarse thread non-metric fasteners are referred to as USS sizes.

Since fasteners of the same size (both standard and metric) may have different strength ratings, be sure to reinstall any bolts, studs or nuts removed from your vehicle in their original locations. Also, when replacing a fastener with a new one, make sure that the new one has a strength rating equal to or greater than the original.

Tightening sequences and procedures

Most threaded fasteners should be tightened to a specific torque value (torque is the twisting force applied to a threaded component such as a nut or bolt). Overtightening the fastener can weaken it and cause it to break, while undertightening can cause it to eventually come loose. Bolts, screws and studs, depending on the material they are made of and their thread diameters, have specific torque values, many of which are noted in the Specifications at the beginning of each Chapter. Be sure to follow the torque recommendations closely. For fasteners not assigned a specific torque, a general torque value chart is presented here as a guide. These torque values are for dry (unlubricated)

fasteners threaded into steel or cast iron (not aluminum). As was previously mentioned, the size and grade of a fastener determine the amount of torque that can safely be applied to it. The figures listed here are approximate for Grade 2 and Grade 3 fasteners. Higher grades can tolerate higher torque values.

Fasteners laid out in a pattern, such as cylinder head bolts, oil pan bolts, differential cover bolts, etc., must be loosened or tightened in sequence to avoid warping the component. This sequence will normally be shown in the appropriate Chapter. If a specific pattern is not given, the following procedures can be used to prevent warping.

Initially, the bolts or nuts should be assembled finger-tight only. Next, they should be tightened one full turn each, in a criss-cross or diagonal pattern. After each one has been tightened one full turn, return to the first one and tighten them all one-half turn, following the same pattern. Finally, tighten each of them one-quarter turn at a time until each fastener has been tightened to the proper torque. To loosen and remove the fasteners, the procedure would be reversed.

Component disassembly

Component disassembly should be done with care and purpose to help ensure that the parts go back together properly. Always keep track of the sequence in which parts are removed. Make note of special characteristics or marks on parts that can be installed more than one way, such as a grooved thrust washer on a shaft. It is a good idea to lay the disassembled parts out on a clean surface in the order that they were removed. It may also be helpful to make sketches or take instant photos of components before removal.

When removing fasteners from a component, keep track of their locations. Sometimes threading a bolt back in a part, or putting the washers and nut back on a stud, can prevent mix-ups later. If nuts and bolts cannot be returned to their original locations, they should be kept in a compartmented box or a series of small boxes. A cupcake or muffin tin is ideal for this purpose, since each cavity can hold the bolts and nuts from a particular area (i.e. oil pan bolts, valve cover bolts, engine mount bolts, etc.). A pan of this type is especially helpful when working on assemblies with very small parts, such as the carburetor, alternator, valve train or interior dash and trim pieces. The cavities can be marked with paint or tape to identify the contents.

Whenever wiring looms, harnesses or connectors are separated, it is a good idea to identify the two halves with numbered pieces of masking tape so they can be easily reconnected.

Gasket sealing surfaces

Throughout any vehicle, gaskets are used to seal the mating surfaces between two parts and keep lubricants, fluids, vacuum or pressure contained in an assembly.

Metric thread sizes

	Ft-lbs	Nm
M-6	6 to 9	9 to 12
M-8	14 to 21	19 to 28
M-10	28 to 40	38 to 54
M-12	50 to 71	68 to 96
M-14	80 to 140	109 to 154

Pipe thread sizes

1/8	5 to 8	7 to 10
1/4	12 to 18	17 to 24
3/8	22 to 33	30 to 44
1/2	25 to 35	34 to 47

U.S. thread sizes

1/4 - 20	6 to 9	9 to 12
5/16 - 18	12 to 18	17 to 24
5/16 - 24	14 to 20	19 to 27
3/8 - 16	22 to 32	30 to 43
3/8 - 24	27 to 38	37 to 51
7/16 - 14	40 to 55	55 to 74
7/16 - 20	40 to 60	55 to 81
1/2 - 13	55 to 80	75 to 108

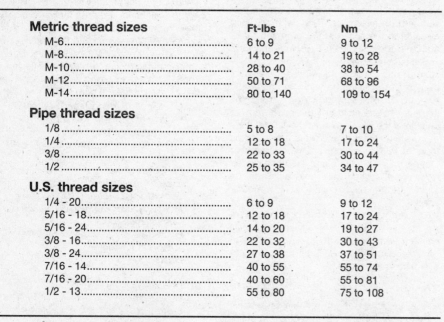

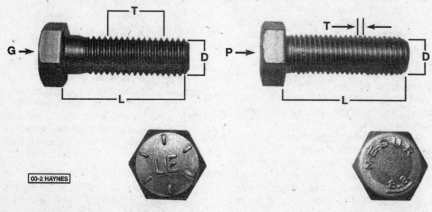

Standard (SAE and USS) bolt dimensions/grade marks

G Grade marks (bolt strength)
L Length (in inches)
T Thread pitch (number of threads per inch)
D Nominal diameter (in inches)

Metric bolt dimensions/grade marks

P Property class (bolt strength)
L Length (in millimeters)
T Thread pitch (distance between threads in millimeters)
D Diameter

00-2 HAYNES

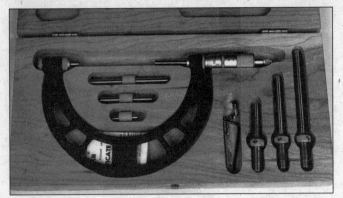

Micrometer set

Dial indicator set

Many times these gaskets are coated with a liquid or paste-type gasket sealing compound before assembly. Age, heat and pressure can sometimes cause the two parts to stick together so tightly that they are very difficult to separate. Often, the assembly can be loosened by striking it with a soft-face hammer near the mating surfaces. A regular hammer can be used if a block of wood is placed between the hammer and the part. Do not hammer on cast parts or parts that could be easily damaged. With any particularly stubborn part, always recheck to make sure that every fastener has been removed.

Avoid using a screwdriver or bar to pry apart an assembly, as they can easily mar the gasket sealing surfaces of the parts, which must remain smooth. If prying is absolutely necessary, use an old broom handle, but keep in mind that extra clean up will be necessary if the wood splinters.

After the parts are separated, the old gasket must be carefully scraped off and the gasket surfaces cleaned. Stubborn gasket material can be soaked with rust penetrant or treated with a special chemical to soften it so it can be easily scraped off. A scraper can be fashioned from a piece of copper tubing by flattening and sharpening one end. Copper is recommended because it is usually softer than the surfaces to be scraped, which reduces the chance of gouging the part. Some gaskets can be removed with a wire brush, but regardless of the method used, the mating surfaces must be left clean and smooth. If for some reason the gasket surface is gouged, then a gasket sealer thick enough to fill scratches will have to be used during reassembly of the components. For most applications, a non-drying (or semi-drying) gasket sealer should be used.

Hose removal tips

Warning: *If the vehicle is equipped with air conditioning, do not disconnect any of the A/C hoses without first having the system depressurized by a dealer service department or a service station.*

Hose removal precautions closely parallel gasket removal precautions. Avoid scratching or gouging the surface that the hose mates against or the connection may leak. This is especially true for radiator hoses.

Because of various chemical reactions, the rubber in hoses can bond itself to the metal spigot that the hose fits over. To remove a hose, first loosen the hose clamps that secure it to the spigot. Then, with slip-joint pliers, grab the hose at the clamp and rotate it around the spigot. Work it back and forth until it is completely free, then pull it off. Silicone or other lubricants will ease removal if they can be applied between the hose and the outside of the spigot. Apply the same lubricant to the inside of the hose and the outside of the spigot to simplify installation.

As a last resort (and if the hose is to be replaced with a new one anyway), the rubber can be slit with a knife and the hose peeled from the spigot. If this must be done, be careful that the metal connection is not damaged.

If a hose clamp is broken or damaged, do not reuse it. Wire-type clamps usually weaken with age, so it is a good idea to replace them with screw-type clamps whenever a hose is removed.

Tools

A selection of good tools is a basic requirement for anyone who plans to maintain and repair his or her own vehicle. For the owner who has few tools, the initial investment might seem high, but when compared to the spiraling costs of professional auto maintenance and repair, it is a wise one.

To help the owner decide which tools are needed to perform the tasks detailed in this manual, the following tool lists are offered: *Maintenance and minor repair, Repair/overhaul* and *Special.*

The newcomer to practical mechanics should start off with the *maintenance and minor repair* tool kit, which is adequate for the simpler jobs performed on a vehicle. Then, as confidence and experience grow, the owner can tackle more difficult tasks, buying additional tools as they are needed. Eventually the basic kit will be expanded into the *repair and overhaul* tool set. Over a period of time, the experienced do-it-yourselfer will assemble a tool set complete enough for most repair and overhaul procedures and will add tools from the special category when it is felt that the expense is justified by the frequency of use.

Maintenance and minor repair tool kit

The tools in this list should be considered the minimum required for performance of routine maintenance, servicing and minor repair work. We recommend the purchase of combination wrenches (box-end and open-end combined in one wrench). While more expensive than open end wrenches, they offer the advantages of both types of wrench.

> *Combination wrench set (1/4-inch to 1 inch or 6 mm to 19 mm)*
> *Adjustable wrench, 8 inch*
> *Spark plug wrench with rubber insert*
> *Spark plug gap adjusting tool*
> *Feeler gauge set*
> *Brake bleeder wrench*
> *Standard screwdriver (5/16-inch x 6 inch)*
> *Phillips screwdriver (No. 2 x 6 inch)*
> *Combination pliers - 6 inch*
> *Hacksaw and assortment of blades*
> *Tire pressure gauge*
> *Grease gun*
> *Oil can*
> *Fine emery cloth*
> *Wire brush*
> *Battery post and cable cleaning tool*
> *Oil filter wrench*
> *Funnel (medium size)*
> *Safety goggles*
> *Jackstands (2)*
> *Drain pan*

Note: *If basic tune-ups are going to be part of routine maintenance, it will be necessary to purchase a good quality stroboscopic timing light and combination tachometer/dwell meter. Although they are included in the list of special tools, it is mentioned here because they are absolutely necessary for tuning most vehicles properly.*

Repair and overhaul tool set

These tools are essential for anyone who plans to perform major repairs and are in addition to those in the maintenance and minor repair tool kit. Included is a comprehensive set of sockets which, though expensive, are invaluable because of their versatility, especially when various extensions and drives are available. We recommend the 1/2-inch drive over the 3/8-inch drive. Although the larger drive is bulky and more expensive, it has the capacity of accepting a very wide

Dial caliper

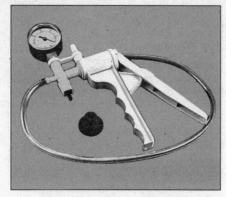

Hand-operated vacuum pump

Timing light

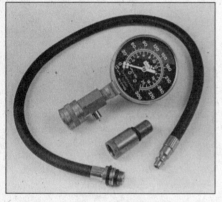

Compression gauge with spark plug
hole adapter

Damper/steering wheel puller

General purpose puller

Hydraulic lifter removal tool

Valve spring compressor

Valve spring compressor

Ridge reamer

Piston ring groove cleaning tool

Ring removal/installation tool

Ring compressor

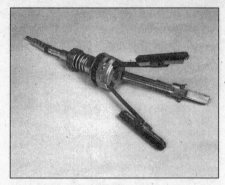

Cylinder hone

Brake hold-down spring tool

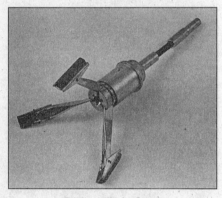

Brake cylinder hone

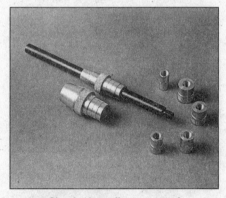

Clutch plate alignment tool

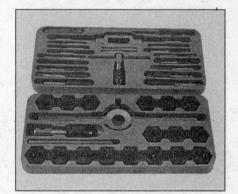

Tap and die set

range of large sockets. Ideally, however, the mechanic should have a 3/8-inch drive set and a 1/2-inch drive set.

> Socket set(s)
> Reversible ratchet
> Extension - 10 inch
> Universal joint
> Torque wrench (same size drive as sockets)
> Ball peen hammer - 8 ounce
> Soft-face hammer (plastic/rubber)
> Standard screwdriver (1/4-inch x 6 inch)
> Standard screwdriver (stubby - 5/16-inch)
> Phillips screwdriver (No. 3 x 8 inch)
> Phillips screwdriver (stubby - No. 2)
> Pliers - vise grip
> Pliers - lineman's
> Pliers - needle nose
> Pliers - snap-ring (internal and external)
> Cold chisel - 1/2-inch
> Scribe
> Scraper (made from flattened copper tubing)
> Centerpunch
> Pin punches (1/16, 1/8, 3/16-inch)
> Steel rule/straightedge - 12 inch
> Allen wrench set (1/8 to 3/8-inch or 4 mm to 10 mm)
> A selection of files
> Wire brush (large)
> Jackstands (second set)
> Jack (scissor or hydraulic type)

Note: Another tool which is often useful is an electric drill with a chuck capacity of 3/8-inch and a set of good quality drill bits.

Special tools

The tools in this list include those which are not used regularly, are expensive to buy, or which need to be used in accordance with their manufacturer's instructions. Unless these tools will be used frequently, it is not very economical to purchase many of them. A consideration would be to split the cost and use between yourself and a friend or friends. In addition, most of these tools can be obtained from a tool rental shop on a temporary basis.

This list primarily contains only those tools and instruments widely available to the public, and not those special tools produced by the vehicle manufacturer for distribution to dealer service departments. Occasionally, references to the manufacturer's special tools are included in the text of this manual. Generally, an alternative method of doing the job without the special tool is offered. However, sometimes there is no alternative to their use. Where this is the case, and the tool cannot be purchased or borrowed, the work should be turned over to the dealer service department or an automotive repair shop.

> Valve spring compressor
> Piston ring groove cleaning tool
> Piston ring compressor
> Piston ring installation tool
> Cylinder compression gauge
> Cylinder ridge reamer
> Cylinder surfacing hone
> Cylinder bore gauge
> Micrometers and/or dial calipers
> Hydraulic lifter removal tool

> Balljoint separator
> Universal-type puller
> Impact screwdriver
> Dial indicator set
> Stroboscopic timing light (inductive pick-up)
> Hand operated vacuum/pressure pump
> Tachometer/dwell meter
> Universal electrical multimeter
> Cable hoist
> Brake spring removal and installation tools
> Floor jack

Buying tools

For the do-it-yourselfer who is just starting to get involved in vehicle maintenance and repair, there are a number of options available when purchasing tools. If maintenance and minor repair is the extent of the work to be done, the purchase of individual tools is satisfactory. If, on the other hand, extensive work is planned, it would be a good idea to purchase a modest tool set from one of the large retail chain stores. A set can usually be bought at a substantial savings over the individual tool prices, and they often come with a tool box. As additional tools are needed, add-on sets, individual tools and a larger tool box can be purchased to expand the tool selection. Building a tool set gradually allows the cost of the tools to be spread over a longer period of time and gives the mechanic the freedom to choose only those tools that will actually be used.

Tool stores will often be the only source of some of the special tools that are needed, but regardless of where tools are bought, try to avoid cheap ones, especially when buying screwdrivers and sockets, because they won't last very long. The expense involved in replacing cheap tools will eventually be greater than the initial cost of quality tools.

Care and maintenance of tools

Good tools are expensive, so it makes sense to treat them with respect. Keep them clean and in usable condition and store them properly when not in use. Always wipe off any dirt, grease or metal chips before putting them away. Never leave tools lying around in the work area. Upon completion of a job, always check closely under the hood for tools that may have been left there so they won't get lost during a test drive.

Some tools, such as screwdrivers, pliers, wrenches and sockets, can be hung on a panel mounted on the garage or workshop wall, while others should be kept in a tool box or tray. Measuring instruments, gauges, meters, etc. must be carefully stored where they cannot be damaged by weather or impact from other tools.

When tools are used with care and stored properly, they will last a very long time. Even with the best of care, though, tools will wear out if used frequently. When a tool is damaged or worn out, replace it. Subsequent jobs will be safer and more enjoyable if you do.

How to repair damaged threads

Sometimes, the internal threads of a nut or bolt hole can become stripped, usually from overtightening. Stripping threads is an all-too-common occurrence, especially when working with aluminum parts, because aluminum is so soft that it easily strips out.

Usually, external or internal threads are only partially stripped. After they've been cleaned up with a tap or die, they'll still work. Sometimes, however, threads are badly damaged. When this happens, you've got three choices:

1) *Drill and tap the hole to the next suitable oversize and install a larger diameter bolt, screw or stud.*
2) *Drill and tap the hole to accept a threaded plug, then drill and tap the plug to the original screw size. You can also buy a plug already threaded to the original size. Then you simply drill a hole to the specified size, then run the threaded plug into the hole with a bolt and jam nut. Once the plug is fully seated, remove the jam nut and bolt.*
3) *The third method uses a patented thread repair kit like Heli-Coil or Slimsert. These easy-to-use kits are designed to repair damaged threads in straight-through holes and blind holes. Both are available as kits which can handle a variety of sizes and thread patterns. Drill the hole, then tap it with the special included tap. Install the Heli-Coil and the hole is back to its original diameter and thread pitch.*

Regardless of which method you use, be sure to proceed calmly and carefully. A little impatience or carelessness during one of these relatively simple procedures can ruin your whole day's work and cost you a bundle if you wreck an expensive part.

Working facilities

Not to be overlooked when discussing tools is the workshop. If anything more than routine maintenance is to be carried out, some sort of suitable work area is essential.

It is understood, and appreciated, that many home mechanics do not have a good workshop or garage available, and end up removing an engine or doing major repairs outside. It is recommended, however, that the overhaul or repair be completed under the cover of a roof.

A clean, flat workbench or table of comfortable working height is an absolute necessity. The workbench should be equipped with a vise that has a jaw opening of at least four inches.

As mentioned previously, some clean, dry storage space is also required for tools, as well as the lubricants, fluids, cleaning solvents, etc. which soon become necessary.

Sometimes waste oil and fluids, drained from the engine or cooling system during normal maintenance or repairs, present a disposal problem. To avoid pouring them on the ground or into a sewage system, pour the used fluids into large containers, seal them with caps and take them to an authorized disposal site or recycling center. Plastic jugs, such as old antifreeze containers, are ideal for this purpose.

Always keep a supply of old newspapers and clean rags available. Old towels are excellent for mopping up spills. Many mechanics use rolls of paper towels for most work because they are readily available and disposable. To help keep the area under the vehicle clean, a large cardboard box can be cut open and flattened to protect the garage or shop floor.

Whenever working over a painted surface, such as when leaning over a fender to service something under the hood, always cover it with an old blanket or bedspread to protect the finish. Vinyl covered pads, made especially for this purpose, are available at auto parts stores.

Booster battery (jump) starting

Observe these precautions when using a booster battery to start a vehicle:

a) *Before connecting the booster battery, make sure the ignition switch is in the Off position.*
b) *Turn off the lights, heater and other electrical loads.*
c) *Your eyes should be shielded. Safety goggles are a good idea.*
d) *Make sure the booster battery is the same voltage as the dead one in the vehicle.*
e) *The two vehicles MUST NOT TOUCH each other!*
f) *Make sure the transaxle is in Neutral (manual) or Park (automatic).*
g) *If the booster battery is not a maintenance-free type, remove the vent caps and lay a cloth over the vent holes.*

Connect the red jumper cable to the positive (+) terminals of each battery **(see illustration)**.

Connect one end of the black jumper cable to the negative (-) terminal of the booster battery. The other end of this cable should be connected to a good ground on the vehicle to be started, such as a bolt or bracket on the body.

Start the engine using the booster battery, then, with the engine running at idle speed, disconnect the jumper cables in the reverse order of connection.

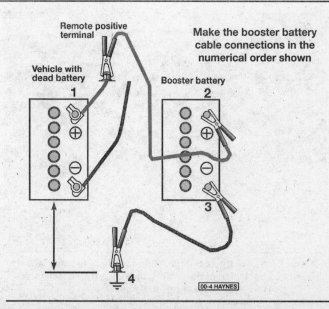

Make the booster battery cable connections in the numerical order shown

Remote positive terminal

Vehicle with dead battery

Booster battery

00-4 HAYNES

Jacking and towing

Jacking

Warning: *The jack supplied with the vehicle should only be used for changing a tire or placing jackstands under the frame. Never work under the vehicle or start the engine while this jack is being used as the only means of support.*

The vehicle should be on level ground. Place the shift lever in Park, if you have an automatic, or Reverse if you have a manual transaxle. Block the wheel diagonally opposite the wheel being changed. Set the parking brake.

Remove the spare tire and jack from stowage. Remove the wheel cover and trim ring (if so equipped) with the tapered end of the lug nut wrench by inserting and twisting the handle and then prying against the back of the wheel cover. Loosen the wheel lug nuts about 1/4-to-1/2 turn each.

Place the scissors-type jack under the side of the vehicle and adjust the jack height until it fits in the notch in the vertical rocker panel flange nearest the wheel to be changed. There is a front and rear jacking point on each side of the vehicle **(see illustrations)**.

Turn the jack handle clockwise until the tire clears the ground. Remove the lug nuts and pull the wheel off. Replace it with the spare.

Install the lug nuts with the beveled edges facing in. Tighten them snugly. Don't attempt to tighten them completely until the vehicle is lowered or it could slip off the jack. Turn the jack handle counterclockwise to lower the vehicle. Remove the jack and tighten the lug nuts in a diagonal pattern.

Install the cover (and trim ring, if used) and be sure it's snapped into place all the way around.

Stow the tire, jack and wrench. Unblock the wheels.

Towing

As a general rule, the vehicle should be towed with the front (drive) wheels off the ground. If they can't be raised, place them on a dolly. The ignition key must be in the ACC position, since the steering lock mechanism isn't strong enough to hold the front wheels straight while towing.

Vehicles equipped with an automatic transaxle can be towed from the front only with all four wheels on the ground, provided that speeds don't exceed 30 mph and the distance is not over 40 miles. Before towing, check the transmission fluid level (see Chapter 1). If the level is below the HOT line on the dipstick, add fluid or use a towing dolly.

Caution: *Never tow a vehicle with an automatic transaxle from the rear with the front wheels on the ground.*

When towing a vehicle equipped with a manual transaxle with all four wheels on the ground, be sure to place the shift lever in neutral and release the parking brake.

Equipment specifically designed for towing should be used. It should be attached to the main structural members of the vehicle, not the bumpers or brackets.

Safety is a major consideration when towing and all applicable state and local laws must be obeyed. A safety chain system must be used at all times.

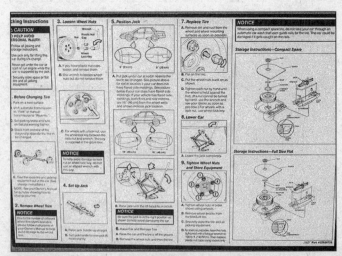

The jack fits over the rocker panel flange (there are two jacking points on each side of the vehicle, indicated by a notch in the rocker panel flange)

The vehicle is equipped with jacking instructions located in the trunk area

Automotive chemicals and lubricants

A number of automotive chemicals and lubricants are available for use during vehicle maintenance and repair. They include a wide variety of products ranging from cleaning solvents and degreasers to lubricants and protective sprays for rubber, plastic and vinyl.

Cleaners

Carburetor cleaner and choke cleaner is a strong solvent for gum, varnish and carbon. Most carburetor cleaners leave a dry-type lubricant film which will not harden or gum up. Because of this film it is not recommended for use on electrical components.

Brake system cleaner is used to remove grease and brake fluid from the brake system, where clean surfaces are absolutely necessary. It leaves no residue and often eliminates brake squeal caused by contaminants.

Electrical cleaner removes oxidation, corrosion and carbon deposits from electrical contacts, restoring full current flow. It can also be used to clean spark plugs, carburetor jets, voltage regulators and other parts where an oil-free surface is desired.

Demoisturants remove water and moisture from electrical components such as alternators, voltage regulators, electrical connectors and fuse blocks. They are non-conductive, non-corrosive and non-flammable.

Degreasers are heavy-duty solvents used to remove grease from the outside of the engine and from chassis components. They can be sprayed or brushed on and, depending on the type, are rinsed off either with water or solvent.

Lubricants

Motor oil is the lubricant formulated for use in engines. It normally contains a wide variety of additives to prevent corrosion and reduce foaming and wear. Motor oil comes in various weights (viscosity ratings) from 0 to 50. The recommended weight of the oil depends on the season, temperature and the demands on the engine. Light oil is used in cold climates and under light load conditions. Heavy oil is used in hot climates and where high loads are encountered. Multi-viscosity oils are designed to have characteristics of both light and heavy oils and are available in a number of weights from 5W-20 to 20W-50.

Gear oil is designed to be used in differentials, manual transmissions and other areas where high-temperature lubrication is required.

Chassis and wheel bearing grease is a heavy grease used where increased loads and friction are encountered, such as for wheel bearings, balljoints, tie-rod ends and universal joints.

High-temperature wheel bearing grease is designed to withstand the extreme temperatures encountered by wheel bearings in disc brake equipped vehicles. It usually contains molybdenum disulfide (moly), which is a dry-type lubricant.

White grease is a heavy grease for metal-to-metal applications where water is a problem. White grease stays soft under both low and high temperatures (usually from -100 to +190-degrees F), and will not wash off or dilute in the presence of water.

Assembly lube is a special extreme pressure lubricant, usually containing moly, used to lubricate high-load parts (such as main and rod bearings and cam lobes) for initial start-up of a new engine. The assembly lube lubricates the parts without being squeezed out or washed away until the engine oiling system begins to function.

Silicone lubricants are used to protect rubber, plastic, vinyl and nylon parts.

Graphite lubricants are used where oils cannot be used due to contamination problems, such as in locks. The dry graphite will lubricate metal parts while remaining uncontaminated by dirt, water, oil or acids. It is electrically conductive and will not foul electrical contacts in locks such as the ignition switch.

Moly penetrants loosen and lubricate frozen, rusted and corroded fasteners and prevent future rusting or freezing.

Heat-sink grease is a special electrically non-conductive grease that is used for mounting electronic ignition modules where it is essential that heat is transferred away from the module.

Sealants

RTV sealant is one of the most widely used gasket compounds. Made from silicone, RTV is air curing, it seals, bonds, waterproofs, fills surface irregularities, remains flexible, doesn't shrink, is relatively easy to remove, and is used as a supplementary sealer with almost all low and medium temperature gaskets.

Anaerobic sealant is much like RTV in that it can be used either to seal gaskets or to form gaskets by itself. It remains flexible, is solvent resistant and fills surface imperfections. The difference between an anaerobic sealant and an RTV-type sealant is in the curing. RTV cures when exposed to air, while an anaerobic sealant cures only in the absence of air. This means that an anaerobic sealant cures only after the assembly of parts, sealing them together.

Thread and pipe sealant is used for sealing hydraulic and pneumatic fittings and vacuum lines. It is usually made from a Teflon compound, and comes in a spray, a paint-on liquid and as a wrap-around tape.

Chemicals

Anti-seize compound prevents seizing, galling, cold welding, rust and corrosion in fasteners. High-temperature ant-seize, usually made with copper and graphite lubricants, is used for exhaust system and exhaust manifold bolts.

Anaerobic locking compounds are used to keep fasteners from vibrating or working loose and cure only after installation, in the absence of air. Medium strength locking compound is used for small nuts, bolts and screws that may be removed later. High-strength locking compound is for large nuts, bolts and studs which aren't removed on a regular basis.

Oil additives range from viscosity index improvers to chemical treatments that claim to reduce internal engine friction. It should be noted that most oil manufacturers caution against using additives with their oils.

Gas additives perform several functions, depending on their chemical makeup. They usually contain solvents that help dissolve gum and varnish that build up on carburetor, fuel injection and intake parts. They also serve to break down carbon deposits that form on the inside surfaces of the combustion chambers. Some additives contain upper cylinder lubricants for valves and piston rings, and others contain chemicals to remove condensation from the gas tank.

Miscellaneous

Brake fluid is specially formulated hydraulic fluid that can withstand the heat and pressure encountered in brake systems. Care must be taken so this fluid does not come in contact with painted surfaces or plastics. An opened container should always be resealed to prevent contamination by water or dirt.

Weatherstrip adhesive is used to bond weatherstripping around doors, windows and trunk lids. It is sometimes used to attach trim pieces.

Undercoating is a petroleum-based, tar-like substance that is designed to protect metal surfaces on the underside of the vehicle from corrosion. It also acts as a sound-deadening agent by insulating the bottom of the vehicle.

Waxes and polishes are used to help protect painted and plated surfaces from the weather. Different types of paint may require the use of different types of wax and polish. Some polishes utilize a chemical or abrasive cleaner to help remove the top layer of oxidized (dull) paint on older vehicles. In recent years many non-wax polishes that contain a wide variety of chemicals such as polymers and silicones have been introduced. These non-wax polishes are usually easier to apply and last longer than conventional waxes and polishes.

Conversion factors

Length (distance)

Inches (in)	X	25.4	= Millimetres (mm)	X 0.0394	= Inches (in)
Feet (ft)	X	0.305	= Metres (m)	X 3.281	= Feet (ft)
Miles	X	1.609	= Kilometres (km)	X 0.621	= Miles

Volume (capacity)

Cubic inches (cu in; in³)	X	16.387	= Cubic centimetres (cc; cm³)	X 0.061	= Cubic inches (cu in; in³)
Imperial pints (Imp pt)	X	0.568	= Litres (l)	X 1.76	= Imperial pints (Imp pt)
Imperial quarts (Imp qt)	X	1.137	= Litres (l)	X 0.88	= Imperial quarts (Imp qt)
Imperial quarts (Imp qt)	X	1.201	= US quarts (US qt)	X 0.833	= Imperial quarts (Imp qt)
US quarts (US qt)	X	0.946	= Litres (l)	X 1.057	= US quarts (US qt)
Imperial gallons (Imp gal)	X	4.546	= Litres (l)	X 0.22	= Imperial gallons (Imp gal)
Imperial gallons (Imp gal)	X	1.201	= US gallons (US gal)	X 0.833	= Imperial gallons (Imp gal)
US gallons (US gal)	X	3.785	= Litres (l)	X 0.264	= US gallons (US gal)

Mass (weight)

Ounces (oz)	X	28.35	= Grams (g)	X 0.035	= Ounces (oz)
Pounds (lb)	X	0.454	= Kilograms (kg)	X 2.205	= Pounds (lb)

Force

Ounces-force (ozf; oz)	X	0.278	= Newtons (N)	X 3.6	= Ounces-force (ozf; oz)
Pounds-force (lbf; lb)	X	4.448	= Newtons (N)	X 0.225	= Pounds-force (lbf; lb)
Newtons (N)	X	0.1	= Kilograms-force (kgf; kg)	X 9.81	= Newtons (N)

Pressure

Pounds-force per square inch (psi; lbf/in²; lb/in²)	X	0.070	= Kilograms-force per square centimetre (kgf/cm²; kg/cm²)	X 14.223	= Pounds-force per square inch (psi; lbf/in²; lb/in²)
Pounds-force per square inch (psi; lbf/in²; lb/in²)	X	0.068	= Atmospheres (atm)	X 14.696	= Pounds-force per square inch (psi; lbf/in²; lb/in²)
Pounds-force per square inch (psi; lbf/in²; lb/in²)	X	0.069	= Bars	X 14.5	= Pounds-force per square inch (psi; lbf/in²; lb/in²)
Pounds-force per square inch (psi; lbf/in²; lb/in²)	X	6.895	= Kilopascals (kPa)	X 0.145	= Pounds-force per square inch (psi; lbf/in²; lb/in²)
Kilopascals (kPa)	X	0.01	= Kilograms-force per square centimetre (kgf/cm²; kg/cm²)	X 98.1	= Kilopascals (kPa)

Torque (moment of force)

Pounds-force inches (lbf in; lb in)	X	1.152	= Kilograms-force centimetre (kgf cm; kg cm)	X 0.868	= Pounds-force inches (lbf in; lb in)
Pounds-force inches (lbf in; lb in)	X	0.113	= Newton metres (Nm)	X 8.85	= Pounds-force inches (lbf in; lb in)
Pounds-force inches (lbf in; lb in)	X	0.083	= Pounds-force feet (lbf ft; lb ft)	X 12	= Pounds-force inches (lbf in; lb in)
Pounds-force feet (lbf ft; lb ft)	X	0.138	= Kilograms-force metres (kgf m; kg m)	X 7.233	= Pounds-force feet (lbf ft; lb ft)
Pounds-force feet (lbf ft; lb ft)	X	1.356	= Newton metres (Nm)	X 0.738	= Pounds-force feet (lbf ft; lb ft)
Newton metres (Nm)	X	0.102	= Kilograms-force metres (kgf m; kg m)	X 9.804	= Newton metres (Nm)

Vacuum

Inches mercury (in. Hg)	X	3.377	= Kilopascals (kPa)	X 0.2961	= Inches mercury
Inches mercury (in. Hg)	X	25.4	= Millimeters mercury (mm Hg)	X 0.0394	= Inches mercury

Power

Horsepower (hp)	X	745.7	= Watts (W)	X 0.0013	= Horsepower (hp)

Velocity (speed)

Miles per hour (miles/hr; mph)	X	1.609	= Kilometres per hour (km/hr; kph)	X 0.621	= Miles per hour (miles/hr; mph)

Fuel consumption*

Miles per gallon, Imperial (mpg)	X	0.354	= Kilometres per litre (km/l)	X 2.825	= Miles per gallon, Imperial (mpg)
Miles per gallon, US (mpg)	X	0.425	= Kilometres per litre (km/l)	X 2.352	= Miles per gallon, US (mpg)

Temperature

Degrees Fahrenheit = ($°C \times 1.8$) + 32   Degrees Celsius (Degrees Centigrade; °C) = ($°F - 32$) x 0.56

*It is common practice to convert from miles per gallon (mpg) to litres/100 kilometres (l/100km), where mpg (Imperial) x l/100 km = 282 and mpg (US) x l/100 km = 235

Safety first!

Regardless of how enthusiastic you may be about getting on with the job at hand, take the time to ensure that your safety is not jeopardized. A moment's lack of attention can result in an accident, as can failure to observe certain simple safety precautions. The possibility of an accident will always exist, and the following points should not be considered a comprehensive list of all dangers. Rather, they are intended to make you aware of the risks and to encourage a safety conscious approach to all work you carry out on your vehicle.

Essential DOs and DON'Ts

DON'T rely on a jack when working under the vehicle. Always use approved jackstands to support the weight of the vehicle and place them under the recommended lift or support points.

DON'T attempt to loosen extremely tight fasteners (i.e. wheel lug nuts) while the vehicle is on a jack - it may fall.

DON'T start the engine without first making sure that the transmission is in Neutral (or Park where applicable) and the parking brake is set.

DON'T remove the radiator cap from a hot cooling system - let it cool or cover it with a cloth and release the pressure gradually.

DON'T attempt to drain the engine oil until you are sure it has cooled to the point that it will not burn you.

DON'T touch any part of the engine or exhaust system until it has cooled sufficiently to avoid burns.

DON'T siphon toxic liquids such as gasoline, antifreeze and brake fluid by mouth, or allow them to remain on your skin.

DON'T inhale brake lining dust - it is potentially hazardous (see *Asbestos* below).

DON'T allow spilled oil or grease to remain on the floor - wipe it up before someone slips on it.

DON'T use loose fitting wrenches or other tools which may slip and cause injury.

DON'T push on wrenches when loosening or tightening nuts or bolts. Always try to pull the wrench toward you. If the situation calls for pushing the wrench away, push with an open hand to avoid scraped knuckles if the wrench should slip.

DON'T attempt to lift a heavy component alone - get someone to help you.

DON'T rush or take unsafe shortcuts to finish a job.

DON'T allow children or animals in or around the vehicle while you are working on it.

DO wear eye protection when using power tools such as a drill, sander, bench grinder, etc. and when working under a vehicle.

DO keep loose clothing and long hair well out of the way of moving parts.

DO make sure that any hoist used has a safe working load rating adequate for the job.

DO get someone to check on you periodically when working alone on a vehicle.

DO carry out work in a logical sequence and make sure that everything is correctly assembled and tightened.

DO keep chemicals and fluids tightly capped and out of the reach of children and pets.

DO remember that your vehicle's safety affects that of yourself and others. If in doubt on any point, get professional advice.

Asbestos

Certain friction, insulating, sealing, and other products - such as brake linings, brake bands, clutch linings, torque converters, gaskets, etc. - may contain asbestos. Extreme care must be taken to avoid inhalation of dust from such products, since it is hazardous to health. If in doubt, assume that they do contain asbestos.

Fire

Remember at all times that gasoline is highly flammable. Never smoke or have any kind of open flame around when working on a vehicle. But the risk does not end there. A spark caused by an electrical short circuit, by two metal surfaces contacting each other, or even by static electricity built up in your body under certain conditions, can ignite gasoline vapors, which in a confined space are highly explosive. Do not, under any circumstances, use gasoline for cleaning parts. Use an approved safety solvent.

Always disconnect the battery ground (-) cable at the battery before working on any part of the fuel system or electrical system. Never risk spilling fuel on a hot engine or exhaust component. It is strongly recommended that a fire extinguisher suitable for use on fuel and electrical fires be kept handy in the garage or workshop at all times. Never try to extinguish a fuel or electrical fire with water.

Fumes

Certain fumes are highly toxic and can quickly cause unconsciousness and even death if inhaled to any extent. Gasoline vapor falls into this category, as do the vapors from some cleaning solvents. Any draining or pouring of such volatile fluids should be done in a well ventilated area.

When using cleaning fluids and solvents, read the instructions on the container carefully. Never use materials from unmarked containers.

Never run the engine in an enclosed space, such as a garage. Exhaust fumes contain carbon monoxide, which is extremely poisonous. If you need to run the engine, always do so in the open air, or at least have the rear of the vehicle outside the work area.

If you are fortunate enough to have the use of an inspection pit, never drain or pour gasoline and never run the engine while the vehicle is over the pit. The fumes, being heavier than air, will concentrate in the pit with possibly lethal results.

The battery

Never create a spark or allow a bare light bulb near a battery. They normally give off a certain amount of hydrogen gas, which is highly explosive.

Always disconnect the battery ground (-) cable at the battery before working on the fuel or electrical systems.

If possible, loosen the filler caps or cover when charging the battery from an external source (this does not apply to sealed or maintenance-free batteries). Do not charge at an excessive rate or the battery may burst.

Take care when adding water to a non maintenance-free battery and when carrying a battery. The electrolyte, even when diluted, is very corrosive and should not be allowed to contact clothing or skin.

Always wear eye protection when cleaning the battery to prevent the caustic deposits from entering your eyes.

Household current

When using an electric power tool, inspection light, etc., which operates on household current, always make sure that the tool is correctly connected to its plug and that, where necessary, it is properly grounded. Do not use such items in damp conditions and, again, do not create a spark or apply excessive heat in the vicinity of fuel or fuel vapor.

Secondary ignition system voltage

A severe electric shock can result from touching certain parts of the ignition system (such as the spark plug wires) when the engine is running or being cranked, particularly if components are damp or the insulation is defective. In the case of an electronic ignition system, the secondary system voltage is much higher and could prove fatal.

Troubleshooting

Contents

This section provides an easy reference guide to the more common problems which may occur during the operation of your vehicle. These problems and their possible causes are grouped under headings denoting various components or systems, such as Engine, Cooling system, etc. They also refer you to the chapter and/or section which deals with the problem.

Remember that successful troubleshooting is not a mysterious black art practiced only by professional mechanics. It is simply the result of the right knowledge combined with an intelligent, systematic approach to the problem. Always work by a process of elimination, starting with the simplest solution and working through to the most complex - and never overlook the obvious. Anyone can run the gas tank dry or leave the lights on overnight, so don't assume that you are exempt from such oversights.

Finally, always establish a clear idea of why a problem has occurred and take steps to ensure that it doesn't happen again. If the electrical system fails because of a poor connection, check the other connections in the system to make sure that they don't fail as well. If a particular fuse continues to blow, find out why - don't just replace one fuse after another. Remember, failure of a small component can often be indicative of potential failure or incorrect functioning of a more important component or system.

Engine

1   Engine will not rotate when attempting to start

1   Battery terminal connections loose or corroded (Chapter 1).
2   Battery discharged or faulty (Chapter 1).
3   Automatic transaxle not completely engaged in Park (Chapter 7) or clutch pedal not completely depressed (Chapter 8).
4   Broken, loose or disconnected wiring in the starting circuit (Chapters 5 and 12).
5   Starter motor pinion jammed in flywheel ring gear (Chapter 5).
6   Starter solenoid faulty (Chapter 5).
7   Starter motor faulty (Chapter 5).
8   Ignition switch faulty (Chapter 12).
9   Starter pinion or flywheel teeth worn or broken (Chapter 5).
10   Defective fusible link (Chapter 12).

2   Engine rotates but will not start

1   Fuel tank empty.
2   Battery discharged (engine rotates slowly) (Chapter 5).
3   Battery terminal connections loose or corroded (Chapter 1).
4   Leaking fuel injector(s), faulty fuel pump, pressure regulator, etc. (Chapter 4).
5   Worn, faulty or incorrectly gapped spark plugs (Chapter 1).
6   Broken, loose or disconnected wiring in the starting circuit (Chapter 5).
7   Broken, loose or disconnected wires at the ignition coils or faulty coils or ignition module (Chapter 5).
8   Defective crankshaft sensor or PCM (Chapter 6).
9   Broken timing chain (Chapter 2).

3   Engine hard to start when cold

1   Battery discharged or low (Chapter 1).
2   Malfunctioning fuel system (Chapter 4).
3   Faulty coolant temperature sensor or intake air temperature sensor (Chapter 6).
4   Fuel injector(s) leaking (Chapter 4).
5   Faulty ignition system (Chapter 5).
6   Defective MAP sensor (see Chapter 6).

4   Engine hard to start when hot

1   Air filter clogged (Chapter 1).
2   Fuel not reaching the fuel injection system (Chapter 4).
3   Corroded battery connections, especially ground (Chapter 1).
4   Faulty coolant temperature sensor or intake air temperature sensor (Chapter 6).

5   Starter motor noisy or excessively rough in engagement

1   Pinion or flywheel gear teeth worn or broken (Chapter 5).
2   Starter motor mounting bolts loose or missing (Chapter 5).

6   Engine starts but stops immediately

1   Loose or faulty electrical connections at ignition coil (Chapter 5).
2   Insufficient fuel reaching the fuel injector(s) (Chapters 4).
3   Vacuum leak at the gasket between the intake manifold/plenum and throttle body (Chapter 4).
4   Fault in the engine control system (Chapter 6).
5   Intake air leaks, broken vacuum lines (see Chapter 4)

7   Oil puddle under engine

1   Oil pan gasket and/or oil pan drain bolt washer leaking (Chapter 2).
2   Oil pressure sending unit leaking (Chapter 2).
3   Valve covers leaking (Chapter 2).
4   Engine oil seals leaking (Chapter 2).

8   Engine lopes while idling or idles erratically

1   Vacuum leakage (Chapters 2 and 4).
2   Leaking EGR valve (Chapter 6).
3   Air filter clogged (Chapter 1).
4   Fuel pump not delivering sufficient fuel to the fuel injection system (Chapter 4).
5   Leaking head gasket (Chapter 2).
6   Camshaft lobes worn (Chapter 2).

9   Engine misses at idle speed

1   Spark plugs worn or not gapped properly (Chapter 1).
2   Faulty spark plug wires (Chapter 1).
3   Vacuum leaks (Chapters 2 and 4).
4   Faulty ignition coil(s) (Chapter 5).
5   Uneven or low compression (Chapter 2).
6   Faulty fuel injector(s) (Chapter 4).

10   Engine misses throughout driving speed range

1   Fuel filter clogged and/or impurities in the fuel system (Chapter 1).
2   Low fuel output at the fuel injector(s) (Chapter 4).
3   Faulty or incorrectly gapped spark plugs (Chapter 1).
4   Leaking spark plug wires (Chapters 1 or 5).
5   Faulty emission system components (Chapter 6).
6   Low or uneven cylinder compression pressures (Chapter 2).
7   Burned valves (Chapter 2).
8   Weak or faulty ignition system (Chapter 5).
9   Vacuum leak in fuel injection system, throttle body, intake manifold or vacuum hoses (Chapter 4).

11   Engine stumbles on acceleration

1   Spark plugs fouled (Chapter 1).
2   Problem with fuel injection system (Chapter 4).
3   Fuel filter clogged (Chapters 1 and 4).

4   Fault in the engine control system (Chapter 6).
5   Intake manifold air leak (Chapters 2 and 4).
6   EGR system malfunction (Chapter 6).

12   Engine surges while holding accelerator steady

1   Intake air leak (Chapter 4).
2   Fuel pump or fuel pressure regulator faulty (Chapter 4).
3   Problem with fuel injection system (Chapter 4).
4   Problem with the emissions control system (Chapter 6).

13   Engine stalls

1   Idle speed incorrect (Chapter 1).
2   Fuel filter clogged and/or water and impurities in the fuel system (Chapters 1 and 4).
3   Ignition components damp or damaged (Chapter 5).
4   Faulty emissions system components (Chapter 6).
5   Faulty or incorrectly gapped spark plugs (Chapter 1).
6   Faulty spark plug wires (Chapter 1).
7   Vacuum leak in the fuel injection system, intake manifold or vacuum hoses (Chapters 2 and 4).

14   Engine lacks power

1   Worn camshaft lobes (Chapter 2).
2   Burned valves or incorrect valve timing (Chapter 2).
3   Faulty spark plug wires or faulty coil (Chapters 1 and 5).
4   Faulty or incorrectly gapped spark plugs (Chapter 1).
5   Problem with the fuel injection system (Chapter 4).
6   Plugged air filter (Chapter 1).
7   Brakes binding (Chapter 9).
8   Automatic transaxle fluid level incorrect (Chapter 1).
9   Clutch slipping (Chapter 8).
10   Fuel filter clogged and/or impurities in the fuel system (Chapters 1 and 4).
11   Emission control system not functioning properly (Chapter 6).
12   Low or uneven cylinder compression pressures (Chapter 2).
13   Restricted exhaust system (Chapters 4).

15   Engine backfires

1   Emission control system not functioning properly (Chapter 6).
2   Faulty spark plug wires or coil(s) (Chapter 5).
3   Problem with the fuel injection system (Chapter 4).
4   Vacuum leak at fuel injector(s), intake manifold or vacuum hoses (Chapters 2 and 4).
5   Burned valves or incorrect valve timing (Chapter 2).

16   Pinging or knocking engine sounds during acceleration or uphill

1   Incorrect grade of fuel.
2   Problem with the engine control system (Chapter 6).
3   Fuel injection system faulty (Chapter 4).
4   Improper or damaged spark plugs or wires (Chapter 1).
5   EGR valve not functioning (Chapter 6).
6   Vacuum leak (Chapters 2 and 4).

17   Engine runs with oil pressure light on

1   Low oil level (Chapter 1).

2   Idle rpm below specification (Chapter 1).
3   Short in wiring circuit (Chapter 12).
4   Faulty oil pressure sender (Chapter 2).
5   Worn engine bearings and/or oil pump (Chapter 2).

18   Engine diesels (continues to run) after switching off

1   Idle speed too high (Chapter 1).
2   Excessive engine operating temperature (Chapter 3).
3   Excessive carbon deposits on valves and pistons (Chapter 2).

Engine electrical system

19   Battery will not hold a charge

1   Alternator drivebelt defective or not adjusted properly (Chapter 1).
2   Battery electrolyte level low (Chapter 1).
3   Battery terminals loose or corroded (Chapter 1).
4   Alternator not charging properly (Chapter 5).
5   Loose, broken or faulty wiring in the charging circuit (Chapter 5).
6   Short in vehicle wiring (Chapter 12).
7   Internally defective battery (Chapters 1 and 5).

20   Alternator light fails to go out

1   Faulty alternator or charging circuit (Chapter 5).
2   Alternator drivebelt defective or out of adjustment (Chapter 1).
3   Alternator voltage regulator inoperative (Chapter 5).

21   Alternator light fails to come on when key is turned on

1   Warning light bulb defective (Chapter 12).
2   Fault in the printed circuit, dash wiring or bulb holder (Chapter 12).

Fuel system

22   Excessive fuel consumption

1   Dirty or clogged air filter element (Chapter 1).
2   Emissions system not functioning properly (Chapter 6).
3   Fuel injection system not functioning properly (Chapter 4).
4   Low tire pressure or incorrect tire size (Chapter 1).

23   Fuel leakage and/or fuel odor

1   Leaking fuel feed or return line (Chapters 1 and 4).
2   Tank overfilled.
3   Evaporative canister filter clogged (Chapters 1 and 6).
4   Problem with fuel injection system (Chapter 4).

Cooling system

24   Overheating

1   Insufficient coolant in system (Chapter 1).
2   Water pump defective (Chapter 3).
3   Radiator core blocked or grille restricted (Chapter 3).
4   Thermostat faulty (Chapter 3).

5   Electric coolant fan inoperative or blades broken (Chapter 3).
6   Radiator cap not maintaining proper pressure (Chapter 3).

25   Overcooling

1   Faulty thermostat (Chapter 3).
2   Inaccurate temperature gauge sending unit (Chapter 3)

26   External coolant leakage

1   Deteriorated/damaged hoses; loose clamps (Chapters 1 and 3).
2   Water pump defective (Chapter 3).
3   Leakage from radiator core or coolant reservoir bottle (Chapter 3).
4   Engine drain or water jacket core plugs leaking (Chapter 2).

27   Internal coolant leakage

1   Leaking cylinder head gasket (Chapter 2).
2   Cracked cylinder bore or cylinder head (Chapter 2).

28   Coolant loss

1   Too much coolant in system (Chapter 1).
2   Coolant boiling away because of overheating (Chapter 3).
3   Internal or external leakage (Chapter 3).
4   Faulty pressure cap (Chapter 3).

29   Poor coolant circulation

1   Inoperative water pump (Chapter 3).
2   Restriction in cooling system (Chapters 1 and 3).
3   Thermostat sticking (Chapter 3).

Clutch

30   Pedal travels to floor - no pressure or very little resistance

1   Leaking clutch hydraulic release system or air in system (Chapter 8).
2   Broken release bearing or fork (Chapter 8).

31   Unable to select gears

1   Faulty transaxle (Chapter 7).
2   Faulty clutch disc or pressure plate (Chapter 8).
3   Faulty release cylinder or release bearing (Chapter 8).
4   Faulty shift lever assembly or rods (Chapter 8).

32   Clutch slips (engine speed increases with no increase in vehicle speed)

1   Clutch plate worn (Chapter 8).
2   Clutch plate is oil soaked by leaking rear main seal (Chapter 8).
3   Clutch plate not seated (Chapter 8).
4   Warped pressure plate or flywheel (Chapter 8).

5   Weak diaphragm springs (Chapter 8).
6   Clutch plate overheated. Allow to cool.
7   Faulty clutch self-adjusting mechanism (Chapter 8).

33   Grabbing (chattering) as clutch is engaged

1   Oil on clutch plate lining, burned or glazed facings (Chapter 8).
2   Worn or loose engine or transaxle mounts (Chapters 2 and 7).
3   Worn splines on clutch plate hub (Chapter 8).
4   Warped pressure plate or flywheel (Chapter 8).
5   Burned or smeared resin on flywheel or pressure plate (Chapter 8).

34   Transaxle rattling (clicking)

1   Release fork loose (Chapter 8).
2   Low engine idle speed (Chapter 1).

35   Noise in clutch area

Faulty bearing (Chapter 8).

36   Clutch pedal stays on floor

1   Broken release bearing or fork (Chapter 8).
2   Broken or disconnected clutch cable (Chapter 8).

37   High pedal effort

1   Binding clutch cable (Chapter 8).
2   Pressure plate faulty (Chapter 8).

Manual transaxle

38   Knocking noise at low speeds

1   Worn driveaxle constant velocity (CV) joints (Chapter 8).
2   Worn side gear shaft counterbore in differential case (Chapter 7A).*

39   Noise most pronounced when turning

Differential gear noise (Chapter 7A).*

40   Clunk on acceleration or deceleration

1   Loose engine or transaxle mounts (Chapters 2 and 7A).
2   Worn differential pinion shaft in case.*
3   Worn side gear shaft counterbore in differential case (Chapter 7A).*
4   Worn or damaged inner CV joints (Chapter 8).

41   Clicking noise in turns

Worn or damaged outer CV joint (Chapter 8).

42 Vibration

1   Rough wheel bearing (Chapters 1 and 10).
2   Damaged driveaxle (Chapter 8).
3   Out of round tires (Chapter 1).
4   Tire out of balance (Chapters 1 and 10).
5   Worn CV joint (Chapter 8).

43 Noisy in neutral with engine running

1   Damaged input gear bearing (Chapter 7A).*
2   Damaged clutch release bearing (Chapter 8).

44 Noisy in one particular gear

1   Damaged or worn constant-mesh gears (Chapter 7A).*
2   Damaged or worn synchronizers (Chapter 7A).*
3   Bent reverse fork (Chapter 7A).*
4   Damaged fourth speed gear or output gear (Chapter 7A).*
5   Worn or damaged reverse idler gear or idler bushing (Chapter 7A).*

45 Noisy in all gears

1   Insufficient lubricant (Chapter 7A).
2   Damaged or worn bearings (Chapter 7A).*
3   Worn or damaged input gear shaft and/or output gear shaft (Chapter 7A).*

46 Slips out of gear

1   Worn or improperly adjusted linkage (Chapter 7A).
2   Transaxle loose on engine (Chapter 7A).
3   Shift linkage does not work freely, binds (Chapter 7A).
4   Input gear bearing retainer broken or loose (Chapter 7A).*
5   Dirt between clutch cover and engine housing (Chapter 7A).
6   Worn shift fork (Chapter 7A).*

47 Leaks lubricant

1   Driveshaft seals worn (Chapter 7A).
2   Excessive amount of lubricant in transaxle (Chapters 1 and 7A).
3   Loose or broken input gear shaft bearing retainer (Chapter 7A).*
4   Input gear bearing retainer O-ring and/or lip seal damaged (Chapter 7A).*
5   Vehicle speed sensor O-ring leaking (Chapter 7A).

48 Hard to shift

Shift linkage loose or worn (Chapter 7A).
Although the corrective action necessary to remedy the symptoms described is beyond the scope of this manual, the above information should be helpful in isolating the cause of the condition so that the owner can communicate clearly with a professional mechanic.

Automatic transaxle

Note: *Due to the complexity of the automatic transaxle, it is difficult for the home mechanic to properly diagnose and service this component. For problems other than the following, the vehicle should be taken to a dealer or transaxle shop.*

49 Fluid leakage

1   Automatic transaxle fluid is a deep red color. Fluid leaks should not be confused with engine oil, which can easily be blown onto the transaxle by air flow.
2   To pinpoint a leak, first remove all built-up dirt and grime from the transaxle housing with degreasing agents and/or steam cleaning. Then drive the vehicle at low speeds so air flow will not blow the leak far from its source. Raise the vehicle and determine where the leak is coming from. Common areas of leakage are:
a) Pan (Chapters 1 and 7)
b) Dipstick tube (Chapters 1 and 7)
c) Transaxle oil lines (Chapter 7)
d) Speed sensor (Chapter 7)
e) Driveaxle oil seals (Chapter 7)

50 Transaxle fluid brown or has a burned smell

Transaxle fluid overheated (Chapter 1).

51 General shift mechanism problems

1   Chapter 7, Part B, deals with checking and adjusting the shift linkage on automatic transaxles. Common problems which may be attributed to poorly adjusted linkage are:
a) Engine starting in gears other than Park or Neutral.
b) Indicator on shifter pointing to a gear other than the one actually being used.
c) Vehicle moves when in Park.
2   Refer to Chapter 7B for the shift linkage adjustment procedure.

52 Transaxle will not downshift with accelerator pedal pressed to the floor

1   On non-electronically-controlled transmissions, check the throttle valve (TV) cable.
2   On electronically-controlled transaxles, this type of problem - which is caused by a malfunction in the control unit, a sensor or solenoid, or the circuit itself - is beyond the scope of this book. Take the vehicle to a dealer service department or a competent automatic transmission shop.

53 Engine will start in gears other than Park or Neutral

Neutral start switch out of adjustment or malfunctioning (Chapter 7B).

54 Transaxle slips, shifts roughly, is noisy or has no drive in forward or reverse gears

There are many probable causes for the above problems, but the home mechanic should be concerned with only one possibility - fluid level. Before taking the vehicle to a repair shop, check the level and condition of the fluid as described in Chapter 1. Correct the fluid level as necessary or change the fluid and filter if needed. If the problem persists, have a professional diagnose the cause.

Driveaxles

55 Clicking noise in turns

Worn or damaged outboard CV joint (Chapter 8).

56 Shudder or vibration during acceleration

1   Excessive toe-in (Chapter 10).
2   Incorrect spring heights (Chapter 10).
3   Worn or damaged inboard or outboard CV joints (Chapter 8).
4   Sticking inboard CV joint assembly (Chapter 8).

57 Vibration at highway speeds

1   Out of balance front wheels and/or tires (Chapters 1 and 10).
2   Out of round front tires (Chapters 1 and 10).
3   Worn CV joint(s) (Chapter 8).

Brakes

Note: *Before assuming that a brake problem exists, make sure that:*
   a) *The tires are in good condition and properly inflated (Chapter 1).*
   b) *The front end alignment is correct (Chapter 10).*
   c) *The vehicle is not loaded with weight in an unequal manner.*

58 Vehicle pulls to one side during braking

1   Incorrect tire pressures (Chapter 1).
2   Front end out of alignment (have the front end aligned).
3   Front, or rear, tire sizes not matched to one another.
4   Restricted brake lines or hoses (Chapter 9).
5   Malfunctioning drum brake or caliper assembly (Chapter 9).
6   Loose suspension parts (Chapter 10).
7   Loose calipers (Chapter 9).
8   Excessive wear of brake shoe or pad material or disc/drum on one side.

59 Noise (high-pitched squeal when the brakes are applied)

   Front and/or rear disc brake pads worn out. The noise comes from the wear sensor rubbing against the disc (does not apply to all vehicles). Replace pads with new ones immediately (Chapter 9).

60 Brake roughness or chatter (pedal pulsates)

1   Excessive lateral runout (Chapter 9).
2   Uneven pad wear (Chapter 9).
3   Defective disc (Chapter 9).

61 Excessive brake pedal effort required to stop vehicle

1   Malfunctioning power brake booster (Chapter 9).
2   Partial system failure (Chapter 9).
3   Excessively worn pads or shoes (Chapter 9).
4   Piston in caliper or wheel cylinder stuck or sluggish (Chapter 9).
5   Brake pads or shoes contaminated with oil or grease (Chapter 9).
6   Brake disc grooved and/or glazed (Chapter 1).
7   New pads or shoes installed and not yet seated. It will take a while for the new material to seat against the disc or drum.

62 Excessive brake pedal travel

1   Partial brake system failure (Chapter 9).
2   Insufficient fluid in master cylinder (Chapters 1 and 9).
3   Air trapped in system (Chapters 1 and 9).

63 Dragging brakes

1   Incorrect adjustment of brake light switch (Chapter 9).
2   Master cylinder pistons not returning correctly (Chapter 9).
3   Restricted brakes lines or hoses (Chapters 1 and 9).
4   Incorrect parking brake adjustment (Chapter 9).

64 Grabbing or uneven braking action

1   Malfunction of proportioning valve (Chapter 9).
2   Malfunction of power brake booster unit (Chapter 9).
3   Binding brake pedal mechanism (Chapter 9).

65 Brake pedal feels spongy when depressed

1   Air in hydraulic lines (Chapter 9).
2   Master cylinder mounting bolts loose (Chapter 9).
3   Master cylinder defective (Chapter 9).

66 Brake pedal travels to the floor with little resistance

1   Little or no fluid in the master cylinder reservoir caused by leaking caliper piston(s) (Chapter 9).
2   Loose, damaged or disconnected brake lines (Chapter 9).

67 Parking brake does not hold

   Parking brake linkage improperly adjusted (Chapters 1 and 9).

Suspension and steering systems

Note: *Before attempting to diagnose the suspension and steering systems, perform the following preliminary checks:*
   a) *Tires for wrong pressure and uneven wear.*
   b) *Steering universal joints from the column to the rack and pinion for loose connectors or wear.*
   c) *Front and rear suspension and the rack and pinion assembly for loose or damaged parts.*
   d) *Out-of-round or out-of-balance tires, bent rims and loose and/or rough wheel bearings.*

68 Vehicle pulls to one side

1   Mismatched or uneven tires (Chapter 10).
2   Broken or sagging springs (Chapter 10).
3   Wheel alignment out-of-specifications (Chapter 10).
4   Front brake dragging (Chapter 9).

69 Abnormal or excessive tire wear

1   Wheel alignment out-of-specifications (Chapter 10).
2   Sagging or broken springs (Chapter 10).
3   Tire out-of-balance (Chapter 10).
4   Worn strut damper (Chapter 10).
5   Overloaded vehicle.
6   Tires not rotated regularly.

70 Wheel makes a thumping noise

1   Blister or bump on tire (Chapter 10).
2   Improper strut damper action (Chapter 10).

71 Shimmy, shake or vibration

1   Tire or wheel out-of-balance or out-of-round (Chapter 10).
2   Loose or worn wheel bearings (Chapters 1, 8 and 10).
3   Worn tie-rod ends (Chapter 10).
4   Worn lower balljoints (Chapters 1 and 10).
5   Excessive wheel runout (Chapter 10).
6   Blister or bump on tire (Chapter 10).

72 Hard steering

1   Lack of lubrication at balljoints, tie-rod ends and rack and pinion assembly (Chapter 10).
2   Front wheel alignment out-of-specifications (Chapter 10).
3   Low tire pressure(s) (Chapters 1 and 10).

73 Poor returnability of steering to center

1   Lack of lubrication at balljoints and tie-rod ends (Chapter 10).
2   Binding in balljoints (Chapter 10).
3   Binding in steering column (Chapter 10).
4   Lack of lubricant in steering gear assembly (Chapter 10).
5   Front wheel alignment out-of-specifications (Chapter 10).

74 Abnormal noise at the front end

1   Lack of lubrication at balljoints and tie-rod ends (Chapters 1 and 10).
2   Damaged strut mounting (Chapter 10).
3   Worn control arm bushings or tie-rod ends (Chapter 10).
4   Loose stabilizer bar (Chapter 10).
5   Loose wheel nuts (Chapters 1 and 10).
6   Loose suspension bolts (Chapter 10)

75 Wander or poor steering stability

1   Mismatched or uneven tires (Chapter 10).
2   Lack of lubrication at balljoints and tie-rod ends (Chapters 1 and 10).
3   Worn strut assemblies (Chapter 10).
4   Loose stabilizer bar (Chapter 10).
5   Broken or sagging springs (Chapter 10).
6   Wheels out of alignment (Chapter 10).

76 Erratic steering when braking

1   Wheel bearings worn (Chapter 10).
2   Broken or sagging springs (Chapter 10).
3   Leaking wheel cylinder or caliper (Chapter 10).
4   Warped rotors or drums (Chapter 10).

77 Excessive pitching and/or rolling around corners or during braking

1   Loose stabilizer bar (Chapter 10).
2   Worn strut dampers or mountings (Chapter 10).
3   Broken or sagging springs (Chapter 10).
4   Overloaded vehicle.

78 Suspension bottoms

1   Overloaded vehicle.
2   Worn strut dampers (Chapter 10).
3   Incorrect, broken or sagging springs (Chapter 10).

79 Cupped tires

1   Front wheel or rear wheel alignment out-of-specifications (Chapter 10).
2   Worn strut dampers (Chapter 10).
3   Wheel bearings worn (Chapter 10).
4   Excessive tire or wheel runout (Chapter 10).
5   Worn balljoints (Chapter 10).

80 Excessive tire wear on outside edge

1   Inflation pressures incorrect (Chapter 1).
2   Excessive speed in turns.
3   Front end alignment incorrect (excessive toe-in). Have professionally aligned.
4   Suspension arm bent or twisted (Chapter 10).

81 Excessive tire wear on inside edge

1   Inflation pressures incorrect (Chapter 1).
2   Front end alignment incorrect (toe-out). Have professionally aligned.
3   Loose or damaged steering components (Chapter 10).

82 Tire tread worn in one place

1   Tires out-of-balance.
2   Damaged or buckled wheel. Inspect and replace if necessary.
3   Defective tire (Chapter 1).

83 Excessive play or looseness in steering system

1   Wheel bearing(s) worn (Chapter 10).
2   Tie-rod end loose (Chapter 10).
3   Steering gear loose (Chapter 10).
4   Worn or loose steering intermediate shaft (Chapter 10).

84 Rattling or clicking noise in steering gear

1   Steering gear loose (Chapter 10).
2   Steering gear defective.

Chapter 1
Tune-up and routine maintenance

Contents

Specifications

Recommended lubricants and fluids

Note: *Listed here are manufacturer recommendations at the time this manual was written. Manufacturers occasionally upgrade their fluid and lubricant specifications, so check with your local auto parts store for current recommendations.*

Engine oil	
Type	API grade SG, SG/CC or SG/CD multigrade and fuel-efficient oil
Viscosity	See accompanying chart
Automatic transaxle fluid	Dexron IIE or III Automatic Transaxle Fluid (ATF)
Manual transaxle fluid	Syncromesh Transmission fluid (GM part number 12345349)
Engine coolant	50/50 mixture of water and the specified ethylene glycol-based (green color) antifreeze or "DEX-COOL, silicate-free (orange-color) coolant - DO NOT mix the two types
Brake and clutch fluid	DOT 3 fluid
Power steering fluid	GM power steering fluid or equivalent
Chassis grease	SAE NLGI no. 2 chassis grease

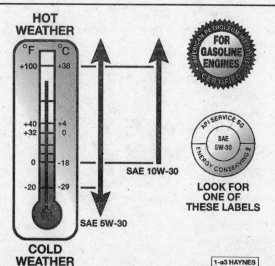

Engine oil viscosity chart - For best fuel economy and cold starting, select the lowest SAE viscosity grade for the expected temperature range

1-a3 HAYNES

Capacities*

Engine oil (with filter change)	4.0 to 4.5 quarts
Fuel tank	15.2 gallons
Cooling system	
2.2L OHV engine	10.3 quarts
2.3 and 2.4L OHC engines	10.7 quarts
Automatic transaxle (fluid and filter replacement (torque converter not included)	4.0 quarts
Manual transaxle	2.0 quarts

All capacities approximate. Add as necessary to bring to appropriate level.

Ignition system

Spark plug type and gap	
2.2L OHV engine	
1998 and earlier	AC type 41-928 or equivalent @ 0.060 in.
1999	AC type 41-928 or equivalent @ 0.050 in.
2.3L and 2.4L OHC engines	
1995 thru 1997	AC type 41-910 or equivalent @ 0.060 in.
1998 and later	AC type 41-942 or equivalent @ 0.050 in.
Firing order	1-3-4-2

General

Coolant reservoir pressure cap rating	15 psi
Brake pad wear limit	1/8 inch

Torque specifications

Ft-lbs (unless otherwise indicated)

Automatic transaxle pan bolts	89 in-lbs
Engine oil drain plug	15 to 20
Spark plugs	11
Throttle body nuts/bolts	16
Wheel lug nuts	100

1995 through 1997 2.2L OHV engine

38016-1-2A HAYNES

1998 and later 2.2L OHV engine

38016-1-2B HAYNES

2.3L and 2.4L OHC engines

38016-1-2C HAYNES

Cylinder and coil terminal location

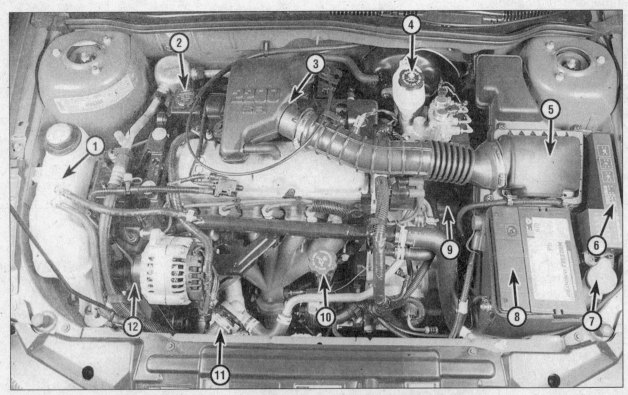

Typical engine compartment component location - 2.2L OHV engine

1 Pressurized engine coolant reservoir
2 Power steering fluid reservoir
3 Throttle body (under air intake assembly)
4 Brake master cylinder reservoir
5 Air filter housing
6 Fuse and relay block
7 Windshield washer reservoir
8 Battery
9 Automatic transaxle check plug (located below master cylinder on transaxle)
10 Engine oil dipstick/fill tube
11 Upper radiator hose
12 Drivebelt

Typical front underside component location

1	Drivebelt	3	Front brake caliper	5	Automatic transaxle fluid pan
2	Strut assembly	4	Tie-rod end	6	Engine oil drain plug

Typical rear underside component location

1	Coil-over-shock assembly	3	Fuel filter	5	Exhaust pipe
2	Axle assembly	4	Fuel tank	6	Muffler

1 Chevrolet Cavalier and Pontiac Sunfire Maintenance schedule

The following maintenance intervals are based on the assumption that the vehicle owner will be doing the maintenance or service work, as opposed to having a dealer service department do the work. Although the time/mileage intervals are loosely based on factory recommendations, most have been shortened to ensure, for example, that such items as lubricants and fluids are checked/changed at intervals that promote maximum engine/driveline service life. Also, subject to the preference of the individual owner interested in keeping his or her vehicle in peak condition at all times, and with the vehicle's ultimate resale in mind, many of the maintenance procedures may be performed more often than recommended in the following schedule. We encourage such owner initiative.

When the vehicle is new it should be serviced initially by a factory authorized dealer service department to protect the factory warranty. In many cases the initial maintenance check is done at no cost to the owner (check with your dealer service department for more information).

Every 250 miles or weekly, whichever comes first

Check the engine oil level (Section 4)
Check the engine coolant level (Section 4)
Check the windshield washer fluid level (Section 4)
Check the brake fluid level (Section 4)
Check the tires and tire pressures (Section 5)

Every 3000 miles or 3 months, whichever comes first

All items listed above plus:
Check the power steering fluid level (Section 6)
Check the automatic transaxle fluid level (Section 7)
Change the engine oil and filter (Section 9)
Lubricate the chassis (Section 10)

Every 6000 miles or 6 months, whichever comes first

All items listed above plus:
Check and service the battery (Section 11)
Check the cooling system (Section 12)
Inspect and replace, if necessary, all underhood hoses (Section 13)
Check the engine drivebelt (Section 14)
Inspect the suspension and steering components (Section 15)
Check the brakes (Section 16)*
Rotate the tires (Section 17)
Inspect the exhaust system (Section 18)

Every 12,000 miles or 12 months, whichever comes first

Check the throttle body nut torque (Section 19)
Inspect and replace, if necessary, the windshield wiper blades (Section 20)
Inspect the seat belts (Section 21)
Replace the air filter and PCV inlet filter (Section 22)

Every 30,000 miles or 24 months, whichever comes first

All items listed above plus:
Inspect and replace, if necessary, the PCV valve (2.2L) or oil/air separator (2.3L and 2.4L) (Section 23)
Inspect and replace if necessary, the spark plug wires (2.2L) (Section 25)
Check the EGR valve (Section 24)
Replace the fuel filter (Section 26)
Inspect the fuel system (Section 27)
Change the automatic transaxle fluid and filter (Section 29)**
Service the cooling system (drain, flush and refill) (green-colored ethylene glycol anti-freeze only) (Section 28)

Every 50,000 miles or 3 years, whichever comes first

Change the fluid in the automatic transaxle (Section 29)
Change the fluid in the manual transaxle (Section 30)

Every 100,000 miles or 5 years, whichever comes first

Replace the spark plugs (platinum-tipped spark plugs) (Section 31)
Service the cooling system (drain, flush and refill) (orange-colored "DEX-COOL" silicate-free coolant only) (Section 28)

* If the vehicle frequently tows a trailer, is operated primarily in stop-and-go conditions or its brakes receive severe usage for any other reason, check the brakes every 3000 miles or three months.
** If operated under one or more of the following conditions, change the automatic transaxle fluid every 15,000 miles:
In heavy city traffic where the outside temperature regularly reaches 90-degrees F (32-degrees C) or higher
In hilly or mountainous terrain
Frequent trailer pulling

2   Introduction

This Chapter is designed to help the home mechanic maintain the Chevrolet Cavalier and the Pontiac Sunfire models with the goals of maximum performance, economy, safety and reliability in mind.

Included is a master maintenance schedule (page 1-5), followed by procedures dealing specifically with each item on the schedule. Visual checks, adjustments, component replacement and other helpful items are included. Refer to the **accompanying illustrations** of the engine compartment and the underside of the vehicle for the locations of various components.

Servicing your vehicle in accordance with the mileage/time maintenance schedule and the step-by-step procedures will result in a planned maintenance program that should produce a long and reliable service life. Keep in mind that it's a comprehensive plan, so maintaining some items but not others at the specified intervals will not produce the same results.

As you service your vehicle, you'll discover that many of the procedures can - and should - be grouped together because of the nature of the particular procedure you're performing or because of the close proximity of two otherwise unrelated components to one another.

For example, if the vehicle is raised, you should inspect the exhaust, suspension, steering and fuel systems while you're under the vehicle. When you're rotating the tires, it makes good sense to check the brakes since the wheels are already removed. Finally, let's suppose you have to borrow or rent a torque wrench. Even if you only need it to tighten the spark plugs, you might as well check the torque of as many critical fasteners as time allows.

The first step in this maintenance program is to prepare yourself before the actual work begins. Read through all the procedures you're planning to do, then gather up all the parts and tools needed. If it looks like you might run into problems during a particular job, seek advice from a mechanic or an experienced do-it-yourselfer.

Caution: *If the vehicle is equipped with a Delco Loc II or Theftlock audio system, make sure you have the correct activation code before disconnecting the battery.*

3   Tune-up general information

The term tune-up is used in this manual to represent a combination of individual operations rather than one specific procedure. If, from the time the vehicle is new, the routine maintenance schedule is followed closely and frequent checks are made of fluid levels and high wear items, as suggested throughout this manual, the engine will be kept in relatively good running condition and the need for additional work will be minimized due to lack of regular maintenance. This is even more likely if a used vehicle, which has not received regular and frequent mainte-

nance checks, is purchased. In such cases, an engine tune-up will be needed outside of the regular routine maintenance intervals.

The first step in any tune-up or diagnostic procedure to help correct a poor running engine is a cylinder compression check. A compression check (see Chapter 2, Part C) will help determine the condition of internal engine components and should be used as a guide for tune-up and repair procedures. If, for instance, a compression check indicates serious internal engine wear, a conventional tune-up won't improve the performance of the engine and would be a waste of time and money. Because of its importance, the compression check should be done by someone with the right equipment and the knowledge to use it properly.

The following procedures are those most often needed to bring a generally poor running engine back into a proper state of tune.

Minor tune-up

Check all engine related fluids (Section 4)
Clean, inspect and test the battery (Section 11)
Check the cooling system (Section 12)
Check all underhood hoses (Section 13)
Check and adjust the drivebelts (Section 14)
Check the air filter (Section 22)
Check the PCV valve (Section 23)
Replace the spark plugs (Section 31)
Inspect the spark plug wires (Section 25)

Major tune-up

All items listed under Minor tune-up plus . . .
Replace the air filter (Section 22)
Replace the spark plug wires (Section 25)
Replace the fuel filter (Section 26)
Check the fuel system (Section 27)
Check the ignition timing (Chapter 5)
Check the charging system (Chapter 5)
Check the EGR system (Chapter 6)

4   Fluid level checks (every 250 miles or weekly)

Note: *The electronic instrument cluster on the dash utilizes several warning lamps to indicate low oil level (CHECK OIL), oil pressure problems (LOW OIL PRESSURE), low coolant level (LOW COOLANT LEVEL) and fuel/emissions systems failure (CHECK ENGINE). Refer to the appropriate Chapters and Sections to check the systems before serious engine trouble develops. Low fluid level checks can be remedied with proper maintenance. Refer to Chapter 2C for oil pressure checks and to Chapter 6 for fuel/emissions systems diagnostic procedures.*

Note: *The following are fluid level checks to be done on a 250 mile or weekly basis. Additional fluid level checks can be found in specific maintenance procedures which follow. Regardless of intervals, be alert to fluid leaks under the vehicle which would indicate a problem to be corrected immediately.*

1   Fluids are an essential part of the lubrication, cooling, brake and windshield washer systems. Because the fluids gradually become depleted and/or contaminated during normal operation of the vehicle, they must be periodically replenished. See *Recommended lubricants and fluids* at the beginning of this Chapter before adding fluid to any of the following components. **Note:** *The vehicle must be on level ground when fluid levels are checked.*

Engine oil

Refer to illustrations 4.2, 4.4 and 4.6

2   The engine oil level is checked with a dipstick **(see illustration)**. The dipstick extends through a metal tube down into the oil pan. **Note:** *2.2L OHV models are equipped with an oil fill cap with a dipstick attached as an assembly. 2.3L and 2.4L OHC engines are equipped with a separate oil fill tube and dipstick.*

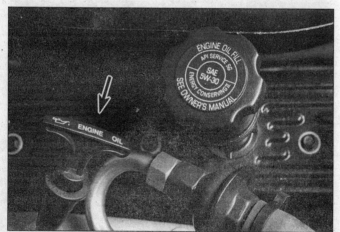

4.2 The engine oil dipstick (arrow) is located on the front side of the engine

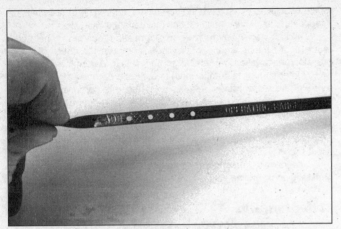

4.4 The oil level should be at or near the upper hole or cross-hatched area on the dipstick - if it's below the ADD line, add enough oil to bring the level into the upper hole or cross-hatched area

3   The oil level should be checked before the vehicle has been driven, or about 15 minutes after the engine has been shut off. If the oil is checked immediately after driving the vehicle, some of the oil will remain in the upper part of the engine, resulting in an inaccurate reading on the dipstick.

4   Pull the dipstick from the tube and wipe all the oil from the end with a clean rag or paper towel. Insert the clean dipstick all the way back into the tube and pull it out again. Note the oil at the end of the dipstick. Add oil as necessary to keep the level above the ADD mark in the cross hatched area of the dipstick **(see illustration)**.

5   Do not overfill the engine by adding too much oil since this may result in oil fouled spark plugs, oil leaks or oil seal failures.

6   Oil is added to the engine after removing a twist-off cap located on the engine **(see illustration)**. A funnel may help to reduce spills.

7   Checking the oil level is an important preventive maintenance step. A consistently low oil level indicates oil leakage through damaged seals, defective gaskets or past worn rings or valve guides. If the oil looks milky in color or has water droplets in it, the cylinder head gasket may be blown or the head or block may be cracked. The engine should be checked immediately. The condition of the oil should also be checked. Whenever you check the oil level, slide your thumb and index finger up the dipstick before wiping off the oil. If you see small dirt or metal particles clinging to the dipstick, the oil should be changed (see Section 9).

Engine coolant

Refer to illustration 4.9

Warning: *Do not allow antifreeze to come in contact with your skin or painted surfaces of the vehicle. Flush contaminated areas immediately with plenty of water. Do not store new coolant or leave old coolant lying around where it's accessible to children or pets - they're attracted by its sweet smell. Ingestion of even a small amount of coolant can be fatal! Wipe up garage floor and drip pan coolant spills immediately. Keep antifreeze containers covered and repair leaks in the cooling system immediately.*

Caution: *The manufacturer recommends using only DEX-COOL coolant for these systems. DEX-COOL is a long-lasting coolant designed for 100,000 miles or 5 years. Never mix green-colored ethylene glycol antifreeze and orange-colored "DEX-COOL" silicate-free coolant because doing so will destroy the efficiency of the "DEX-COOL".*

8   The models covered by this manual are equipped with a pressurized coolant recovery system. A plastic coolant reservoir located at the front of the engine compartment is connected by a hose to the radiator assembly. As the engine warms up and the coolant expands, it escapes through a valve in the radiator cap and travels from the radiator through the hose into the reservoir. As the engine cools, the coolant is automatically drawn back into the cooling system to maintain the correct level.

4.6 The engine oil filler cap/dipstick assembly (arrow) is clearly marked and threads into the tube on the front of the engine - turn it counter clockwise to remove it (2.2L OHV engine shown)

9   The coolant level in the reservoir should be checked regularly. **Warning:** *Do not remove the reservoir cap to check the coolant level when the engine is warm. The level in the reservoir varies with the temperature of the engine. When the engine is cold, the coolant level should be at or slightly above the FULL COLD mark on the reservoir* **(see illustration)**. *Once the engine has warmed up, the level should be at or near the FULL HOT mark. If it isn't, add coolant to the reservoir. Unscrew the cap from the reservoir and add a 50/50 mixture of ethylene glycol based green-colored antifreeze or orange-colored "DEX-COOL" silicate-free coolant and water* (see **Caution** above).

10   Drive the vehicle and recheck the coolant level. If only a small amount of coolant is required to bring the system up to the proper level, water can be used. However, repeated additions of water will dilute the antifreeze and water solution. In order to maintain the proper ratio of antifreeze and water, always top up the coolant level with the correct mixture. An empty plastic milk jug or bleach bottle makes an excellent container for mixing coolant. Do not use rust inhibitors or additives.

11   If the coolant level drops consistently, there may be a leak in the system. Inspect the radiator, hoses, filler cap, drain plugs and water pump (see Section 12). If no leaks are noted, have the coolant reservoir cap pressure tested by a service station.

12   If you have to remove the reservoir cap, wait until the engine has cooled completely, then wrap a thick cloth around the cap and turn it to the first stop. If coolant or steam escapes, let the engine cool down longer, then remove the cap.

13   Check the condition of the coolant as well. It should be relatively

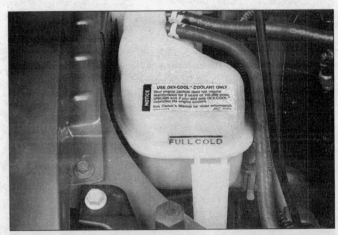

4.9 The coolant reservoir is located in the right (passenger's side) front corner of the engine compartment. The coolant level can be checked by observing the level through the translucent plastic

clear. If it is brown or rust colored, the system should be drained, flushed and refilled. Even if the coolant appears to be normal, the corrosion inhibitors wear out, so it must be replaced at the specified intervals.

Windshield washer fluid

Refer to illustration 4.14

14   Fluid for the windshield washer system is located in a plastic reservoir on the left side of the engine compartment **(see illustration)**. In milder climates, plain water can be used in the reservoir, but it should be kept no more than two-thirds full to allow for expansion if the water freezes. In colder climates, use windshield washer system antifreeze, available at any auto parts store, to lower the freezing point of the fluid. Mix the antifreeze with water in accordance with the manufacturer's directions on the container. **Caution:** *Do not use cooling system antifreeze - it will damage the vehicle's paint.*

15   To help prevent icing in cold weather, warm the windshield with the defroster before using the washer.

Battery electrolyte

Refer to illustration 4.16

16   All vehicles covered by this manual are equipped with a battery which is permanently sealed (except for vent holes) and has no filler caps. Water does not have to be added to these batteries at any time; however, if a maintenance-type battery has been installed on the vehicle since it was new, remove all the cell caps on top of the battery **(see illustration)** (usually there are two caps that cover three cells each). If the electrolyte level is low, add distilled water until the level is above the plates. There is usually a split-ring indicator in each cell to help you judge when enough water has been added. Add water until the electrolyte level is just up to the bottom of the split ring indicator. Do not overfill the battery or it will spew out electrolyte when it is charging.

Brake fluid

Refer to illustration 4.18

17   The brake fluid level is checked by looking through the plastic reservoir mounted on the master cylinder. The master cylinder is mounted on the front of the power booster unit in the left (driver's side) rear corner of the engine compartment.

18   The fluid level should be between the MAX and MIN lines on the side of the reservoir **(see illustration)**. If the fluid level is low, wipe the top of the reservoir and the lid with a clean rag to prevent contamination of the system as the lid is pried off.

19   When adding fluid, pour it carefully into the reservoir to avoid spilling it on surrounding painted surfaces. Be sure the specified fluid is used, since mixing different types of brake fluid can cause damage to the system. See *Recommended lubricants and fluids* at the front of this Chapter or your owner's manual. **Warning:** *Brake fluid can harm your eyes and damage painted surfaces, so use extreme caution when handling or pouring it. Do not use brake fluid that has been standing open or*

4.14 Flip the windshield washer fluid cap up to add fluid

is more than one year old. Brake fluid absorbs moisture from the air. Excess moisture can cause a dangerous loss of braking effectiveness.

20   At this time the fluid and master cylinder can be inspected for contamination. The system should be drained and refilled if deposits, dirt particles or water droplets are seen in the fluid.

21   After filling the reservoir to the proper level, make sure the lid completely snaps in place to prevent fluid leakage.

22   The brake fluid level in the master cylinder will drop slightly as the pads at each wheel wear down during normal operation. If the master cylinder requires repeated replenishing to keep it at the proper level, this is an indication of leakage in the brake system, which should be corrected immediately. Check all brake lines and connections (see Section 16 for more information).

23   If, when checking the master cylinder fluid level, you discover one or both reservoirs empty or nearly empty, the brake system should be bled (see Chapter 9).

5   Tire and tire pressure checks (every 250 miles or weekly)

Refer to illustrations 5.2, 5.3, 5.4a, 5.4b and 5.8

1   Periodic inspection of the tires may spare you the inconvenience of being stranded with a flat tire. It can also provide you with vital information regarding possible problems in the steering and suspension systems before major damage occurs.

2   The original tires on this vehicle are equipped with 1/2-inch wide bands that appear when tread depth reaches 1/16-inch, indicating the tires are worn out. Tread wear can be monitored with a simple, inexpensive device known as a tread depth indicator **(see illustration)**.

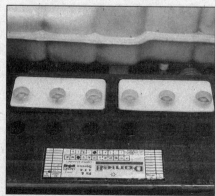

4.16 If equipped with a maintenance type battery, remove the cell caps to check the battery electrolyte level

4.18 The fluid level inside the brake fluid reservoir can easily be checked by observing the level from the outside - fluid can be added to the reservoir after the cover is removed by prying up on the tabs

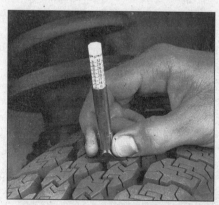

5.2 Use a tire tread depth gauge to monitor tire wear - they are available at auto parts stores and service stations and cost very little

UNDERINFLATION

CUPPING

Cupping may be caused by:

- **Underinflation and/or mechanical irregularities such as out-of-balance condition of wheel and/or tire, and bent or damaged wheel.**
- **Loose or worn steering tie-rod or steering idler arm.**
- **Loose, damaged or worn front suspension parts.**

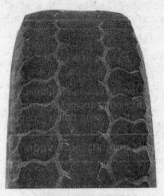

OVERINFLATION

INCORRECT TOE-IN OR EXTREME CAMBER

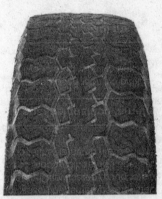

FEATHERING DUE TO MISALIGNMENT

5.3 This chart will help you determine the condition of the tires, the probable cause(s) of abnormal wear and the corrective action necessary

3   Note any abnormal tread wear **(see illustration)**. Tread pattern irregularities such as cupping, flat spots and more wear on one side than the other are indications of front end alignment and/or balance problems. If any of these conditions are noted, take the vehicle to a tire shop or service station to correct the problem.

4   Look closely for cuts, punctures and embedded nails or tacks. Sometimes a tire will hold air pressure for short time or leak down very slowly after a nail has embedded itself in the tread. If a slow leak persists, check the valve stem core to make sure it's tight **(see illustration)**. Examine the tread for an object that may have embedded itself in the tire or for a "plug" that may have begun to leak (radial tire punctures are repaired with a plug that's installed in a puncture). If a puncture is suspected, it can be easily verified by spraying a solution of soapy water onto the suspected area **(see illustration)**. The soapy solution will bubble if there's a leak. Unless the puncture is unusually large, a tire shop or service station can usually repair the tire.

5   Carefully inspect the inner sidewall of each tire for evidence of brake fluid. If you see any, inspect the brakes immediately.

6   Correct air pressure adds miles to the lifespan of the tires, improves mileage and enhances overall ride quality. Tire pressure cannot be accurately estimated by looking at a tire, especially if it's a radial. A tire pressure gauge is essential. Keep an accurate gauge in the vehicle. The pressure gauges attached to the nozzles of air hoses at gas stations are often inaccurate.

7   Always check tire pressure when the tires are cold. Cold, in this case, means the vehicle has not been driven over a mile in the three hours preceding a tire pressure check. A pressure rise of four to eight

pounds is not uncommon once the tires are warm.

8   Unscrew the valve cap protruding from the wheel or hubcap and push the gauge firmly onto the valve stem **(see illustration)**. Note the reading on the gauge and compare the figure to the recommended tire pressure shown on the label attached to the inside of the glove compartment door. Be sure to reinstall the valve cap to keep dirt and moisture out of the valve stem mechanism. Check all four tires and, if necessary, add enough air to bring them up to the recommended pressure.

9   Don't forget to keep the spare tire inflated to the specified pressure (refer to your owner's manual or the tire sidewall).

6   Power steering fluid level check (every 3000 miles or 3 months)

Refer to illustrations 6.2 and 6.6

1   Unlike manual steering, the power steering system relies on fluid which may, over a period of time, require replenishing.

2   The fluid reservoir for the power steering pump is located on the passenger side of the engine compartment **(see illustration)**.

3   For the check, the front wheels should be pointed straight ahead and the engine should be off.

4   Use a clean rag to wipe off the reservoir cap and the area around the cap. This will help prevent any foreign matter from entering the reservoir during the check.

5   Twist off the cap and check the temperature of the fluid at the end of the dipstick with your finger.

5.4a If a tire loses air on a steady basis, check the valve core first to make sure it's snug (special inexpensive wrenches are commonly available at auto parts stores)

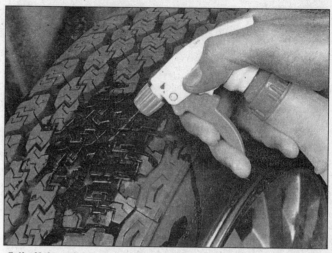

5.4b If the valve core is tight, raise the corner of the vehicle with the low tire and spray a soapy water solution onto the tread as the tire is turned slowly - leaks will cause small bubbles to appear

6   Wipe off the fluid with a clean rag, reinsert it, then withdraw it and read the fluid level. The level should be at the HOT mark if the fluid was hot to the touch **(see illustration)**. It should be at the COLD mark if the fluid was cool to the touch.

7   If additional fluid is required, pour the specified type directly into the reservoir, using a funnel to prevent spills.

8   If the reservoir requires frequent fluid additions, all power steering hoses, hose connections, the power steering pump and the rack and pinion assembly should be carefully checked for leaks.

7   Automatic transaxle fluid level check (every 3000 miles or 3 months)

1   The automatic transaxle fluid level should be carefully maintained. Low fluid level can lead to slipping or loss of drive, while overfilling can cause foaming and loss of fluid.

2   With the parking brake set, start the engine, then move the shift lever through all the gear ranges, ending in Park. The fluid level must be checked with the transaxle at operating temperature and sitting level. **Note:** *Incorrect fluid level readings will result if the vehicle has just been driven at high speeds for an extended period, in hot weather in city traffic, or if it has been pulling a trailer. If any of these conditions apply, wait until the fluid has cooled (about 30 minutes).*

5.8 To extend the life of the tires, check the air pressure at least once a week with an accurate gauge (don't forget the spare)

6.2 The power steering fluid reservoir is located on the right (passenger's side) of the engine compartment - turn the cap counterclockwise for removal

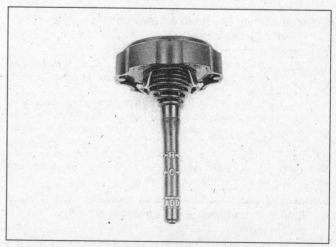

6.6 The marks on the dipstick indicate the safe fluid range

7.6 With the automatic transaxle at normal operating temperature, the fluid level must be maintained within the cross-hatched area on the dipstick

3-speed automatic transaxle

Refer to illustration 7.6

3   With the engine idling and the transaxle at normal operating temperature (driven 15 miles or idling five minutes), remove the dipstick from the filler tube. The dipstick is located at the rear section of the transaxle.
4   Wipe the fluid from the dipstick with a clean rag and push it back into the filler tube until the cap seats.
5   Pull the dipstick out again and note the fluid level.
6   The level should be within the cross-hatched upper areas on the dipstick **(see illustration)**. If additional fluid is required, pour it directly into the tube using a funnel. It takes about one pint to raise the level from the lower mark to the upper edge of the cross-hatched area with the transaxle at normal operating temperature, so add the fluid a little at a time and keep checking the level until it's correct.
7   The condition of the fluid should also be checked along with the level. If the fluid at the end of the dipstick is a dark reddish-brown color, or if the fluid has a burned smell, the fluid should be changed. If you're in doubt about the condition of the fluid, purchase some new fluid and compare the two for color and smell.

4-speed automatic transaxle

On these models, it is not necessary to check the fluid level at regular intervals. If leaks or shifting problems lead you to suspect a low fluid level, refer to Section 29.

---

8   Manual transaxle lubricant level check (3,000 miles or 3 months)

1   The manual transaxle fluid level should be carefully maintained. Low fluid level can lead to gear and synchromesh scalding, chipping and gear damage.
2   With the engine OFF, the vehicle level and the engine temperature cold, remove the indicator from the fluid level vent plug and check the increments. If the oil level reads "ADD", it will be necessary to add the correct amount.
3   Use a narrow funnel designed to fit into the fill vent tube and slowly add the necessary transaxle fluid. Refer to the Specifications listed in this Chapter for the correct fluid type.

---

9   Engine oil and filter change (every 3000 miles or 3 months)

Refer to illustrations 9.2, 9.7, 9.12 and 9.14
1   Frequent oil changes are the best preventive maintenance the home mechanic can give the engine, because aging oil becomes

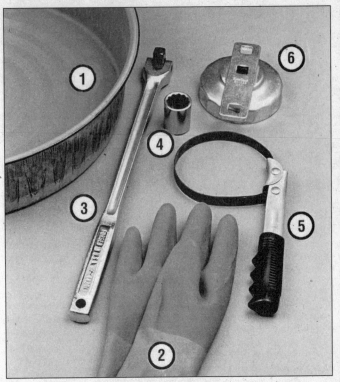

9.2 These tools are required when changing the engine oil and filter

1   *Drain pan - It should be fairly shallow in depth, but wide to prevent spills*
2   *Rubber gloves - When removing the drain plug and filter, you will get oil on your hands (the gloves will prevent burns)*
3   *Breaker bar - Sometimes the oil drain plug is tight, and a long breaker bar is needed to loosen it*
4   *Socket - To be used with the breaker bar or a ratchet (must be the correct size to fit the drain plug)*
5   *Filter wrench - This is a metal band-type wrench, which requires clearance around the filter to be effective*
6   *Filter wrench - This type fits on the bottom of the filter and can be turned with a ratchet or breaker bar (different-size wrenches are available for different types of filters)*

diluted and contaminated, which leads to premature engine wear.
2   Make sure you have all the necessary tools before you begin this procedure **(see illustration)**. You should also have plenty of rags or newspapers handy for mopping up any spills.
3   Access to the underside of the vehicle is greatly improved if the vehicle can be lifted on a hoist, driven onto ramps or supported by jackstands. **Warning:** *Do not work under a vehicle which is supported only by a hydraulic or scissors-type jack.*
4   If this is your first oil change, get under the vehicle and familiarize yourself with the locations of the oil drain plug and the oil filter. The engine and exhaust components will be warm during the actual work, so try to anticipate any potential problems before the engine and accessories are hot.
5   Park the vehicle on a level spot. Start the engine and allow it to reach its normal operating temperature. Warm oil and sludge will flow out more easily. Turn off the engine when it's warmed up. Remove the filler cap from the valve cover.
6   Raise the vehicle and support it securely on jackstands. **Warning:** *Never get beneath the vehicle when it is supported only by a jack. The jack provided with your vehicle is designed solely for raising the vehicle to remove and replace the wheels. Always use jackstands to support the vehicle when it becomes necessary to place your body underneath the vehicle.*

9.7 The engine oil drain plug is located at the rear of the oil pan - it is usually very tight, so use a box-end wrench to avoid rounding off the hex

9.12 The oil filter is usually on very tight as well and will require a special wrench for removal - DO NOT use the wrench to tighten the new filter!

9.14 Lubricate the oil filter gasket with clean engine oil before installing the filter on the engine

1

7   Being careful not to touch the hot exhaust components, place the drain pan under the drain plug in the bottom of the pan and remove the plug **(see illustration)**. You may want to wear gloves while unscrewing the plug the final few turns if the engine is hot.

8   Allow the old oil to drain into the pan. It may be necessary to move the pan farther under the engine as the oil flow slows to a trickle. Inspect the old oil for the presence of metal shavings and chips.

9   After all the oil has drained, wipe off the drain plug with a clean rag. Even minute metal particles clinging to the plug would immediately contaminate the new oil.

10   Clean the area around the drain plug opening, reinstall the plug and tighten it securely, but do not strip the threads.

11   Move the drain pan into position under the oil filter.

12   Loosen the oil filter **(see illustration)** by turning it counterclockwise with the filter wrench. Use a quality filter wrench of the correct size and be careful not to collapse the canister as you apply pressure. Once the filter is loose, use your hands to unscrew it from the block. Just as the filter is detached from the block, immediately tilt the open end up to prevent the oil inside the filter from spilling out. **Warning:** *The exhaust system may still be hot, so be careful.*

13   With a clean rag, wipe off the mounting surface on the block. If a residue of old oil is allowed to remain, it will smoke when the block is heated up. Also make sure that none of the old gasket remains stuck to the mounting surface. It can be removed with a scraper if necessary.

14   Compare the old filter with the new one to make sure they are the same type. Smear some clean engine oil on the rubber gasket of the new filter and screw it into place **(see illustration)**. Because over-tightening the filter will damage the gasket, do not use a filter wrench to tighten the filter. Tighten it by hand until the gasket contacts the seating surface. Then seat the filter by giving it an additional 3/4-turn.

15   Remove all tools, rags, etc. from under the vehicle, being careful not to spill the oil in the drain pan, then lower the vehicle.

16   Add new oil to the engine through the oil filler cap in the valve cover. Use a funnel, if necessary, to prevent oil from spilling onto the top of the engine. Pour three quarts of fresh oil into the engine. Wait a few minutes to allow the oil to drain into the pan, then check the level on the oil dipstick (see Section 4 if necessary). If the oil level is at or near the upper hole on the dipstick, install the filler cap hand tight, start the engine and allow the new oil to circulate.

17   Allow the engine to run for about a minute. While the engine is running, look under the vehicle and check for leaks at the oil pan drain plug and around the oil filter. If either is leaking, stop the engine and tighten the plug or filter.

18   Wait a few minutes to allow the oil to trickle down into the pan, then recheck the level on the dipstick and, if necessary, add enough oil to bring the level to the upper hole.

19   During the first few trips after an oil change, make it a point to check frequently for leaks and proper oil level.

20   The old oil drained from the engine cannot be re-used in its present state and should be discarded. Check with your local refuse disposal company, disposal facility or environmental agency to see whether they will accept the oil for recycling. Don't pour used oil into drains or onto the ground. After the oil has cooled, it can be drained into a suitable container (capped plastic jugs, topped bottles, milk cartons, etc.) for transport to one of these disposal sites.

10 Chassis lubrication (every 3000 miles or 3 months)

Refer to illustration 10.1

1   Refer to *Recommended lubricants and fluids* at the front of this Chapter to obtain the necessary grease, etc. You'll also need a grease gun **(see illustration)**. Occasionally plugs will be installed rather than grease fittings. If so, grease fittings will have to be purchased and installed.

2   Look under the vehicle and see if grease fittings or plugs are

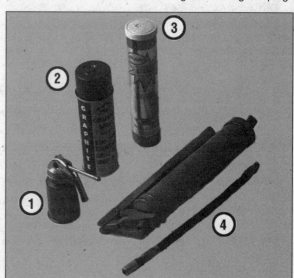

10.1 Materials required for chassis and body lubrication

1   *Engine oil - Light engine oil in a can like this can be used for door and hood hinges*

2   *Graphite spray - Used to lubricate lock cylinders*

3   *Grease - Grease, in a variety of types and weights, is available for use in a grease gun. Check the Specification for your requirements*

4   *Grease gun - A common grease gun, shown here with a detachable hose and nozzle, is needed for chassis lubrication. After use, clean it thoroughly!*

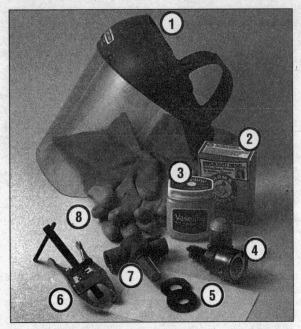

11.1 Tools and materials required for battery maintenance

1   **Face shield/safety goggles -** *When removing corrosion with a brush, the acidic particles can easily fly up into your eyes*
2   **Rubber gloves -** *Another safety item to consider when servicing the battery - remember that's acid inside the battery!*
3   **Battery terminal/cable cleaner -** *This wire brush cleaning tool will remove all traces of corrosion from the battery and cable*
4   **Treated felt washers -** *Placing one of these on each terminal, directly under the cable end, will help prevent corrosion (be sure to get the correct type for side-terminal batteries)*
5   **Baking soda -** *A solution of baking soda and water can be used to neutralize corrosion*
6   **Petroleum jelly -** *A layer of this on the battery terminal bolts will help prevent corrosion*

installed. If there are plugs, remove them and buy grease fittings, which will thread into the component. A dealer or auto parts store will be able to supply the correct fittings. Straight, as well as angled, fittings are available.
3   For easier access under the vehicle, raise it with a jack and place jackstands under the frame. Make sure it's safely supported by the stands. If the wheels are to be removed at this interval for tire rotation or brake inspection, loosen the lug nuts slightly while the vehicle is still on the ground.
4   Before beginning, force a little grease out of the nozzle to remove any dirt from the end of the gun. Wipe the nozzle clean with a rag.
5   With the grease gun and plenty of clean rags, crawl under the vehicle and begin lubricating the components.
6   Wipe one of the grease fittings clean and push the nozzle firmly over it. Pump the gun until the balljoint rubber seal is firm to the touch. Do not pump too much grease into the fitting as it could rupture the seal.
7   Wipe the excess grease from the components and the grease fitting. Repeat the procedure for the remaining fitting.
8   Clean and lubricate the parking brake cable, along with the cable guides and levers. This can be done by smearing some of the chassis grease onto the cable and its related parts with your fingers.
9   Open the hood and smear a little chassis grease on the hood latch mechanism. Have an assistant pull the hood release lever from inside the vehicle as you lubricate the cable at the latch.
10   Lubricate all the hinges (door, hood, etc.) with engine oil to keep them in proper working order.
11   The key lock cylinders can be lubricated with spray graphite or silicone lubricant, which is available at auto parts stores.
12   Lubricate the door weatherstripping with silicone spray. This will reduce chafing and retard wear.

11.4 Check the tightness of the battery cable terminal bolts (arrows)

---

11   Battery check, maintenance and charging (every 6000 miles or 6 months)

Refer to illustrations 11.1, 11.4, 11.5a, 11.5b and 11.5c
Warning: *Hydrogen gas is produced by the battery, so keep open flames and lighted tobacco away from it at all times. Always wear eye protection when working around the battery. Rinse off spilled electrolyte immediately with large amounts of water. When removing the battery cables, always detach the negative cable first and hook it up last!*
Caution: *If the vehicle is equipped with a Delco Loc II or Theftlock audio system, make sure you have the correct activation code before disconnecting the battery.*
1   Battery maintenance is an important procedure which will help ensure you aren't stranded because of a dead battery. Several tools are required for this procedure **(see illustration)**.
2   A sealed battery is standard equipment on all vehicles covered by this manual. Although this type of battery has many advantages over the older, capped cell type, and never requires the addition of water, it should still be routinely maintained according to the procedures which follow.

Check
3   The battery is located in the right front corner of the engine compartment. The exterior of the battery should be inspected periodically for damage such as a cracked case or cover.
4   Check the tightness of the battery cable terminals and connections to ensure good electrical connections and check the entire length of each cable for cracks and frayed conductors **(see illustration)**.
5   If corrosion (visible as white, fluffy deposits) is evident, remove the cables from the terminals, clean them with a battery brush and reinstall the cables **(see illustrations)**. Corrosion can be kept to a minimum by using special treated fiber washers available at auto parts stores or by applying a layer of petroleum jelly to the terminals and cables after they are assembled.
6   Make sure that the battery tray is in good condition and the hold-down clamp bolt is tight. If the battery is removed from the tray, make sure no parts remain in the bottom of the tray when the battery is reinstalled. When reinstalling the hold-down clamp bolt, do not overtighten it.
7   Information on removing and installing the battery can be found in Chapter 5. Information on jump starting can be found at the front of this manual. For more detailed battery checking procedures, refer to the *Haynes Automotive Electrical Manual*.

Cleaning
8   Corrosion on the hold-down components, battery case and surrounding areas can be removed with a solution of water and baking soda. Thoroughly rinse all cleaned areas with plain water.
9   Any metal parts of the vehicle damaged by corrosion should be covered with a zinc-based primer, then painted.

11.5a A tool like this one (available at auto parts stores) is used to clean the side terminal type battery contact area

11.5b Use the brush to finish the cleaning job

1

Charging

Warning: *When batteries are being charged, hydrogen gas, which is very explosive and flammable, is produced. Do not smoke or allow open flames near a charging or a recently charged battery. Wear eye protection when near the battery during charging. Also, make sure the charger is unplugged before connecting or disconnecting the battery from the charger.*

10   Slow-rate charging is the best way to restore a battery that's discharged to the point where it will not start the engine. It's also a good way to maintain the battery charge in a vehicle that's only driven a few miles between starts. Maintaining the battery charge is particularly important in the winter when the battery must work harder to start the engine and electrical accessories that drain the battery are in greater use.

11   It's best to use a one or two-amp battery charger (sometimes called a "trickle" charger). They are the safest and put the least strain on the battery. They are also the least expensive. For a faster charge, you can use a higher amperage charger, but don't use one rated more than 1/10th the amp/hour rating of the battery. Rapid boost charges that claim to restore the power of the battery in one to two hours are hardest on the battery and can damage batteries that aren't in good condition. This type of charging should only be used in emergency situations.

12   The average time necessary to charge a battery should be listed in the instructions that come with the charger. As a general rule, a trickle charger will charge a battery in 12 to 16 hours.

13   Remove all of the cell caps (if equipped) and cover the holes with a clean cloth to prevent spattering electrolyte. Disconnect the negative battery cable and hook the battery charger leads to the battery posts (positive to positive, negative to negative), then plug in the charger. Make sure it is set at 12-volts if it has a selector switch.

14   If you're using a charger with a rate higher than two amps, check the battery regularly during charging to make sure it doesn't overheat. If you're using a trickle charger, you can safely let the battery charge overnight after you've checked it regularly for the first couple of hours.

15   If the battery has removable cell caps, measure the specific gravity with a hydrometer every hour during the last few hours of the charging cycle. Hydrometers are available inexpensively from auto parts stores - follow the instructions that come with the hydrometer. Consider the battery charged when there's no change in the specific gravity reading for two hours and the electrolyte in the cells is gassing (bubbling) freely. The specific gravity reading from each cell should be very close to the others. If not, the battery probably has a bad cell(s).

16   Some batteries with sealed tops have built-in hydrometers on the top that indicate the state of charge by the color displayed in the hydrometer window. Normally, a bright-colored hydrometer indicates a full charge and a dark hydrometer indicates the battery still needs charging. Check the battery manufacturer's instructions to be sure you know what the colors mean.

17   If the battery has a sealed top and no built-in hydrometer, you can

hook up a digital voltmeter across the battery terminals to check the charge. A fully charged battery should read 12.5-volts or higher.

12   Cooling system check (every 6000 miles or 6 months)

Refer to illustration 12.4

Caution: *The manufacturer recommends using only DEX-COOL coolant for these systems. DEX-COOL is a long-lasting coolant designed for 100,000 miles or 5 years. Never mix green-colored ethylene glycol antifreeze and orange-colored "DEX-COOL" silicate-free coolant because doing so will destroy the efficiency of the "DEX-COOL".*

1   Many major engine failures can be attributed to a faulty cooling system. If the vehicle is equipped with an automatic transaxle, the cooling system also cools the transaxle fluid and plays an important role in prolonging transaxle life.

2   The cooling system should be checked with the engine cold. Do this before the vehicle is driven for the day or after the engine has been shut off for at least three hours.

3   Remove the reservoir cap by turning it to the left until it reaches a stop. If you hear any hissing sounds (indicating there is still pressure in the system), wait until it stops. Now continue turning to the left until the cap can be removed. Thoroughly clean the cap, inside and out, with clean water. Also clean the filler neck on the radiator. All traces of corrosion should be removed. The coolant inside the radiator should be relatively transparent. If it is rust colored, the system should be drained and refilled (see Section 28). If the coolant level is not up to the top, add additional antifreeze/coolant mixture (see Section 4).

4   Carefully check the large upper and lower radiator hoses along

11.5c The result should be a clean, shiny terminal area 12.4 Hoses, like drivebelts, have a habit of failing at the worst possible time - to prevent the inconvenience of a blown radiator or heater hose, inspect them carefully as shown here

**Check for a chafed area that
could fail prematurely.**

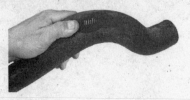

**Check for a soft area indicating
the hose has deteriorated inside.**

**Overtightening the clamp on a
hardened hose will damage the
hose and cause a leak.**

**Check each hose for swelling and
oil-soaked ends. Cracks and breaks
can be located by squeezing the hose.**

**12.4 Hoses, like drivebelts, have a habit of failing at the worst
possible time - to prevent the inconvenience of a blown radiator
or heater hose, inspect them carefully as shown here**

with any smaller diameter heater hoses which run from the engine to
the firewall. Inspect each hose along its entire length, replacing any
hose which is cracked, swollen or shows signs of deterioration. Cracks
may become more apparent if the hose is squeezed **(see illustration)**.
5   Make sure all hose connections are tight. A leak in the cooling
system will usually show up as white or rust colored deposits on the
areas adjoining the leak. If wire-type clamps are used at the ends of
the hoses, it may be wise to replace them with more secure screw-
type clamps.
6   Use compressed air or a soft brush to remove bugs, leaves, etc.
from the front of the radiator or air conditioning condenser. Be careful
not to damage the delicate cooling fins or cut yourself on them.
7   Every other inspection, or at the first indication of cooling system
problems, have the cap and system pressure tested. If you don't have
a pressure tester, most gas stations and repair shops will do this for a
minimal charge.

13   Underhood hose check and replacement
   (every 6000 miles or 6 months)

General

1   **Warning:** *Replacement of air conditioning hoses must be left to a
dealer service department or air conditioning shop that has the equipment to depressurize the system safely. Never remove air conditioning
components or hoses until the system has been depressurized.*

2   High temperatures under the hood can cause the deterioration of
the rubber and plastic hoses used for engine, accessory and emission
systems operation. Periodic inspection should be made for cracks,
loose clamps, material hardening and leaks. Information specific to the
cooling system hoses can be found in Section 12.
3   Some, but not all, hoses are secured to the fittings with clamps.
Where clamps are used, check to be sure they haven't lost their tension, allowing the hose to leak. If clamps aren't used, make sure the
hose hasn't expanded and/or hardened where it slips over the fitting,
allowing it to leak.

Vacuum hoses

4   It's quite common for vacuum hoses, especially those in the
emissions system, to be color coded or identified by colored stripes
molded into each hose. Various systems require hoses with different
wall thickness, collapse resistance and temperature resistance. When
replacing hoses, be sure the new ones are made of the same material.
5   Often the only effective way to check a hose is to remove it completely from the vehicle. If more than one hose is removed, be sure to
label the hoses and fittings to ensure correct installation.
6   When checking vacuum hoses, be sure to include any plastic T-
fittings in the check. Inspect the fittings for cracks and the hose where
it fits over the fitting for distortion, which could cause leakage.
7   A small piece of vacuum hose (1/4-inch inside diameter) can be
used as a stethoscope to detect vacuum leaks. Hold one end of the
hose to your ear and probe around vacuum hoses and fittings, listening for the "hissing" sound characteristic of a vacuum leak. **Warning:**
*When probing with the vacuum hose stethoscope, be careful not to
allow your body or the hose to come into contact with moving engine
components such as the drivebelt, cooling fan, etc.*

Fuel lines

Warning: *Gasoline is extremely flammable, so take extra precautions
when you work on any part of the fuel system. Don't smoke or allow
open flames or bare light bulbs near the work area, and don't work in a
garage where a natural gas-type appliance (such as a water heater or
clothes dryer) with a pilot light is present. If you spill any fuel on your
skin, rinse it off immediately with soap and water. When you perform
any kind of work on the fuel system, wear safety glasses and have a
Class B type fire extinguisher on hand. The fuel system is under pressure, so if any lines must be disconnected, the pressure in the system
must be relieved first (see Chapter 4 for more information).*
8   Check all fuel lines for deterioration and chafing. Check especially
for cracks in areas where the tubing bends and just before fittings,
such as where a line attaches to the fuel filter and fuel injection unit.
Modular nylon fuel lines are used in conjunction with fuel system components. The nylon lines are flexible and are formed around bends, but
will restrict fuel flow if kinked. Check the quick-connect fittings for
damage or leakage.
9   High quality fuel line should be used for fuel line replacement.
Never, under any circumstances, use unreinforced vacuum line, clear
plastic tubing or water hose for fuel lines. Replace defective nylon lines
with components meeting original equipment specifications.
10   Spring-type clamps are commonly used on neoprene fuel lines.
These clamps often lose their tension over a period of time, and can be
"sprung" during the removal process. As a result spring-type clamps
should be replaced with screw-type clamps whenever a hose is
replaced.

Metal lines

11   Sections of steel tubing often used for fuel line between the fuel
pump and fuel injection unit. Check carefully for cracks, kinks and flat
spots in the line.
12   If a section of metal fuel line must be replaced, only seamless
steel tubing should be used, since copper and aluminum tubing do not
have the strength necessary to withstand normal engine vibration.
13   Check the metal brake lines where they enter the master cylinder
and brake proportioning unit (if used) for cracks in the lines and loose
fittings. Any sign of brake fluid leakage calls for an immediate thorough
inspection of the brake system.

14 Drivebelt check and replacement (every 6000 miles or 6 months)

Refer to illustrations 14.2, 14.5 and 14.7

1   A single serpentine drivebelt is located at the front of the engine and plays an important role in the overall operation of the engine and its components. Due to its function and material make up, the belt is prone to wear and should be periodically inspected. The serpentine belt drives the alternator, power steering pump, water pump and air conditioning compressor.

2   With the engine off, open the hood and use your fingers (and a flashlight, if necessary), to move along the belt checking for cracks and separation of the belt plies. Also check for fraying and glazing, which gives the belt a shiny appearance **(see illustration)**. Both sides of the belt should be inspected, which means you will have to twist the belt to check the underside.

3   Check the ribs on the underside of the belt. They should all be the same depth, with none of the surface uneven.

4   The tension of the belt is maintained by the tensioner assembly and isn't adjustable. The belt should be checked at the mileage specified in the maintenance schedule at the front of this chapter, if the belt shows noticeable damage or wear during these checks it should be replaced.

5   To replace the belt, rotate the tensioner pulley clockwise to release belt tension **(see illustration)**.

6   Remove the belt from the auxiliary components and slowly release the tensioner.

7   Route the new belt over the various pulleys, again rotating the tensioner to allow the belt to be installed, then release the belt tensioner. **Note:** *These models have a drivebelt routing decal on the radiator shroud to help during drivebelt installation* **(see illustration)**.

15 Steering, suspension and driveaxle boot check (every 6000 miles or 6 months)

Refer to illustrations 15.3 and 15.4

Steering and suspension check

1   Indications of a fault in these systems are excessive play in the steering wheel before the front wheels react, excessive sway around corners, body movement over rough roads or binding at some point as the steering wheel is turned.

2   Raise the front of the vehicle periodically and visually check the suspension and steering components for wear. Because of the work to be done, make sure the vehicle cannot fall from the stands.

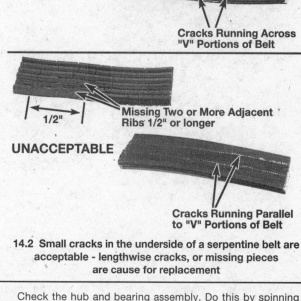

ACCEPTABLE

Cracks Running Across "V" Portions of Belt

1/2"

Missing Two or More Adjacent Ribs 1/2" or longer

UNACCEPTABLE

Cracks Running Parallel to "V" Portions of Belt

14.2 Small cracks in the underside of a serpentine belt are acceptable - lengthwise cracks, or missing pieces are cause for replacement

3   Check the hub and bearing assembly. Do this by spinning the front wheels. Listen for any abnormal noises and watch to make sure the wheel spins true (doesn't wobble). Grab the top and bottom of the tire and pull in-and-out on it. Notice any movement which would indicate a loose bearing assembly **(see illustration)**. Often times, it will be necessary to drive the vehicle slowly and listen for grinding noises from the front end. The weight and load of the vehicle will make the symptoms pronounced. Refer to Chapter 10 for additional information.

4   From under the vehicle check for loose bolts, broken or disconnected parts and deteriorated rubber bushings on all suspension and steering components. Check the power steering hoses and connections for leaks. Check the shock absorbers or leaking fluid or damage.

5   Have an assistant turn the steering wheel from side-to-side and check the steering components for free movement, chafing and binding. If the steering doesn't react with the movement of the steering wheel, try to determine where the slack is located.

Driveaxle boot check

Refer to illustration 15.7

6   The driveaxle boots are very important because they prevent dirt, water and foreign material from entering and damaging the constant

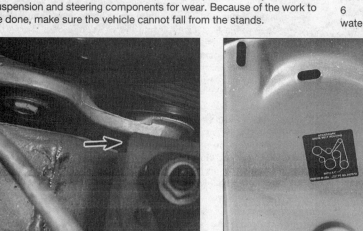

14.5 Using a wrench placed on the tensioner pulley bolt, rotate the tensioner clockwise to remove or install the belt

14.7 The serpentine drivebelt routing diagram is located on the radiator shroud (later model diagram shown)

15.3 Grab the top and bottom of the tire and pull in-and-out on it - Check for any noticeable movement which would indicate a loose wheel bearing assembly

15.7 Check the driveaxle boots for wear or damage

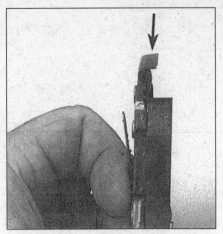

16.3 The disc brake pads are equipped with wear indicators that contact the disc and make a squealing sound when the pad has worn to its limit

16.5 With the wheels removed, the brake pad lining (arrows) can be inspected through the caliper window and at each end of the caliper

velocity joints. Oil and grease can cause the boot material to deteriorate prematurely, so it's a good idea to wash the boots with soap and water.

7   Inspect the boots for tears and cracks as well as loose clamps **(see illustration)**. If there is any evidence of cracks or leaking lubricant, they must be replaced as described in Chapter 8.

16   Brake check (every 6000 miles or 6 months)

Note: *For detailed information of the brake system,* refer to Chapter 9.
Warning: *Brake system dust may contain asbestos, which is hazardous to your health. DO NOT blow it out with compressed air, inhale it or use gasoline or solvents to remove it. Use brake system cleaner only.*
1   In addition to the specified intervals, the brakes should be inspected every time the wheels are removed or whenever a defect is suspected. Raise the vehicle and place it securely on jackstands. Remove the wheels (see *Jacking and towing* at the front of the manual, if necessary).

Disc brakes

Refer to illustrations 16.3, 16.5 and 16.7
2   Disc brakes can be checked without removing any parts except the wheels. Extensive disc damage can occur if the pads are not replaced when needed.
3   The disc brake pads have built-in wear indicators **(see illustration)** which make a high-pitched squealing sound when the pads are worn. **Caution:** *Expensive damage to the disc can result if the pads are not replaced soon after the wear indicators start squealing.*
4   The disc brake calipers, which contain the brake pads, have an inner pad and outer pad in each caliper. All pads should be inspected.
5   Each caliper has a "window" to inspect the pads **(see illustration)**. If the pad material has worn to about 1/8-inch thick or less, the pads should be replaced.
6   If you're unsure about the exact thickness of the remaining lining material, remove the pads for further inspection or replacement (see Chapter 9).
7   Before installing the wheels, check for leakage and/or damage at the brake hoses and connections **(see illustration)**. Replace the hose or fittings as necessary, (see Chapter 9).
8   Check the condition of the brake disc. Look for score marks, deep scratches and overheated areas (they will appear blue or discolored). If damage or wear is noted, the disc can be removed and resurfaced by an automotive machine shop or replaced with a new one. See Chapter 9 for more detailed inspection and repair procedures.

Drum brakes

Refer to illustrations 16.14 and 16.16
9   Raise the vehicle and support it securely on jackstands. Block the front tires to prevent the vehicle from rolling; however, don't apply the parking brake or it will lock the drums in place.
10   Remove the wheels, referring to *Jacking and towing* at the front of this manual if necessary.
11   Mark the hub so it can be reinstalled in the same position. Use a scribe, chalk, etc. on the drum, hub and backing plate.
12   Remove the brake drum.
13   With the drum removed, carefully clean the brake assembly with brake system cleaner. **Warning:** *Don't blow the dust out with compressed air and don't inhale any of it (it may contain asbestos, which is harmful to your health).*
14   Note the thickness of the lining material on both front and rear brake shoes. If the material has worn away to within 1/8-inch of the recessed rivets or metal backing, the shoes should be replaced **(see illustration)**. The shoes should also be replaced if they're cracked, glazed (shiny areas), or covered with brake fluid.
15   Make sure all the brake assembly springs are connected and in good condition.
16   Check the brake components for signs of fluid leakage. With your finger or a small screwdriver, carefully pry back the rubber cups on the wheel cylinder located at the top of the brake shoes **(see illustration)**. Any leakage here is an indication that the wheel cylinders should be replaced immediately (see Chapter 9). Also, check all hoses and connections for signs of leakage.
17   Wipe the inside of the drum with a clean rag and denatured alcohol or brake cleaner. Again, be careful not to breathe the dangerous asbestos dust.
18   Check the inside of the drum for cracks, score marks, deep scratches and "hard spots" which will appear as small discolored areas. If imperfections cannot be removed with fine emery cloth, the drum must be taken to an automotive machine shop for resurfacing.
19   Repeat the procedure for the remaining wheel. If the inspection reveals that all parts are in good condition, reinstall the brake drums, install the wheels and lower the vehicle to the ground.

Parking brake

20   The parking brake is operated by a hand lever and locks the rear brake system. The easiest, and perhaps most obvious, method of periodically checking the operation of the parking brake assembly is to park the vehicle on a steep hill with the parking brake set and the transaxle in Neutral (be sure to stay in the vehicle during this check!). If the parking brake cannot prevent the vehicle from rolling, it needs service (see Chapter 9).

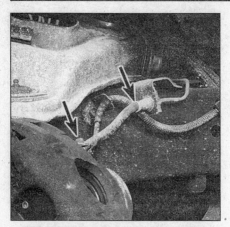

16.7 Check for any sign of brake fluid leakage at the line fittings and the brake hoses

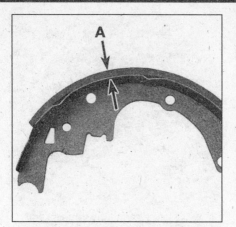

16.14 If the lining is bonded to the brake shoe, measure the lining thickness (A) from the outer surface to the metal shoe, as shown here; if the lining is riveted to the shoe, measure from the lining outer surface to the rivet head

16.16 Check for fluid leakage at both ends of the wheel cylinder dust covers (arrow)

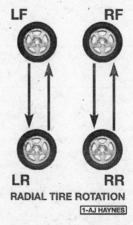

17.2 Tire rotation diagram

18.2a Check the flange connections (arrow) for exhaust leaks - also check that the retaining nuts are securely tightened

18.2b Check the exhaust system hangers (arrows) for damage and cracks

17 Tire rotation (every 6000 miles or 6 months)

Refer to illustration 17.2

1   The tires should be rotated at the specified intervals and whenever uneven wear is noticed.

2   Refer to the **accompanying illustration** for the preferred tire rotation pattern.

3   Refer to the information in *Jacking and towing* at the front of this manual for the proper procedures to follow when raising the vehicle and changing a tire. If the brakes are to be checked, don't apply the parking brake as stated. Make sure the tires are blocked to prevent the vehicle from rolling as it's raised.

4   Preferably, the entire vehicle should be raised at the same time. This can be done on a hoist or by jacking up each corner and then lowering the vehicle onto jackstands placed under the frame rails. Always use four jackstands and make sure the vehicle is safely supported.

5   After rotation, check and adjust the tire pressures as necessary and be sure to properly tighten the lug nuts.

18 Exhaust system check (every 6000 miles or 6 months)

Refer to illustrations 18.2a and 18.2b

1   With the engine cold (at least three hours after the vehicle has been driven), check the complete exhaust system from the engine to the end of the tailpipe. Ideally, the inspection should be done with the vehicle on a hoist to permit unrestricted access. If a hoist is not available, raise the vehicle and support it securely on jackstands.

2   Check the exhaust pipes and connections for evidence of leaks, severe corrosion and damage. Make sure that all brackets and hangers are in good condition and tight **(see illustrations)**.

3   At the same time, inspect the underside of the body for holes, corrosion, open seams, etc. which may allow exhaust gases to enter the interior. Seal all body openings with silicone or body putty.

4   Rattles and other noises can often be traced to the exhaust system, especially the mounts and hangers. Try to move the pipes, muffler and catalytic converter. If the components can come in contact with the body or suspension parts, secure the exhaust system with new mounts.

19 Throttle body mounting nut torque check (every 12,000 miles or 12 months)

Refer to illustration 19.4

1   The throttle body unit is attached to the top of the intake manifold by several bolts or nuts. These fasteners can sometimes work loose from vibration and temperature changes during normal engine operation and cause a vacuum leak.

2   If you suspect that a vacuum leak exists at the bottom of the throttle body, obtain a length of hose about the diameter of fuel hose. Start the engine and place one end of the hose next to your ear as you probe around the base with the other end. You will hear a hissing sound if a leak exists (be careful of hot or moving engine components).

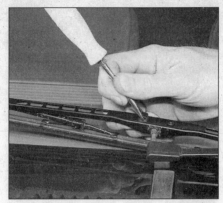

19.4 Tighten the throttle body mounting bolts (arrows) to the torque listed in this Chapter's Specifications

20.3 Gently pry off the trim cap and check the tightness of the wiper arm retaining nut

20.5 Lift the release lever with a flat-bladed screwdriver, then slide the blade assembly off the pin on the end of the wiper arm

3   Remove the air cleaner assembly, tagging each hose to be disconnected with a piece of numbered tape to make reassembly easier.
4   Locate the mounting nuts or bolts at the base of the throttle body **(see illustration)**. Decide what special tools or adapters will be necessary, if any, to tighten the fasteners.
5   Tighten the nuts or bolts to the torque listed in this Chapter's Specifications. Do not overtighten them, as the threads could strip.
6   If, after the nuts or bolts are properly tightened, a vacuum leak still exists, the throttle body must be removed and a new gasket installed. See Chapter 4 for more information.
7   After tightening the fasteners, reinstall the air cleaner and return all hoses to their original positions.

20   Windshield wiper blade inspection and replacement (every 12,000 miles or 12 months)

Refer to illustrations 20.3, 20.5 and 20.7
1   The windshield wiper and blade assemblies should be inspected periodically for damage, loose components and cracked or worn blade elements.
2   Road film can build up on the wiper blades and affect their efficiency, so they should be washed regularly with a mild detergent solution.
3   The action of the wiping mechanism can loosen the bolts, nuts and fasteners, so they should be checked and tightened, as necessary, at the same time the wiper blades are checked **(see illustration)**.
4   If the wiper blade elements (sometimes called inserts) are cracked, worn or warped, they should be replaced with new ones.
5   Remove the wiper blade assembly from the wiper arm by inserting a screwdriver and lifting the release lever while pulling on the blade to release it **(see illustration)**.
6   With the blade removed from the vehicle, you can remove the rubber element from the blade.
7   Grasp the end of the wiper bridge securely with one hand and the element with the other. Detach the end of the element from the bridge claw and slide to free it, then slide the element out **(see illustration)**.
8   Compare the new element with the old for length, design, etc.
9   Slide the new element into the claw into place, notched end last and secure the claw into the notches.
10   Reinstall the blade assembly on the arm, wet the windshield and test for proper operation.

21   Seat belt check (every 12,000 miles or 12 months)

1   Check the seat belts, buckles, latch plates and guide loops for obvious damage and signs of wear.
2   See if the seat belt reminder light comes on when the key is turned to the Run or Start position. A chime should also sound.
3   The seat belts are designed to lock up during a sudden stop or

impact, yet allow free movement during normal driving. Make sure the retractors return the belt against your chest while driving and rewind the belt fully when the buckle is unlatched.
4   If any of the above checks reveal problems with the seat belt system, replace parts as necessary.

22   Air filter replacement (every 12,000 miles or 12 months)

Refer to illustrations 22.2a and 22.2b
1   At the specified intervals, the air filter should be replaced with a new one. The filter should be inspected between changes.
2   The air filter is located inside the air cleaner housing which is mounted in the left, front corner of the engine compartment. Remove the mounting screws, separate the housing halves and lift the filter out **(see illustrations)**.
3   While the filter housing cover is off, be careful not to drop anything down into the air cleaner housing.
4   Wipe out the inside of the air cleaner housing with a clean rag.
5   Place the new filter in the air cleaner housing. Make sure it seats properly.
6   The remainder of installation is the reverse of removal.

23   Positive Crankcase Ventilation (PCV) system check and replacement (every 30,000 miles or 24 months)

Note: *These models are equipped with three different crankcase ventilating systems. On 1997 and earlier 2.2L OHV models, the PCV valve is located on the right side of the valve cover. 1998 and later 2.2L OHV engines are equipped with an oil/air separator located inside the valve rocker arm cover that requires no maintenance. 2.3L and 2.4L OHC engines are equipped with an oil/air separator that is attached to the side of the engine block above the oil filter.*

2.2L OHV engine
Refer to illustrations 23.1, 23.3a and 23.3b
1   The PCV valve is located on the right side of the valve cover **(see illustration)**.
2   To inspect the PCV valve, remove the hose from the nipple.
3   Unscrew the metal retainer and then pull it out of the valve cover using a pair of needlenose pliers **(see illustrations)**. Note its installed position and direction.
4   Make sure the PCV valve is not plugged with oil/dirt residue and that vacuum is allowed to pass through without obstruction.
5   Shake the PCV valve, listening for a rattle. If the valve doesn't rattle, replace it with a new one.
6   When purchasing a replacement PCV valve, make sure it's for

20.7 Squeeze the end of the wiper element to free it from the bridge claw, then slide the element out

22.2a Remove the mounting screws (arrows) from the air cleaner housing and . . .

22.2b . . . lift the air filter from the housing

your particular vehicle, model year and engine size. Compare the old valve with the new one to make sure they are the same.

7   Push the valve into the end into the valve cover until it's seated.

8   Inspect the metal retainer for damage and replace it with a new one if necessary.

9   Push the PCV valve hose securely into position.

2.3L and 2.4L OHC engines

10   The oil/air separator on 2.3L and 2.4L OHC engines, separates the oil suspended in the blow-by gasses and allows it to drain back into the crankcase. Crankcase engine blow-by gasses are also allowed to pass through the separator and back into the air resonator to be burned in the normal combustion process. The oil/air separator may clog due to the heavy driving loads or lack of timely oil changes during the course of operation.

11   Because the oil/air separator is difficult to access, checking will require operating knowledge of the engine. If excess pressure is building up inside the engine then most likely there is a problem with the crankcase recirculation system. Some of the symptoms associated with a clogged oil/air separator are leaky engine seals, excess oil and sludge deposits on the oil dipstick and filler cap, dirty oil and excess crankcase pressure when the oil filler cap is removed. It is a good idea to consult a dealer service department or other qualified technician concerning this system and trouble areas.

12   To change the oil/air separator, remove the air cleaner outlet resonator. The air cleaner outlet resonator is part of the air cleaner housing assembly (see Chapter 4).

13   Remove the intake manifold (see Chapter 2B).

14   Disconnect the oil/air separator hoses and mounting bolts.

15   Remove the assembly from the engine.

16   Installation is the reverse of removal.

---

24   Exhaust Gas Recirculation (EGR) system check (1995 2.2L OHV engine only) (every 30,000 miles or 24 months)

Note: *The 1995 2.2L OHV engine is equipped with a negative back-pressure EGR valve mounted on the side of the cylinder head. All other models are equipped with a linear EGR valve, except the 1995 2.3L OHC engine which is not equipped with an EGR system. The negative backpressure type EGR valve may be checked as described below. The linear EGR valves on the other systems will require a SCAN tool for checking. Refer to Chapter 6 for additional information.*

1   The EGR valve is located on the side of the cylinder head. Most of the time when a problem develops in this emissions system, it's due to a stuck or corroded EGR valve.

2   With the engine cold, to prevent burns, push on the EGR valve diaphragm. Using moderate pressure, you should be able to press the diaphragm in and out within the housing.

3   If the diaphragm doesn't move or moves only with much effort, replace the EGR valve with a new one. If in doubt about the condition of the valve, compare the free movement of your EGR valve with a new valve.

4   Refer to Chapter 6 for more information on the EGR system.

23.1 On 2.2L OHV models, the PCV valve is located on the right side of the valve cover

23.3a Unscrew the metal retainer from the valve cover

23.3b Remove the PCV valve using a pair of needlenose pliers

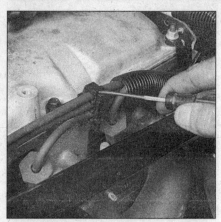

25.3a Use a small screwdriver or awl to release the locking tab on the ignition wire hold-down bracket

25.3b Using a special spark plug wire removal tool, grab the spark plug boot and pull carefully to separate the end from the spark plug

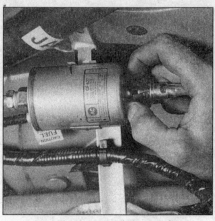

26.5a Squeeze the plastic quick-disconnect tabs together and pull the fuel line away from the filter

25 Spark plug wire check and replacement (2.2L OHV engine) (every 30,000 miles or 24 months)

Refer to illustrations 25.3a and 25.3b

1   The spark plug wires should be checked at the recommended intervals and whenever new spark plugs are installed in the engine.

2   The wires should be inspected one at a time to prevent mixing up the order, which is essential for proper engine operation.

3   Disconnect the plug wire from the spark plug. To do this, grab the rubber boot, twist slightly and pull the wire off. Do not pull on the wire itself, only on the rubber boot **(see illustrations)**.

4   Check inside the boot for corrosion, which will look like a white crusty powder. Push the wire and boot back onto the end of the spark plug. It should be a tight fit on the plug. If it isn't, remove the wire and use pliers to carefully crimp the metal connector inside the boot until it fits securely on the end of the spark plug.

5   Using a clean rag, wipe the entire length of the wire to remove any built-up dirt and grease. Once the wire is clean, check for burns, cracks and other damage. Do not bend the wire excessively or pull the wire lengthwise - the conductor inside might break.

6   Check the remaining spark plug wires one at a time, making sure they are securely fastened at the coil packs and the spark plug when the check is complete.

7   If new spark plug wires are required, purchase a set for your specific engine model. Wire sets are available pre-cut, with the rubber boots already installed. Remove and replace the wires one at a time to avoid mix-ups in the firing order. **Note:** *If an accidental mix-up occurs, refer to the firing order diagrams at the beginning of this Chapter.*

26 Fuel filter replacement (every 30,000 miles or 24 months)

Refer to illustrations 26.5a and 26.5b

Warning: *Gasoline is extremely flammable, so take extra precautions when you work on any part of the fuel system. Don't smoke or allow open flames or bare light bulbs near the work area, and don't work in a garage where a natural gas-type appliance (such as a water heater or clothes dryer) with a pilot light is present. Since gasoline is carcinogenic, wear latex gloves when there's a possibility of being exposed to fuel, and, if you spill any fuel on your skin, rinse it off immediately with soap and water. Mop up any spills immediately and do not store fuel-soaked rags where they could ignite. The fuel system is under constant pressure, so, if any fuel lines are to be disconnected, the fuel pressure in the system must be relieved first (see Chapter 4 for more information). When you perform any kind of work on the fuel system, wear*
safety glasses and have a Class B type fire extinguisher on hand.

1   Relieve the fuel system pressure (see Chapter 4).

2   Raise the vehicle and support it securely on jackstands.

3   The fuel filter is located on the inside of right frame rail in front of the fuel tank.

4   Clean any dirt surrounding the fuel inlet and outlet line fittings, especially around the inside of the quick-connect fitting.

5   Twist the quick-connect fitting 1/4-turn in each direction to loosen the seal. Push the fuel line into the filter, depress the white plastic tabs and pull the fuel line from the fuel filter. Using a flare-nut wrench, loosen the flare-nut fitting **(see illustrations)**. **Note:** *Have spare rags or a small container to catch or wipe up extra gasoline which will spill from the filter assembly.*

6   Remove the fuel filter from the clamp.

7   Install the new fuel filter into the clamp. **Note:** *Make sure the directional arrow on the fuel filter points in the direction of fuel flow.*

8   Apply a few drops of clean engine oil into the quick-connect fitting and press it onto the filter until it snaps into place. Pull in-an-out several times to ensure the fitting is securely installed. Install the flare nut fitting and tighten it securely. Be sure to use a flare-nut wrench and a back-up to prevent damage to the filter or fitting.

9   Lower the vehicle, start the engine and check for fuel leaks at the filter.

27 Fuel system check (every 30,000 miles or 24 months)

Refer to illustration 27.4

Warning: *Gasoline is extremely flammable, so take extra precautions when you work on any part of the fuel system. Don't smoke or allow open flames or bare light bulbs near the work area, and don't work in a garage where a natural gas-type appliance (such as a water heater or clothes dryer) with a pilot light is present. Since gasoline is carcinogenic, wear latex gloves when there's a possibility of being exposed to fuel, and, if you spill any fuel on your skin, rinse it off immediately with soap and water. Mop up any spills immediately and do not store fuel-soaked rags where they could ignite. The fuel system is under constant pressure, so, if any fuel lines are to be disconnected, the fuel pressure in the system must be relieved first (see Chapter 4 for more information). When you perform any kind of work on the fuel system, wear safety glasses and have a Class B type fire extinguisher on hand.*

1   The fuel system is most easily checked with the vehicle raised on a hoist so the components underneath the vehicle are readily visible and accessible.

2   If the smell of gasoline is noticed while driving or after the vehicle has been in the sun, the system should be thoroughly inspected immediately.

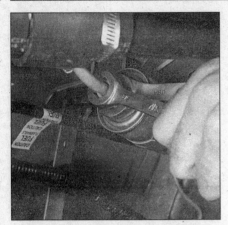

26.5b Loosen the fuel line fitting with a flare-nut wrench - be sure to use an open-end wrench on the filter as a back-up

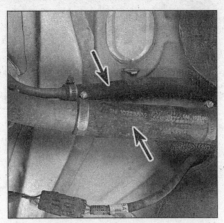

27.4 Check the fuel filler lines (arrows) for cracks and deterioration and the hose clamps for tightness

28.6a The drain plug (arrow) is located at the lower right corner of the radiator

1

28.6b Use a small socket to open the bleeder screw (arrow)

28.6c Location of the heater core bleeder plug (1998 2.2L OHV model shown)

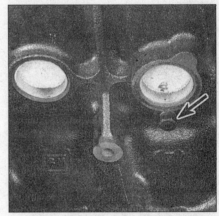

28.7 Location of the block drain plug (arrow) on the 2.2L OHV engine

3   Remove the gas tank cap and check for damage, corrosion and an unbroken sealing imprint on the gasket. Replace the cap with a new one if necessary.

4   With the vehicle raised, inspect the gas tank and filler neck for cracks and other damage **(see illustration)**. The connection between the filler neck and tank is especially critical. Sometimes a filler neck will leak due to cracks, problems a home mechanic can't repair. **Warning:** *Do not, under any circumstances, try to repair a fuel tank yourself (except rubber components). A welding torch or any open flame can easily cause the fuel vapors to explode if the proper precautions are not taken.*

5   Carefully check all hoses and lines leading away from the fuel tank. Check for loose connections, deteriorated hoses, crimped lines and other damage. Follow the lines to the front of the vehicle, carefully inspecting them all the way. Repair or replace damaged sections as necessary.

28 Cooling system servicing (draining, flushing and refilling) (see maintenance schedule for service intervals)

Refer to illustrations 28.6a, 28.6b, 28.6c and 28.7
Warning: *Make sure the engine is completely cool before performing this procedure.*
Caution: *The manufacturer recommends using only DEX-COOL coolant for these systems. DEX-COOL is a long-lasting coolant designed for 100,000 miles or 5 years. Never mix green-colored ethylene glycol anti-freeze and orange-colored "DEX-COOL" silicate-free coolant because doing so will destroy the efficiency of the "DEX-COOL".*

1   Periodically, the cooling system should be drained, flushed and refilled to replenish the antifreeze mixture and prevent formation of rust and corrosion, which can impair the performance of the cooling system and cause engine damage.

2   At the same time the cooling system is serviced, all hoses and the radiator cap should be inspected and replaced if defective (see Section 12).

3   Since antifreeze is a corrosive and poisonous solution, be careful not to spill any of the coolant mixture on the vehicle's paint or your skin. If this happens, rinse it off immediately with plenty of clean water. Consult local authorities about the dumping of antifreeze before draining the cooling system. In many areas, reclamation centers have been set up to collect automobile oil and drained antifreeze/water mixtures, rather than allowing them to be added to the sewage system.

4   With the engine cold, remove the reservoir pressure cap.

5   Move a large container under the radiator to catch the coolant as it's drained.

6   Drain the radiator by opening the drain plug at the bottom on the left side **(see illustration)**. If the drain plug is corroded and can't be turned easily, or if the radiator isn't equipped with a plug, disconnect the lower radiator hose to allow the coolant to drain. Be careful not to get antifreeze on your skin or in your eyes. Some cooling systems are equipped with a bleeder screw, use a screwdriver to open the screw two or three turns **(see illustrations)**.

7   After the coolant stops flowing out of the radiator, move the container under the engine block drain plug and remove the plug, or knock sensor (which double as a plug on some models) **(see illustration)**.

8   Disconnect the hose from the coolant reservoir and remove the reservoir (see Chapter 3). Flush it out with clean water.

9   Remove the upper radiator hose from the radiator, then place a

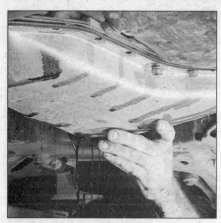

29.6 With the rear bolts in place but loose, pull the front of the pan down to drain the transaxle fluid

29.9a Remove the filter mounting bolts/nuts (arrows) and . . .

29.9b . . . lower it from the transaxle

garden hose in the upper radiator opening and flush the system until the water runs clear at all drain points.

10   In severe cases of contamination or clogging of the radiator, remove it (see Chapter 3) and reverse flush it. This involves inserting the hose in the bottom radiator outlet to allow the water to run against the normal flow, draining through the top. A radiator repair shop should be consulted if further cleaning or repair is necessary.

11   When the coolant is regularly drained and the system refilled with the correct antifreeze/water mixture, there should be no need to use chemical cleaners or descalers.

12   To refill the system, install the block plugs or knock sensors, reconnect any radiator hoses and install the reservoir and the overflow hose.

13   On later models, make sure to use the proper coolant (see **Caution** above). The manufacturer recommends adding a cooling system sealer any time the coolant is changed. Slowly fill the radiator with the recommended mixture of antifreeze and water to the base of the filler neck. On models without a radiator cap, add coolant to the coolant reservoir. Wait two minutes and recheck the coolant level, adding if necessary. Close the bleed screws (if equipped) when the coolant issuing from them is free of bubbles. **Note:** *The low coolant light may illuminate after the draining and flushing procedure has been completed. Start the engine and let it run until it reaches normal operating temperature, then let it completely cool down and add coolant as necessary. Repeat this procedure until the light goes out.*

14   Keep a close watch on the coolant level and the cooling system hoses during the first few miles of driving. Tighten the hose clamps and/or add more coolant as necessary. The coolant level should be a little above the HOT mark on the reservoir with the engine at normal operating temperature.

29   Automatic transaxle fluid and filter change (every 50,000 miles)

Refer to illustrations 29.6, 29.9a, 29.9b and 29.10

1   At the specified intervals, the transaxle fluid should be drained and replaced. Since the fluid will remain hot long after driving, perform this procedure only after the engine has cooled down completely.

2   Before beginning work, purchase the specified transaxle fluid (see *Recommended lubricants and fluids* at the front of this Chapter) and a new filter.

3   Other tools necessary for this job include a floor jack, jackstands to support the vehicle in a raised position, a drain pan capable of holding at least eight quarts, newspapers and clean rags.

4   Raise the vehicle and support it securely on jackstands.

5   Place the drain pan underneath the transaxle pan. Remove the front and side pan mounting bolts, but only loosen the rear pan bolts

approximately four turns.

6   Carefully pry the transaxle pan loose with a screwdriver, allowing the fluid to drain **(see illustration)**.

7   Remove the remaining bolts, pan and gasket. Carefully clean the gasket surface of the transaxle to remove all traces of the old gasket and sealant.

8   Drain the fluid from the transaxle pan, clean it with solvent and dry it with compressed air.

9   Remove the filter from the mount inside the transaxle **(see illustrations)**.

10   If the seal did not come out with the filter, remove it from the transaxle **(see illustration)**. Install a new filter and seal.

11   Make sure the gasket surface on the transaxle pan is clean, then install a new gasket on the pan. Put the pan in place against the transaxle and, working around the pan, tighten each bolt a little at a time until the final torque figure is reached.

12   Lower the vehicle and add approximately 3-1/2 quarts of the specified type of automatic transaxle fluid (Section 7).

13   If you are working on a three-speed transaxle, refer to Section 7 for the fluid level checking procedure. If you're working on a four-speed automatic transaxle, procede to step 15.

14   Check under the vehicle for leaks during the first few trips.

Fluid level check (4-speed automatic transaxle)

Refer to illustration 29.17 and 29.19

15   The automatic transaxle fluid level should be carefully maintained. Low fluid level can lead to slipping or loss of drive, while overfilling can cause foaming and loss of fluid. **Warning:** *This procedure is potentially dangerous and is best left to a professional shop with a safe lifting apparatus. The vehicle must be kept level while being safely raised high enough for access to the check plug on the transaxle.*

16   With the vehicle safely raised and supported, start the engine, then move the shift lever through all the gear ranges, ending in Park. **Note:** *Incorrect fluid level readings will result if the vehicle has just been driven at high speeds for an extended period, in hot weather in city traffic, or if it has been pulling a trailer. If any of these conditions apply, wait until the fluid has cooled (about 30 minutes).*

17   With the engine running and the transaxle at normal operating temperature (having idled for 3 to 5 minutes), locate the check plug on the transaxle. The check plug is located near the pan, adjacent to the engine oil drain plug **(see illustration)**.

18   Place an oil container under the check plug and remove it. Observe the fluid as it drips into the pan, indicating correct fluid level.

19   The fluid level should be at the bottom of the check hole. If fluid pours out excessively, the transaxle may have been overfilled. Double-check to make sure the vehicle is level. If no fluid drips from the check hole, add small amounts of fluid through the vent/fill cap at the top of the transaxle until the level is at the bottom of the check hole **(see illustration)**. A long-necked funnel will be necessary to add fluid.

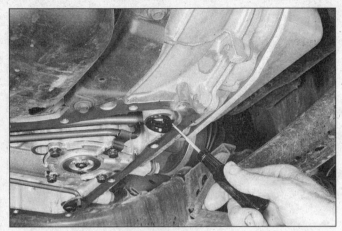

29.10 If necessary, use a screwdriver to remove the seal from the transaxle - be careful not to gouge the aluminum housing

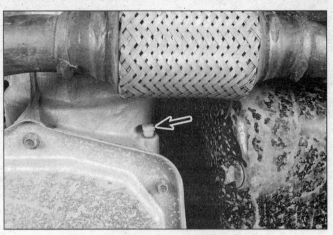

29.17 Location of the check plug on the 4-speed automatic transaxle

1

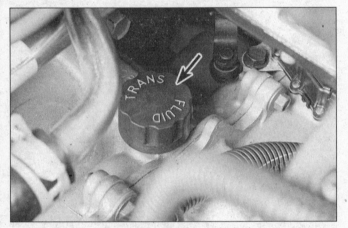

29.19 Location of the vent/fill cap on the 4-speed automatic transaxle

20   The condition of the fluid should also be checked along with the level. If the fluid in the drain pan is a dark reddish-brown color, or if the fluid has a burned smell, the fluid should be changed (see above). If you're in doubt about the condition of the fluid, purchase some new fluid and compare the two for color and smell.

21   Be sure to install the check plug and tighten it securely when you're done.

30   Manual transaxle lubricant change (every 50,000 miles)

1   Remove the drain plug and drain the fluid.
2   Reinstall the drain plugs securely.
3   Use a narrow funnel designed to fit into the fill vent tube and slowly add the necessary transaxle fluid. Refer to the Specifications listed in this Chapter for the correct fluid type.

31   Spark plug check and replacement (see maintenance schedule for service intervals)

Refer to illustrations 31.2, 31.5a, 31.5b, 31.6, 31.8, 31.9 and 31.10
Note: The engines covered by this manual are equipped with either a distributorless Direct Ignition System (DIS) (2.2L OHV engine) or an Integrated Direct Ignition system (IDI) (2.3L and 2.4L OHC engine). On the DIS system, the coil packs and ignition module are mounted on the side of the engine block with spark plug wires connecting the spark plugs to the coil packs. The IDI system on 2.3L and 2.4L OHC engines

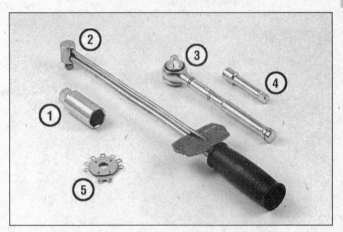

31.2 Tools required for changing spark plugs

1   **Spark plug socket** - This will have special padding inside to protect the spark plug's porcelain insulator
2   **Torque wrench** - Although not mandatory, using this tool is the best way to ensure the plugs are tightened properly
3   **Ratchet** - Standard hand tool to fit the spark plug socket
4   **Extension** - Depending on model and accessories, you may need special extensions and universal joints to reach one or more of the plugs
5   **Spark plug gap gauge** - This gauge for checking the gap comes in a variety of styles. Make sure the gap for your engine is included

incorporates the ignition module, coils and spark plug boots as one complete unit mounted directly on top of the cylinder head.

1   The spark plugs are located at the side of the cylinder head on 2.2L OHV models or in the center of the cylinder head between the intake and exhaust camshaft housings on the 2.3L and 2.4L OHC engines. The spark plugs on the 2.2L OHV engine can be reached easily in the engine compartment. The spark plugs on the 2.3L and 2.4L OHC engine will require removal of the ignition cover, module, coil pack and spark plug boot assembly (see Chapter 5).

2   In most cases, the tools necessary for spark plug replacement include a spark plug socket which fits onto a ratchet (spark plug sockets are padded inside to prevent damage to the porcelain insulators on the new spark plugs), various extensions and a gap gauge to check and adjust the gaps on the new spark plugs **(see illustration)**. A special spark plug wire removal tool is available for separating the wire boots from the spark plugs, and is a good idea on these models because the boots fit very tightly. A torque wrench should be used to tighten the new spark plugs. It is a good idea to allow the engine to cool before removing or installing the spark plugs.

31.5a Spark plug manufacturers recommend using a wire type gauge when checking the gap - if the wire does not slide between the electrodes with a slight drag, adjustment is required

31.5b To change the gap, bend the *side* electrode only, as indicated by the arrows, and be very careful not to crack or chip the porcelain insulator surrounding the center electrode

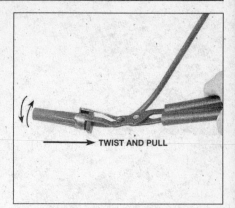

31.6 When removing the spark plug wires, pull only on the boot and use a twisting, pulling motion

31.8 Use an extension and socket to remove the spark plugs from the cylinder head

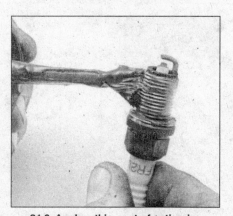

31.9 Apply a thin coat of anti-seize compound to the spark plug threads

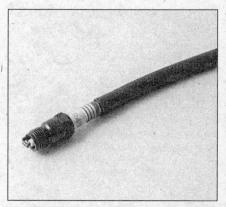

31.10 A length of snug-fitting rubber hose will save time and prevent damaged threads when installing the spark plugs

3   The best approach when replacing the spark plugs is to purchase the new ones in advance, adjust them to the proper gap and replace the spark plugs one at a time. When buying the new spark plugs, be sure to obtain the correct spark plug type for your particular engine. The spark plug type can be found in the Specifications at the front of this Chapter and on the Emission Control Information label located under the hood. If these two sources list different spark plug types, consider the emission control label correct.

4   Allow the engine to cool completely before attempting to remove any of the spark plugs. While you are waiting for the engine to cool, check the new spark plugs for defects and adjust the gaps.

5   Check the gap by inserting the proper thickness gauge between the electrodes at the tip of the spark plug **(see illustration)**. The gap between the electrodes should be the same as the one specified on the Emissions Control Information label or in Chapter 5. The wire should slide between the electrodes with a slight amount of drag. If the gap is incorrect, use the adjuster on the gauge body to bend the curved side electrode slightly until the proper gap is obtained **(see illustration)**. If the side electrode is not exactly over the center electrode, bend it with the adjuster until it is. Check for cracks in the porcelain insulator (if any are found, the spark plug should not be used).

6   With the engine cool, remove the spark plug wire from one spark plug. Pull only on the boot at the end of the wire - do not pull on the wire. A spark plug wire removal tool should be used if available **(see illustration)**. On 2.3L and 2.4L OHC engines it will be necessary to remove the ignition cover, the module, the coil pack assemblies and the spark plug boots as one complete assembly. Refer to Chapter 5 for additional information and illustrations.

7   If compressed air is available, use it to blow any dirt or foreign material away from the spark plug hole. A common bicycle pump will also work. The idea here is to eliminate the possibility of debris falling into the cylinder as the spark plug is removed.

8   The spark plugs on 2.3L and 2.4L OHC engines are, for the most part, difficult to reach so a spark plug socket incorporating a universal joint will be necessary. Place the spark plug socket over the spark plug and remove it from the engine by turning it in a counterclockwise direction **(see illustration)**.

9   Compare the spark plug with the chart shown on the inside back cover of this manual to get an indication of the general running condition of the engine. Before installing the new spark plugs, it is a good idea to apply a thin coat of anti-seize compound to the threads **(see illustration)**.

10   Thread one of the new spark plugs into the hole until you can no longer turn it with your fingers, then tighten it with a torque wrench (if available) or the ratchet. It's a good idea to slip a short length of rubber hose over the end of the spark plug to use as a tool to thread it into place **(see illustration)**. The hose will grip the spark plug well enough to turn it, but will start to slip if the spark plug begins to cross-thread in the hole - this will prevent damaged threads and the accompanying repair costs.

11   Before pushing the spark plug wire onto the end of the spark plug, inspect it following the procedures outlined in Section 25.

12   Attach the spark plug wire to the new spark plug, again using a twisting motion on the boot until it's seated on the spark plug.

13   Repeat the procedure for the remaining spark plugs, replacing them one at a time to prevent mixing up the spark plug wires.

Chapter 2 Part A
Overhead valve (OHV) engine

Contents

2A

Specifications

General

Cylinder numbers (drivebelt end-to-transaxle end)	1-2-3-4
Firing order	1-3-4-2

Torque specifications

Ft-lbs (unless otherwise indicated)

Camshaft sprocket bolt	96
Crankshaft pulley-to-hub bolts	37
Crankshaft pulley center bolt	77
Cylinder head bolts **(in sequence - see illustration 9.23)**	
Step 1	
Short bolts	43
Long bolts	46
Step 2	Tighten an additional 90-degrees
Exhaust manifold fasteners	
1995 through 1997	115 in-lbs
1998 and later	118 in-lbs
Flywheel bolts	55
Driveplate bolts	52
Intake manifold fasteners	
1995 through 1997	22
1998 and later	17
Timing chain cover bolts	97 in-lbs
Timing chain tensioner bolts	18
Oil pan bolts	89 in-lbs
Oil pump mounting bolt	32
Rocker arm nuts/bolts	
1995 through 1997	22
1998 and later	19
Valve cover bolts	89 in-lbs
Valve lifter anti-rotation bracket bolts	97 in-lbs

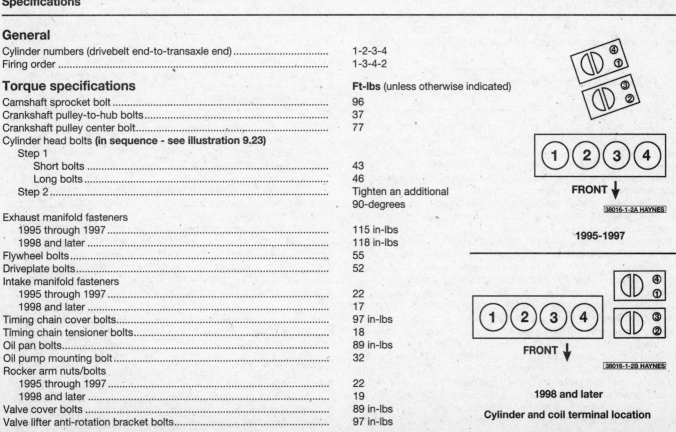

FRONT ↓

38016-1-2A HAYNES

1995-1997

FRONT ↓

38016-1-2B HAYNES

1998 and later

Cylinder and coil terminal location

4.6 Remove the valve cover bolts (arrows) and . . .

4.7 . . . lift the valve cover from the cylinder head

1   General information

Caution: *If the vehicle is equipped with a Delco Loc II or Theftlock audio system, make sure you have the correct activation code before disconnecting the battery. See the information at the front of this manual for the radio re-activation procedure.*

This Part of Chapter 2 is devoted to in-vehicle repair procedures for the 2.2 liter, four-cylinder overhead valve engines. These engines have cast iron blocks and cast aluminum pistons. The aluminum cylinder head has replaceable valve seats and guides. Stamped steel rocker arms and tubular pushrods actuate the valves. All information concerning engine removal and installation and engine block and cylinder head overhaul can be found in Part C of this Chapter.

The following repair procedures are based on the assumption the engine is in the vehicle. If the engine has been removed from the vehicle and mounted on a stand, many of the steps outlined in this Part of Chapter 2 will not apply.

The Specifications included in this Part of Chapter 2 apply only to the procedures contained in this Part. Part C of Chapter 2 contains the Specifications necessary for cylinder head and engine block rebuilding.

2   Repair operations possible with the engine in the vehicle

Many major repair operations can be accomplished without removing the engine from the vehicle. Clean the engine compartment and the exterior of the engine with some type of degreaser before any work is done. It'll make the job easier and help keep dirt out of the internal areas of the engine. Depending on the components involved, it may be helpful to remove the hood to improve access to the engine as repairs are performed (refer to Chapter 11 if necessary). Cover the fenders to prevent damage to the paint. Special pads are available, but an old bedspread or blanket will also work.

If vacuum, exhaust, oil or coolant leaks develop, indicating a need for gasket or seal replacement, the repairs can generally be made with the engine in the vehicle. The intake and exhaust manifold gaskets, timing chain cover gasket, oil pan gasket, crankshaft oil seals and cylinder head gasket are all accessible with the engine in place.

Exterior engine components, such as the intake and exhaust manifolds, the oil pan (and the oil pump), the water pump, the starter motor, the alternator and the fuel system components can be removed for repair with the engine in place. Since the cylinder head can be removed without pulling the engine, valve component servicing can also be accomplished with the engine in the vehicle. Replacement of the timing chain and sprockets is also possible with the engine in the vehicle. In extreme cases caused by a lack of necessary equipment, repair or replacement of piston rings, pistons, connecting rods and rod bearings is possible with the engine in the vehicle. However, this prac-

tice is not recommended because of the cleaning and preparation work that must be done to the components involved.

3   Top Dead Center (TDC) for number one piston - locating

1   Top Dead Center (TDC) is the highest point in the cylinder each piston reaches as it travels up-and-down when the crankshaft turns. Each piston reaches TDC on the compression stroke and again on the exhaust stroke, but TDC generally refers to piston position on the compression stroke.

2   Positioning the piston(s) at TDC is an essential part of certain procedures such as camshaft removal and timing chain/sprocket removal.

3   Before beginning this procedure, be sure to place the transaxle in Neutral (or Park on automatic transaxle models), apply the parking brake and block the rear wheels. Disconnect the cable from the negative terminal of the battery. **Caution:** *If the vehicle is equipped with a Delco Loc II or Theftlock audio system, make sure you have the correct activation code before disconnecting the battery. See the information at the front of this manual for the radio re-activation procedure.*

4   Remove the spark plugs (see Chapter 1).

5   When looking at the drivebelt end of the engine, normal crankshaft rotation is clockwise. In order to bring any piston to TDC, the crankshaft must be turned with a socket and ratchet attached to the bolt threaded into the center of the lower drivebelt pulley on the crankshaft.

6   Have an assistant turn the crankshaft with a socket and ratchet as described above while you hold a finger over the number one spark plug hole. **Note:** *See the Specifications for the number one cylinder location.*

7   When the piston approaches TDC, pressure will be felt at the spark plug hole. Have your assistant stop turning the crankshaft when the timing marks are aligned.

8   If the timing marks are bypassed, turn the crankshaft two complete revolutions clockwise until the timing marks are properly aligned.

9   After the number one piston has been positioned at TDC on the compression stroke, TDC for any of the remaining pistons can be located by turning the crankshaft one-half turn (180-degrees) to get to TDC for the next cylinder in the firing order.

4   Valve cover - removal and installation

Removal

Refer to illustrations 4.6 and 4.7

1   Disconnect the cable from the negative terminal of the battery. **Caution:** *If the vehicle is equipped with a Delco Loc II or Theftlock audio system, make sure you have the correct activation code before*

5.2 Remove the rocker arms one at a time and place them in a marked container to avoid assembly mistakes

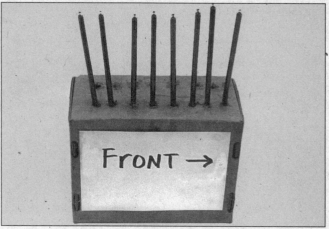

5.4 If more than one pushrod is being removed, store them in a perforated cardboard box to prevent mix-ups during installations (note the label indicating the front of the engine)

2A

disconnecting the battery. See the information at the front of this manual for the radio re-activation procedure.

2   Remove the air cleaner outlet duct from the throttle body to the air cleaner assembly.

3   On 1998 and later models, remove the resonator from the throttle body and the resonator bracket that shields the fuel rail and harness connectors.

4   Remove the PCV hose from the valve cover.

5   Disconnect the accelerator cable, the throttle valve cable (automatic transaxle) and cruise control cable, if equipped.

6   Remove the valve cover bolts **(see illustration)**.

7   Detach the valve cover from the cylinder head **(see illustration)**.
Note: *If the cover is stuck to the cylinder head, use a block of wood and hammer to dislodge it. If that doesn't work, try to slip a flexible putty knife between the cylinder head and cover to break the gasket seal. Don't pry at the cover-to-cylinder head joint or damage to the sealing surfaces may occur (leading to oil leaks in the future).*

Installation

8   The mating surfaces of the cylinder head and valve cover must be perfectly clean when the cover is installed. Use a gasket scraper to remove all traces of sealant or old gasket, then clean the mating surfaces with lacquer thinner or acetone (if there's sealant or oil on the mating surfaces when the cover is installed, oil leaks may develop). The cylinder head and cover are made of aluminum, so be extra careful not to nick or gouge the mating surfaces with the scraper.

9   Clean the mounting bolt threads with a die if necessary to remove any corrosion and restore damaged threads. Make sure the threaded holes in the cylinder head are clean - run a tap into them if necessary to remove corrosion and restore damaged threads.

10   Apply a thin coat of RTV-type sealant to the sealing flange on the cover and install a new gasket.

11   Place the valve cover on the cylinder head and install the mounting bolts. Tighten the bolts a little at a time to the torque listed in this Chapter's Specifications. Work from the center out in a spiral pattern.

12   Complete the installation by reversing the removal procedure.

5   Rocker arms and pushrods - removal, inspection and installation

Removal

Refer to illustrations 5.2 and 5.4

1   Refer to Section 5 and detach the valve cover from the cylinder head.

2   Beginning at the front of the cylinder head, loosen the rocker arm nuts **(see illustration)**.

3   Remove the nuts, the rocker arms and the pivot balls and store them in marked containers (they must be reinstalled in their original locations).

4   Remove the pushrods and store them in order to make sure they don't get mixed up during installation **(see illustration)**.

5   If the pushrod guides must be removed for any reason, make sure they're marked so they can be reinstalled in their original locations.

Inspection

6   Check each rocker arm for wear, cracks and other damage, especially where the pushrods and valve stems contact the rocker arm faces.

7   Make sure the hole at the pushrod end of each rocker arm is open.

8   Check each rocker arm pivot area for wear, cracks and galling. If the rocker arms are worn or damaged, replace them with new ones and use new pivot balls as well.

9   Inspect the pushrods for cracks and excessive wear at the ends. Roll each pushrod across a piece of plate glass to see if it's bent (if it wobbles, it's bent).

Installation

10   Lubricate the lower ends of the pushrods with clean engine oil or moly-base grease and install them in their original locations. Make sure each pushrod seats completely in the lifter socket.

11   Apply moly-base grease to the ends of the valve stems and the upper ends of the pushrods before positioning the rocker arms and installing the nuts.

12   Apply moly-base grease to the pivot balls to prevent damage to the mating surfaces before engine oil pressure builds up. Set the rocker arms in place, then install the pivot balls and nuts. Tighten the nuts to the torque listed in this Chapter's Specifications.

6   Valve springs, retainers and seals - replacement

Refer to illustrations 6.4, 6.9 and 6.17
Note: *Broken valve springs and defective valve stem seals can be replaced without removing the cylinder head. Two special tools and a compressed air source are normally required to perform this operation, so read through this Section carefully and rent or buy the tools before beginning the job. If compressed air isn't available, a length of nylon rope can be used to keep the valves from falling into the cylinder during this procedure.*

1   Refer to Section 4 and remove the valve cover.

2   Remove the spark plug from the cylinder which has the defective component. If all of the valve stem seals are being replaced, all of the spark plugs should be removed.

3   Turn the crankshaft until the piston in the affected cylinder is at top dead center on the compression stroke (see Section 3 for instruc-

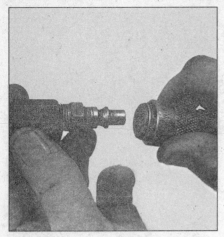

6.4 This is what the air hose adapter that threads into the spark plug hole looks like - they're commonly available from auto part stores

6.9 Once the spring is depressed, the keepers can be removed with a small magnet or needle-nose pliers (a magnet is preferred to prevent dropping the keepers)

6.17 Apply a small dab of grease to the keepers before installation-it will hold them in place on the valve stem as the spring is released

tions). If you're replacing all of the valve stem seals, begin with cylinder number one and work on the valves for one cylinder at a time. Move from cylinder-to-cylinder following the firing order sequence (see this Chapter's Specifications).

4   Thread an adapter into the spark plug hole **(see illustration)** and connect an air hose from a compressed air source to it. Most auto parts stores can supply the air hose adapter. **Note:** *Many cylinder compression gauges utilize a screw-in fitting that may work with your air hose quick-disconnect fitting.*

5   Remove the nut, pivot ball and rocker arm for the valve with the defective part and pull out the pushrod. If all of the valve stem seals are being replaced, all of the rocker arms and pushrods should be removed (refer to Section 5).

6   Apply compressed air to the cylinder. **Warning:** *The piston may be forced down by compressed air, causing the crankshaft to turn suddenly. If the wrench used when positioning the number one piston at TDC is still attached to the bolt in the crankshaft nose, it could cause damage or injury when the crankshaft moves.*

7   The valves should be held in place by the air pressure.

8   If you don't have access to compressed air, an alternative method can be used. Position the piston at a point approximately 45-degrees before TDC on the compression stroke, then feed a long piece of nylon rope through the spark plug hole until it fills the combustion chamber. Be sure to leave the end of the rope hanging out of the engine so it can be removed easily. Use a large ratchet and socket to rotate the crankshaft in the normal direction of rotation (clockwise) until slight resistance is felt.

9   Stuff shop rags into the cylinder head holes above and below the valves to prevent parts and tools from falling into the engine, then use a valve spring compressor to compress the spring. Remove the keepers with small needle-nose pliers or a magnet **(see illustration)**. **Note:** *A couple of different types of tools are available for compressing the valve springs with the cylinder head in place. The type shown here utilizes the rocker arm stud and nut for leverage, while the other type grips the lower spring coils and presses on the retainer as the knob is turned. Both types work very well, although the lever type is usually less expensive.*

10   Remove the spring retainer and valve spring, then remove the valve guide seal. **Note:** *If air pressure fails to hold the valve in the closed position during this operation, the valve face or seat is probably damaged. If so, the cylinder head will have to be removed for additional repair operations.*

11   Wrap a rubber band or tape around the top of the valve stem so the valve won't fall into the combustion chamber, then release the air pressure. **Note:** *If a rope was used instead of air pressure, turn the crankshaft slightly in the direction opposite normal rotation.*

12   Inspect the valve stem for damage. Rotate the valve in the guide and check the end for eccentric movement, which would indicate the

valve stem is bent.

13   Move the valve up-and-down in the guide and make sure it doesn't bind. If the valve stem binds, either the valve is bent or the guide is damaged. In either case, the cylinder head will have to be removed for repair.

14   Reapply air pressure to the cylinder to retain the valve in the closed position, then remove the tape or rubber band from the valve stem. If a rope was used instead of air pressure, rotate the crankshaft in the normal direction of rotation until slight resistance is felt.

15   Lubricate the valve stem with engine oil and install a new valve guide seal. **Note:** *Intake and exhaust valve seals are different.*

16   Install the spring in position over the valve.

17   Install the valve spring retainer. Compress the valve spring and carefully install the keepers in the groove. Apply a small dab of grease to the inside of each keeper to hold it in place if necessary **(see illustration)**. Remove the pressure from the spring tool and make sure the keepers are seated.

18   Disconnect the air hose and remove the adapter from the spark plug hole. If a rope was used in place of air pressure, turn the crankshaft counterclockwise and pull it out of the cylinder.

19   Refer to Section 5 and install the rocker arm(s) and pushrod(s).

20   Install the spark plug(s) and hook up the wire(s).

21   Refer to Section 4 and install the valve cover.

22   Start and run the engine, then check for oil leaks and unusual sounds coming from the valve cover area.

---

7   Intake manifold - removal and installation

Removal

Refer to illustrations 7.10a, 7.10b, 7.11 and 7.12

1   Relieve the fuel pressure (see Chapter 4), then disconnect the negative battery cable from the battery. **Caution:** *If the vehicle is equipped with a Delco Loc II or Theftlock audio system, make sure you have the correct activation code before disconnecting the battery. See the information at the front of this manual for the radio re-activation procedure.*

2   Remove the air intake ducts (see Chapter 4).

3   On 1998 and later models, remove the resonator from the throttle body and the resonator bracket that shields the fuel rail and harness connectors.

4   Remove the PCV hose from the valve cover.

5   Remove the accelerator cable, the cruise control cable and TV cables and brackets (see Chapter 4).

6   Remove the serpentine drivebelt from the engine (see Chapter 1).

7   Remove the power steering pump (if equipped) and tie it aside in

7.10a EGR tube location (arrow) - 1998 models

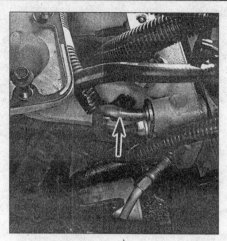

7.10b EGR tube location (arrow) - 1997 and earlier models

7.11 Remove the intake manifold mounting bracket (arrow) bolt and separate the bracket from the engine/intake manifold

2A

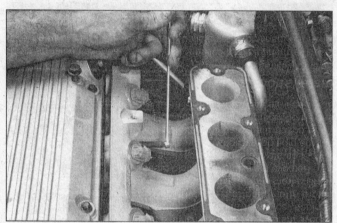

7.12 Remove the intake manifold bolts without dropping them into the engine compartment by using a magnet

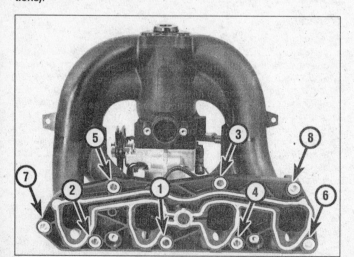

7.15a Intake manifold tightening sequence - 1997 and earlier models

an upright position (see Chapter 10).

8   Raise the front of the vehicle and support it securely on jackstands. Drain the coolant (refer to Chapter 1).

9   Label and disconnect any wires (MAP, TPS, IAC, EGR, etc.) and vacuum hoses which will interfere with manifold removal.

10   Remove the EGR pipe from the intake manifold **(see illustrations)**.

11   Remove the intake manifold mounting bracket and bolt **(see illustration)** from the engine.

12   Remove the mounting nuts from the intake manifold **(see illustration)**.

13   On 1998 and later models, remove the fuel rail and injectors as a complete assembly (see Chapter 4).

14   Separate the intake manifold and gasket from the engine. Scrape all traces of gasket material off the intake manifold and cylinder head gasket mating surfaces.

Installation

Refer to illustrations 7.15a and 7.15b

15   Install the intake manifold using a new gasket. Tighten the nuts/bolts, in the recommended sequence, to the torque listed in this Chapter's Specifications **(see illustrations)**.

16   The remainder of installation is the reverse of removal.

17   Add coolant, run the engine and check for leaks and proper operation.

---

8   Exhaust manifold - removal and installation

Removal

Refer to illustrations 8.8, 8.9 and 8.10

1   Disconnect the negative battery cable from the battery. **Caution:** *If the vehicle is equipped with a Delco Loc II or Theftlock audio system, make sure you have the correct activation code before disconnecting the battery.*

7.15b Intake manifold tightening sequence - 1998 and later models

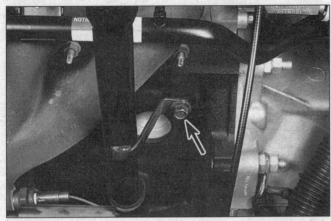

8.8 Remove the oil filler tube mounting bracket bolt (arrow) and remove the tube

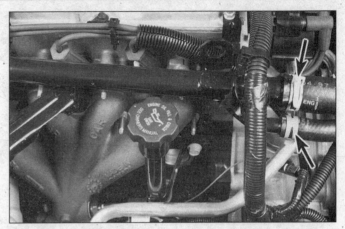

8.9 Remove the mounting clamps (arrows) and separate the heater outlet hoses from the pipes

8.10 Remove the exhaust manifold mounting nuts (arrows) from the cylinder head

9.6 Support the engine and remove the bolts (arrows) from the front engine mount

See the information at the front of this manual for the radio re-activation procedure.

2   Remove the serpentine drivebelt from the engine (see Chapter 1).

3   Partially drain the coolant system (see Chapter 1).

4   Unplug the oxygen sensor lead.

5   Raise the front of the vehicle, support it securely on jackstands and apply the parking brake. Block the rear wheels to keep the vehicle from rolling off the jackstands. Unbolt the exhaust pipe from the manifold. Lower the vehicle.

6   Remove the alternator (see Chapter 5).

7   Remove the bolts from the exhaust manifold flange.

8   Remove the oil fill tube assembly **(see illustration)**.

9   Remove the heater outlet hose assembly **(see illustration)**.

10   Remove the exhaust manifold-to-cylinder head nuts/bolts **(see illustration)**, pull the manifold off the engine and lift it out of the exhaust pipe flange. Remove and discard the gasket.

11   Scrape all traces of gasket material off the exhaust manifold and cylinder head mating surfaces.

12   Clean all bolt and stud threads before installation. A wire brush can be used on the manifold mounting studs, while a tap works well when cleaning the bolt holes.

Installation

13   If a new manifold is being installed, transfer the oxygen sensor from the old manifold to the new one.

14   Install the exhaust manifold using a new gasket. Tighten the nuts/bolts to the torque listed in this Chapter's Specifications. Work in a spiral pattern from the center out.

15   The remainder of installation is the reverse of removal. Refill the cooling system (see Chapter 1).

9   Cylinder head - removal and installation

Caution: *Allow the engine to cool completely before loosening the cylinder head bolts.*

Note: *On vehicles with high mileage or during an engine overhaul, camshaft lobe height should be checked prior to cylinder head removal (see Chapter 2, Part C, Section 13 for instructions).*

Removal

Refer to illustrations 9.6, 9.13a and 9.13b

1   Relieve the fuel pressure (see Chapter 4), then disconnect the negative battery cable from the battery. **Caution:** *If the vehicle is equipped with a Delco Loc II or Theftlock audio system, make sure you have the correct activation code before disconnecting the battery. See the information at the front of this manual for the radio re-activation procedure.*

2   Remove the alternator and brackets as described in Chapter 5. Remove the power steering pump and position the assembly to the side while keeping the power steering lines attached to the pump assembly.

3   Remove the intake manifold as described in Section 7.

4   Remove the exhaust manifold as described in Section 8.

5   Unbolt the drivebelt tensioner.

6   Support the engine from above using an engine support fixture (available at auto parts stores or equipment rental yards) or from below with a floor jack. Use a block of wood between the floor jack and the engine when raising it to prevent damage to the oil pan. Remove the front engine mount from the cylinder head **(see illustration)**.

7   Remove the front engine accessory bracket.

9.13a To avoid mixing up the cylinder head bolts, use a new gasket to transfer the bolt hole pattern to a piece of cardboard, then punch holes to accept the bolts

9.13b Start with the outer bolts and work inward in a circular pattern to prevent warping the cylinder head

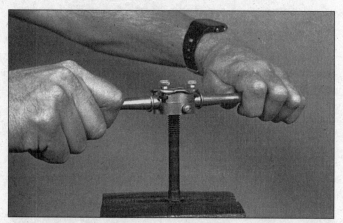

9.19 A die should be used to remove sealant and corrosion from the cylinder head bolt threads prior to installation

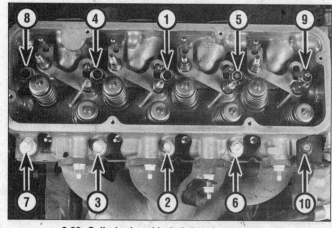

9.23 Cylinder head bolt tightening sequence

8   Disconnect any remaining wires, hoses, fuel and vacuum lines from the cylinder head. Be sure to label them to simplify reinstallation.

9   Disconnect the spark plug wires and remove the spark plugs. Be sure the plug wires are labeled to simplify reinstallation.

10   Remove the valve cover (see Section 4).

11   Remove the rocker arms and pushrods (see Section 5).

12   Remove the ignition coil assembly (coil packs, wires, ignition module) (see Chapter 5).

13   Using the new cylinder head gasket, outline the cylinders and bolt pattern on a piece of cardboard (see illustration). Be sure to indicate the front of the engine for reference. Punch holes at the bolt locations. Loosen each of the cylinder head mounting bolts 1/4-turn at a time until they can be removed by hand (see illustration). Store the bolts in the cardboard holder as they're removed - this will ensure they are reinstalled in their original locations, which is absolutely essential.

14   Lift the cylinder head off the engine. If it's stuck, don't attempt to pry it off - you could damage the sealing surfaces. Instead, use a hammer and block of wood to tap the cylinder head and break the gasket seal. Place the cylinder head on a block of wood to prevent damage to the gasket surface.

15   Remove the cylinder head gasket.

16   Refer to Chapter 2, Part C, for cylinder head disassembly and valve service procedures.

Installation

Refer to illustrations 9.19 and 9.23

17   If a new cylinder head is being installed, transfer all external parts from the old cylinder head to the new one.

18   If not already done, thoroughly clean the gasket surfaces on the cylinder head and the engine block. Do not gouge or otherwise damage the soft aluminum gasket surfaces.

19   To get the proper torque readings, the threads of the cylinder head bolts must be clean (see illustration). This also applies to the threaded holes in the engine block. Run a tap through the holes to ensure they are clean.

20   Place the gasket in position over the engine block dowel pins. Note any marks like "THIS SIDE UP" and install the gasket accordingly.

21   Carefully lower the cylinder head onto the engine, over the dowel pins and the gasket.

22   Install the bolts finger tight. Don't tighten any of the bolts at this time.

23   Tighten each of the bolts in 1/4-turn increments in the recommended sequence (see illustration). Note that the different length bolts have different torque specifications. Continue tightening in the recommended sequence until the torque (and angle of rotation) specified in this Chapter's Specifications is reached.

24   The remaining installation steps are the reverse of removal.

25   Be sure to refill the cooling system and change the oil and filter (see Chapter 1).

10   Valve lifters - removal, inspection and installation

Removal

Refer to illustrations 10.4a, 10.4b, 10.5 and 10.6

1   A noisy valve lifter can be isolated when the engine is idling. Place a length of hose near the location of each valve while listening at the other end of the hose. Another method is to remove the valve cover and, with the engine idling, place a finger on each of the valve spring retainers, one at a time. If a valve lifter is defective it will be evident from the shock felt at the retainer as the valve seats. The most likely cause of a noisy valve lifter is a piece of dirt trapped inside the lifter.

2A

10.4a Remove the anti-rotation bracket bolts (arrows) and . . .

10.4b . . . lift the anti-rotation brackets from the engine block

10.5 Use a magnet to lift the hydraulic lifter from the engine block

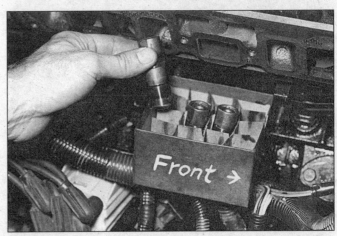

10.6 If you're removing more than one lifter, keep them in order in a clearly labeled box

2   Disconnect the cable from the negative battery terminal. **Caution:** *If the vehicle is equipped with a Delco Loc II or Theftlock audio system, make sure you have the correct activation code before disconnecting the battery. See the information at the front of this manual for the radio re-activation procedure.*

3   Remove the valve cover, the rocker arms and pushrods and then remove the cylinder head (see Section 9).

4   Remove the anti-rotation brackets from the engine block **(see illustrations)**.

5   A magnetic pick-up tool or scribe can be positioned at the top of the lifter and used to raise it up and out of the bore **(see illustration)**. To remove a stuck lifter, a special hydraulic lifter removal tool may be used. Do not use pliers or other tools on the outside of the lifter body - they will damage the machined surface and render the lifter useless.

6   If you're removing more than one lifter at a time, store them in a marked container **(see illustration)**. They must be returned to the same locations. **Note:** *Refer to Chapter 2, Part C, for camshaft removal and inspection procedures.*

Inspection

Refer to illustrations 10.8a and 10.8b

7   Clean the lifters with solvent and dry them thoroughly without mixing them up.

8   Check each lifter wall and pushrod seat for scuffing, score marks and uneven wear **(see illustrations)**. Each roller (the surface that rides on the cam lobe) must be free of nicks, score marks or damage. If the lifter walls are damaged or worn (which isn't very likely), inspect the lifter bores in the engine block as well. If the pushrod seats are worn, check the pushrod ends.

9   On the roller lifters, check the rollers carefully for wear and damage and make sure they turn freely without excessive play.

Installation

10   If new lifters are being installed, a new camshaft must also be installed. If the camshaft is replaced, then install new lifters as well (see Chapter 2, Part C). Never install used lifters unless the original camshaft is used and the lifters can be installed in their original locations! When installing lifters, make sure they're coated with moly-base grease or engine assembly lube.

11   The remaining installation steps are the reverse of removal. There are several steps that must be performed carefully to avoid damaging the valvetrain components.

12   When installing the hydraulic lifters back into the bores, make sure that the flat sides of the lifters are aligned with the flat sides of the anti-rotation brackets. The roller will align parallel with the camshaft also.

13   Install the anti-rotation brackets onto the lifters, making sure they align correctly.

14   Tighten the anti-rotation bracket bolts to the torque listed in this Chapter's Specifications.

11   Crankshaft front oil seal - replacement

Refer to illustrations 11.5a and 11.5b

1   Disconnect the negative battery cable from the battery. **Caution:** *If the vehicle is equipped with a Delco Loc II or Theftlock audio system, make sure you have the correct activation code before disconnecting the battery. See the information at the front of this manual for the radio re-activation procedure.*

2   Remove the engine drivebelt (see Chapter 1).

3   Loosen the right front wheel lug nuts. Raise the vehicle and sup-

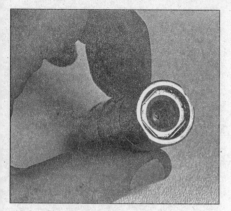

10.8a Check each lifter wall and pushrod seat for scuffing, score marks and uneven wear

10.8b The roller on roller lifters must turn freely - check for wear and excessive play as well

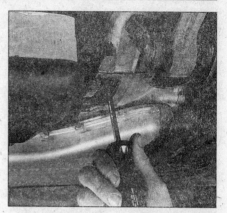

11.5a Have an assistant hold the ring gear with a large screwdriver as the pulley-to-crankshaft bolt is loosened or tightened

2A

port it securely on jackstands.

4   Remove the right front wheel and the splash shield from the fenderwell. Remove the flywheel/driveplate inspection cover.

5   Have an assistant hold the ring gear on the flywheel (manual transaxle) or driveplate (automatic transaxle) to prevent the engine from rotating when loosening the pulley hub bolt **(see illustration)**. Remove the crankshaft pulley bolts and remove the pulley. Remove the pulley hub using a special crankshaft balancer/hub removal tool **(see illustration)**.

6   Pry the old oil seal out with a seal removal tool or a screwdriver. Be very careful not to nick or otherwise damage the crankshaft in the process. Wrap the screwdriver tip with vinyl tape to protect the crankshaft.

7   Apply a thin coat of RTV-type sealant to the outer edge of the new seal. Lubricate the seal lip with multi-purpose grease or clean engine oil.

8   Place the seal squarely in position in the bore and drive it into place with a special seal driver tool.

9   If you don't have the special tool, carefully tap the seal into place with a large socket or piece of pipe and a hammer. The outer diameter of the socket or pipe should be the same size as the seal outer diameter. Make sure the seal is seated completely in the bore.

10   Install the crankshaft hub using a special crankshaft balancer/hub installation tool. If the special tool is not available, press the hub on using a large socket or section of pipe, washers and a long bolt of the correct size and thread pitch to thread into the crankshaft.

11   Install the pulley. Have an assistant hold the ring gear on the flywheel (manual transaxle) or driveplate (automatic transaxle) to prevent the engine from rotating when tightening the pulley hub bolt **(see illustration 11.5a)**.

12   Reinstall the remaining parts in the reverse order of removal.

13   Start the engine and check for oil leaks at the seal.

12   Timing chain cover, chain and sprockets - removal, inspection and installation

Cover removal

Refer to illustration 12.10

1   Disconnect the negative battery cable from the battery. **Caution:** *If the vehicle is equipped with a Delco Loc II or Theftlock audio system, make sure you have the correct activation code before disconnecting the battery. See the information at the front of this manual for the radio re-activation procedure.*

2   Remove the engine drivebelt (see Chapter 1). Remove drivebelt tensioner.

3   Remove the power steering pump and position it to the side while keeping the power steering lines attached to the pump.

4   Remove the alternator (see Chapter 5) and the alternator brace from the engine.

5   Support the engine from above using an engine support fixture (available at auto parts stores or rental yards). Remove the front engine mount **(see illustration 9.6)**.

6   Remove the water pump pulley. Remove the engine accessory bracket.

7   Raise the vehicle and support it securely on jackstands.

8   Remove the oil pan (see Section 13).

9   Remove the crankshaft pulley and hub from the engine (see Section 11).

10   Remove the timing chain cover bolts and separate the cover from the engine **(see illustration)**.

11   Use a putty knife to break the cover loose from the engine, if necessary. Don't strike or pry on the cover, since it is made of plastic.

11.5b Use a puller to remove the crankshaft pulley and/or hub

12.10 Timing chain cover bolt locations

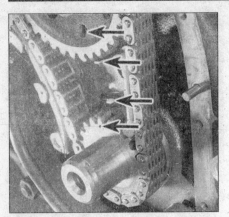

12.13 Align the camshaft and crankshaft sprocket timing marks with the tabs on the timing chain tensioner (arrows)

12.14 Press the tensioner in and insert an appropriate size drill bit through the hole to retain the tensioner in the retracted position

12.15 Wedge a prybar against two bolts in the temporarily installed crankshaft hub to hold the crankshaft while loosening the camshaft sprocket bolt

12   Use a gasket scraper to remove all traces of old gasket material and sealant from the cover and engine block. Clean the gasket sealing surfaces with lacquer thinner or acetone.

Timing chain removal

Refer to illustrations 12.13, 12.14 and 12.15

13   Temporarily install the crankshaft pulley hub and bolt. Rotate the crankshaft until the timing marks on the crankshaft and camshaft sprockets align with the tabs on the chain tensioner housing **(see illustration)**. **Note:** *Before removing the timing chain tensioner, check it carefully. Measure the distance from the hole in the bracket to the unworn surface of the timing chain tensioner shoe. It should not exceed 5/16 inch (8 mm). If out of limits, replace the tensioner, timing chain and both sprockets. This excessive play in the gears and chain can only be removed by installing new parts.*

14   Push the spring back on the timing chain tensioner and insert an appropriate size drill bit into the hole to retain it in the retracted position **(see illustration)**.

15   Use a prybar against two bolts in the crankshaft balancer hub (installed temporarily) to keep the engine from turning while removing the camshaft bolt **(see illustration)**. Do not turn the camshaft in the process (if you do, realign the timing marks before the sprocket is removed).

16   Use two large screwdrivers to carefully pry the camshaft sprocket off the camshaft dowel pin (it may come off easily with no tools required), and remove the sprocket and chain.

Inspection

17   Timing chains and sprockets should be replaced in sets. If you intend to install a new timing chain, remove the crankshaft sprocket with a puller and install a new one. Be sure to align the key in the crankshaft with the keyway in the sprocket during installation.

18   Clean the timing chain and sprockets with solvent and dry them with compressed air (if available). **Warning:** *Wear eye protection when using compressed air.*

19   Inspect the components for wear and damage. Look for teeth that are deformed, chipped, pitted and cracked.

20   The timing chain should be replaced with a new one if the engine has high mileage, the chain has visible damage, or total freeplay (without the tensioner) midway between the sprockets exceeds one inch. Failure to replace a worn timing chain may result in erratic engine performance, backfiring, loss of power and decreased fuel mileage. Loose chains can "jump" timing and in the worst case, will result in severe engine damage.

Installation

21   Mesh the timing chain with the camshaft sprocket, then engage it with the crankshaft sprocket. The timing marks should be aligned as shown in **illustration 12.13**. **Note:** *If the crankshaft has been disturbed, turn it until the mark stamped on the crankshaft sprocket is pointing at*

the projection on the tensioner. If the camshaft was turned, install the sprocket temporarily and turn the camshaft until the timing marks align.

22   Install the camshaft sprocket bolt and tighten it to the torque listed in this Chapter's Specifications.

23   Press the timing chain against the tensioner, pull out the pin retaining the spring and release the tensioner.

24   Lubricate the chain and sprocket with clean engine oil. Rotate the engine through two complete revolutions and check the alignment of the timing marks again.

25   Install the timing chain cover, using a new gasket and tighten the cover bolts to the torque listed in this Chapter's Specifications.

26   Install the oil pan.

27   The remaining installation steps are the reverse of removal.

13   Oil pan - removal and installation

Removal

Refer to illustration 13.10

1   Warm up the engine, then drain the oil and remove the oil filter (see Chapter 1).

2   Disconnect the negative battery cable from the battery. **Caution:** *If the vehicle is equipped with a Delco Loc II or Theftlock audio system, make sure you have the correct activation code before disconnecting the battery. See the information at the front of this manual for the radio re-activation procedure.*

3   Raise the vehicle and support it securely on jackstands.

4   Remove the right engine splash shield and the exhaust pipe shield.

5   On air-conditioned models, remove the air conditioner brace at the starter and compressor bracket.

6   Remove the starter and bracket (see Chapter 5).

7   Remove the flywheel/driveplate inspection cover (see Chapter 7).

8   Remove the engine mount strut and remove the support bolts from the engine mount strut bracket. Lower the bracket slightly to gain clearance for oil pan removal.

9   Remove the oil filter extension (automatic transaxle equipped models only).

10   Remove the bolts and nuts securing the oil pan to the engine block **(see illustration)**.

11   Tap on the pan with a soft-face hammer to break the gasket seal, then detach the oil pan from the engine.

Installation

12   Using a gasket scraper, remove all traces of old gasket and/or sealant from the engine block and oil pan. Make sure the threaded bolt holes in the block are clean. Wash the oil pan with solvent and dry it thoroughly.

13.11 Remove the oil pan mounting bolts
(arrows; not all the bolts are visible in
this photo)

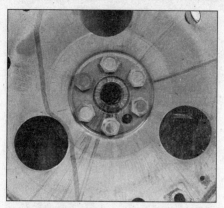

15.3 Most flywheels and driveplates have
locating dowels - if the one you're
working on doesn't, make some marks to
ensure correct installation

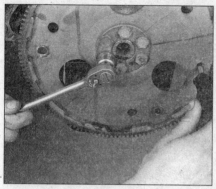

15.4 A large screwdriver wedged in the
starter ring gear teeth or one of the holes
in the driveplate can be used to keep the
flywheel/driveplate from turning as the
mounting bolts are removed

2A

13   Check the gasket flanges for distortion, particularly around the bolt holes. If necessary, place the pan on a block of wood and use a hammer to flatten and restore the gasket surfaces. Clean the mating surfaces with lacquer thinner or acetone.

14   Place a 1/8-inch diameter bead of RTV sealant on the oil pan-to-block sealing flanges and the oil pan-to-front cover surface.

15   Apply a thin coat of RTV sealant to the ends of the rear oil pan seal down to the ears. Press the oil pan seal into position.

16   Carefully place the oil pan against the block.

17   Install the bolts/nuts and tighten them evenly to the torque listed in this Chapter's Specifications. Start with the bolts closest to the center of the pan and work out in a spiral pattern. Don't overtighten them or leakage may occur.

18   Reinstall components removed for access to the oil pan.

19   Add oil and install a new filter (see Chapter 1), run the engine and check for oil leaks.

14   Oil pump - removal and installation

1   Remove the oil pan (see Section 13).

2   Place a large drain pan under the engine.

3   Unbolt the pump from the rear main bearing cap.

4   Lower the pump and extension shaft from the engine.

5   Before installation, prime the pump with engine oil. Pour oil into the pick-up while the pump extension shaft is turned.

6   Attach the pump, extension shaft and retainer to the main bearing cap. While aligning the pump with the dowel pins at the bottom of the main bearing cap, align the top end of the extension shaft with the lower end of the oil pump drive. When aligned properly, it should slip into place easily.

7   Install the pump mounting bolt and tighten it to the torque listed in this Chapter's Specifications.

8   Install the oil pan (see Section 13) and add oil (see Chapter 1).

15   Flywheel/driveplate - removal and installation

Refer to illustrations 15.3 and 15.4

1   Raise the vehicle and support it securely on jackstands, then refer to Chapter 7 and remove the transaxle. If it's leaking, now would be a very good time to have the front pump seal/O-ring replaced (automatic transaxle only).

2   Remove the pressure plate and clutch disc (Chapter 8 - manual transaxle equipped vehicles). Now is a good time to check/replace the clutch components and pilot bearing.

3   If there is no dowel pin, make some marks on the flywheel/driveplate and crankshaft to ensure correct alignment during reinstallation (see illustration).

4   Remove the bolts that secure the flywheel/driveplate to the crankshaft (see illustration). If the crankshaft turns, wedge a screwdriver through the openings in the driveplate (automatic transaxle) or against the flywheel ring gear teeth (manual transaxle). Since the flywheel is fairly heavy, be sure to support it while removing the last bolt.

5   Remove the flywheel/driveplate from the crankshaft.

6   Clean the flywheel to remove grease and oil. Inspect the friction surface for cracks, rivet grooves, burned areas and score marks. Light scoring can be removed with emery cloth. Check for cracked and broken ring gear teeth. Lay the flywheel on a flat surface and use a straightedge to check for warpage.

7   Clean and inspect the mating surfaces of the flywheel/driveplate and the crankshaft. If the crankshaft rear seal is leaking, replace it before reinstalling the flywheel/driveplate.

8   Position the flywheel/driveplate against the crankshaft. Be sure to align the dowel or marks made during removal. Before installing the bolts, apply thread locking compound to the threads.

9   Keep the flywheel/driveplate from turning as described above while you tighten the bolts to the torque listed in this Chapter's Specifications.

10   The remainder of installation is the reverse of the removal procedure.

16   Rear main oil seal - replacement

Refer to illustration 16.2

1   Remove the flywheel/driveplate (see Section 15).

2   Using a flat-bladed screwdriver or seal removal tool, carefully remove the oil seal from the engine block (see illustration). Be very

16.2 Carefully pry the old oil seal out

17.7a Strut mounting bolt locations (arrows)

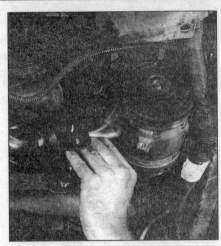

17.7b Removing the engine mount/strut bracket from the bottom of the engine compartment

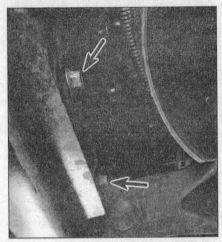

17.7c Remove the transaxle mounting bolts (arrows) to separate the mount

careful not to damage the crankshaft surface while prying the seal out.

3   Clean the bore in the block and the seal contact surface on the crankshaft. Check the seal contact surface on the crankshaft for scratches and nicks that could damage the new seal lip and cause oil leaks - if the crankshaft is damaged, the only alternative is a new or different crankshaft. Inspect the seal bore for nicks and scratches. Carefully smooth it with a fine file if necessary, but don't nick the crankshaft in the process.

4   A special tool is recommended to install the new oil seal. Lubricate the oil seal lips with clean engine oil or multi-purpose grease. Slide the seal onto the mandril until the dust lip bottoms squarely against the collar of the tool. **Note:** *If the special tool isn't available, carefully work the seal lip over the crankshaft and tap it into place with a hammer and punch.*

5   Align the dowel pin on the tool with the dowel pin hole in the crankshaft and attach the tool to the crankshaft by hand-tightening the bolts.

6   Turn the tool handle until the collar bottoms against the case, seating the seal.

7   Loosen the tool handle and remove the bolts. Remove the tool.

8   Check the seal and make sure it's seated squarely in the bore.

9   Install the flywheel/driveplate (see Section 15).

10   Install the transaxle.

---

17   Engine mounts - check and replacement

Refer to illustrations 17.7a, 17.7b, 17.7c and 17.7d

1   Engine mounts seldom require attention, but broken or deteriorated mounts should be replaced immediately or the added strain placed on the driveline components may cause damage or wear.

Check

2   During the check, the engine must be raised slightly to remove the weight from the mounts.

3   Raise the vehicle and support it securely on jackstands, then position a jack under the engine oil pan. Place a large block of wood between the jack head and the oil pan, then carefully raise the engine just enough to take the weight off the mounts. **Warning:** *DO NOT place any part of your body under the engine when it's supported only by a jack!*

4   Check the mounts to see if the rubber is cracked, hardened or

17.7d Remove the bracket from the front engine mount

separated from the metal plates. Sometimes the rubber will split right down the center.

5   Check for relative movement between the mount plates and the engine or frame (use a large screwdriver or pry bar to attempt to move the mounts). If movement is noted, lower the engine and tighten the mount fasteners.

Replacement

6   Disconnect the negative battery cable from the battery, then raise the vehicle and support it securely on jackstands (if not already done). **Caution:** *If the vehicle is equipped with a Delco Loc II or Theftlock audio system, make sure you have the correct activation code before disconnecting the battery. See the information at the front of this manual for the radio re-activation procedure.*

7   Raise the engine slightly with a jack or hoist. Remove the fasteners and detach the mount from the frame bracket **(see illustrations)**.

8   Remove the mount-to-block bracket bolts/nuts and detach the mount.

9   Installation is the reverse of removal. Use thread locking compound on the threads and be sure to tighten everything securely.

10   Rubber preservative should be applied to the mounts to slow deterioration.

Chapter 2 Part B
Overhead camshaft (OHC) engines

Contents

Specifications

General

Cylinder numbers (drivebelt end-to-transaxle end) 1-2-3-4
Firing order 1-3-4-2

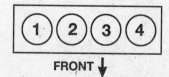

Cylinder numbering on overhead camshaft (OHC) engines

Camshafts and housings

Lobe lift (intake and exhaust)	
1995	0.3750 inch (9.525 mm)
1996 and later	
Intake	0.3540 inch (9.0 mm)
Exhaust	0.3460 inch (8.8 mm)
Lobe taper limit	0.0018 to 0.0033 inch per 0.50 inch (0.046 to 0.083 mm per 14.0 mm)
Endplay	0.0009 to 0.0088 inch (0.025 to 0.225 mm)
Journal diameter	
No. 1	1.5720 to 1.5728 inch (39.93 to 39.95 mm)
All others	1.3751 to 1.3760 inch (34.93 to 34.95 mm)
Bearing oil clearance	0.0019 to 0.0043 inch (0.050 to 0.110 mm)
Lifters	
Bore diameter	1.33381 to 1.3393 inch (33.989 to 34.019 mm)
Outside diameter	1.3369 to 1.3375 inch (33.959 to 33.975 mm)
Lifter-to-bore clearance	0.0006 to 0.0024 inch (0.014 to 0.060 mm)
Camshaft housing warpage limit	0.001 inch per 4.0 inch (0.025 mm per 100 mm)

Balance shafts and housing

Housing warpage maximum	0.0030 inch (0.076 mm)
Chain slack (with 3 lb. applied to chain guide)	0.040 inch (1.0 mm)
Thrust plate thickness	0.1159 to 0.1199 inch (2.945 to 3.045 mm)

Torque specifications

Ft-lbs (unless otherwise indicated)

Balance shaft assembly-to-block bolts **(see illustration 14.19)**	
Step 1	
Bolts 9, 10 and 12	18
Bolt 11	30
Bolt 13	39
Step 2	
Bolts 9, 10 and 12	Tighten an additional 70-degrees
Bolt 11	Tighten an additional 60-degrees
Balance shaft housing bolts **(see illustration 14.19)**	
Step 1	
Bolts 1, 2, 4, 5, 6 and 7	89 in-lbs
Bolts 3 and 8	132 in-lbs
Step 2	Tighten all bolts an additional 40-degrees
Balance shaft sprocket bolt (left hand thread)	
1995 through 1997	
Step 1	22
Step 2	Tighten an additional 45-degrees
1998 and later	
Step 1	30
Step 2	Tighten an additional 45-degrees
Balance shaft chain tensioner bolt	115 in-lbs
Balance shaft chain cover nut and bolt	115 in-lbs
Balance shaft thrust plate bolts	115 in-lbs
Balance shaft chain guide bolt	115 in-lbs
Camshaft housing-to-cylinder head bolts	
1995 and 1996	
Step 1	132 in-lbs
Step 2	Tighten an additional 90-degrees
1997 and later	
Step 1	16
Step 2	Tighten an additional 90-degrees
Camshaft cover-to-camshaft housing bolts (rear two on the intake camshaft housing)	
1995 and 1996	
Step 1	132 in-lbs
Step 2	Tighten an additional 30-degrees
1997 and later	
Step 1	16
Step 2	Tighten an additional 30-degrees
Camshaft sprocket-to-camshaft bolt	52
Crankshaft balancer-to-crankshaft bolt	
Step 1	129
Step 2	Tighten an additional 90-degrees
Crankshaft rear main oil seal housing bolts	106 in-lbs

Cylinder head bolts **(in sequence - see illustration 11.15)**
    1995
        Step 1
            Bolts 1 thru 8 30
            Bolts 9 and 10 26
        Step 2 .... Tighten all bolts an additional 90-degrees
    1996 and later models
        Step 1
            Bolts 1 thru 8 40
            Bolts 9 and 10 30
        Step 2 .... Tighten all bolts an additional 90-degrees
Engine mount-to-bracket bolts 96
Engine mount-to-body nuts 55
Engine mount bracket-to-engine bolts *
    Step 1 44
    Step 2 .... Tighten an additional 90-degrees
Exhaust manifold nuts-to-cylinder head 31
Exhaust manifold brace
    Bolts 41
    Nuts 19
Flywheel-to-crankshaft bolts
    Step 1 22
    Step 2 .... Tighten an additional 45-degrees
Intake manifold-to-cylinder head nuts/bolts 18
Oil pan bolts
    6 mm .... 106 in-lbs
    8 mm 18
Oil pump-to-balance shaft housing .... 106 in-lbs
Timing chain cover-to-housing bolts .... 106 in-lbs
Timing chain housing bolts 19
Timing chain housing-to-block stud 21
Timing chain tensioner .... 89 in-lbs

* Replace with new bolts anytime they are removed.

1 General information

Caution: *If the vehicle is equipped with a Delco Loc II or Theftlock audio system, make sure you have the correct activation code before disconnecting the battery.*

This Part of Chapter 2 is devoted to in-vehicle repair procedures for the 2.3L and 2.4L four-cylinder (Quad-4) engine. All information concerning engine removal and installation and engine block and cylinder head overhaul can be found in Part C of this Chapter. The following repair procedures are based on the assumption the engine is installed in the vehicle. If the engine has been removed from the vehicle and mounted on a stand, many of the steps outlined in this Part of Chapter 2 will not apply. The Specifications included in this Part of Chapter 2 apply only to the procedures contained in this Part. Part C of Chapter 2 contains the Specifications necessary for cylinder head and engine block rebuilding.

These engines utilize a number of advanced design features to increase power output and improve durability. The aluminum cylinder head contains four valves per cylinder. A double-row timing chain drives two overhead camshafts - one for intake and one for exhaust. Lightweight bucket-type hydraulic lifters actuate the valves. Rotators are used on all valves for extended service life.

A balance shaft assembly has been added to smooth power pulsations. This assembly is bolted to the bottom of the main bearing webs and is chain driven from the rear of the crank. The gerotor oil pump is driven by the balance shaft trailing shaft and is mounted to the rear of the balance shaft housing.

2 Repair operations possible with the engine in the vehicle

Many major repair operations can be accomplished without removing the engine from the vehicle. Clean the engine compartment and the exterior of the engine with some type of degreaser before any work is done. It'll make the job easier and help keep dirt out of the internal areas of the engine.

Depending on the components involved, it may be helpful to remove the hood to improve access to the engine as repairs are performed (refer to Chapter 11 if necessary). Cover the fenders to prevent damage to the paint. Special pads are available, but an old bedspread or blanket will also work.

If vacuum, exhaust, oil or coolant leaks develop, indicating a need for gasket or seal replacement, the repairs can generally be made with the engine in the vehicle. The intake and exhaust manifold gaskets, timing chain housing gasket, oil pan gasket, crankshaft oil seals and cylinder head gasket are all accessible with the engine in place.

Exterior engine components, such as the intake and exhaust manifolds, the oil pan (and the oil pump), the water pump, the starter motor, the alternator and the fuel system components can be removed for repair with the engine in place.

Since the cylinder head can be removed without pulling the engine, camshaft and valve component servicing can also be accomplished with the engine in the vehicle. Replacement of the timing chain and sprockets is also possible with the engine in the vehicle.

In extreme cases caused by a lack of necessary equipment,

3.1 Timing marks (arrows) on the OHC engine

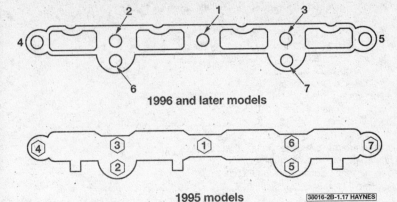

1996 and later models

1995 models   `38016-2B-1.17 HAYNES`

4.17 Intake manifold fastener tightening sequence

repair or replacement of piston rings, pistons, connecting rods and rod bearings is possible with the engine in the vehicle. However, this practice is not recommended because of the cleaning and preparation work that must be done to the components involved.

3 Top Dead Center (TDC) for number one piston - locating

Refer to illustration 3.1

1   Refer to Chapter 2A, Section 3 for the TDC locating procedure **(see illustration)**. The TDC locating procedure for these engines is the same as for the 2.2L engines.

4 Intake manifold - removal and installation

Removal

1   Relieve the fuel system pressure as described in Chapter 4.
2   Detach the cable from the negative terminal of the battery. **Caution:** *If the vehicle is equipped with a Delco Loc II or Theftlock audio system, make sure you have the correct activation code before disconnecting the battery.*
3   Remove the accelerator cable (see Chapter 4), the cruise control cable (if equipped) and the TV cable (automatic transaxle).
4   Label and disconnect the vacuum and breather hoses and electrical wires (MAP, IAT, EVAP, fuel injectors, etc.).
5   Remove the PCV oil/air separator (see Chapter 1).
6   Remove the oil fill cap and dipstick assembly. Unbolt the oil fill tube and detach it from the engine block, rotating it as necessary to gain clearance between the intake tubes.
7   Remove the alternator bracket bolt closest to the engine block with the stud end pointing UP.
8   Remove the EGR pipe from the EGR adapter (see Chapter 6).
9   Remove the intake manifold support brace.
10   Loosen the manifold mounting nuts/bolts in 1/4-turn increments until they can be removed by hand.
11   The manifold will probably be stuck to the cylinder head and force may be required to break the gasket seal. If necessary, dislodge the manifold with a soft-face hammer. **Caution:** *Don't pry between the cylinder head and manifold or damage to the gasket sealing surfaces will result and vacuum leaks could develop.*

Installation

Refer to illustration 4.17
Note: *The mating surfaces of the cylinder head and manifold must be perfectly clean when the manifold is installed. Gasket removal solvents in aerosol cans are available at most auto parts stores and may be helpful when removing old gasket material stuck to the cylinder head*

and manifold (since the components are made of aluminum, aggressive scraping can cause damage). Be sure to follow the directions printed on the container.
12   Use a gasket scraper to remove all traces of sealant and old gasket material, then clean the mating surfaces with lacquer thinner or acetone. If there's old sealant or oil on the mating surfaces when the manifold is reinstalled, vacuum leaks may develop.
13   Use a tap of the correct size to chase the threads in the bolt holes, then use compressed air (if available) to remove the debris from the holes. **Warning:** *Wear safety glasses or a face shield to protect your eyes when using compressed air. Use a die to clean and restore the stud threads.*
14   Position the gasket on the cylinder head. Make sure all intake port openings, coolant passage holes and bolt holes are aligned correctly.
15   Install the manifold, taking care to avoid damaging the gasket.
16   Thread the nuts/bolts into place by hand.
17   Tighten the nuts/bolts to the torque listed in this Chapter's Specifications following the recommended sequence **(see illustration)**. Work up to the final torque in three steps.
18   The remaining installation steps are the reverse of removal. Start the engine and check carefully for leaks at the intake manifold joints.

5 Exhaust manifold - removal and installation

Removal

Warning: *Allow the engine to cool completely before performing this procedure.*
1   Detach the cable from the negative terminal of the battery. **Caution:** *If the vehicle is equipped with a Delco Loc II or Theftlock audio system, make sure you have the correct activation code before disconnecting the battery.*
2   Unplug the oxygen sensor (see Chapter 6).
3   Remove the manifold heat shields.
4   Raise the vehicle and support it securely on jackstands, then remove the exhaust pipe-to-manifold nuts. The nuts are usually rusted in place, so penetrating oil should be applied to the stud threads before attempting to remove them. Loosen them a little at a time, working from side-to-side to prevent the flange from jamming.
5   Remove the exhaust manifold brace.
6   Separate the exhaust pipe flange from the manifold studs, then pull the pipe down slightly to break the seal at the manifold joint.
7   Loosen the exhaust manifold mounting nuts 1/4-turn at a time each, working from the inside out, until they can be removed by hand.
8   Separate the manifold from the cylinder head and remove it.

Installation

Refer to illustration 5.10
9   The manifold and cylinder head mating surfaces must be clean when the manifold is reinstalled. Use a gasket scraper to remove all

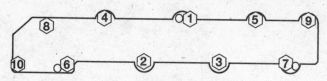

1996 and later models

1995 models

38016-2B-5.10 HAYNES

5.10 Exhaust manifold fastener tightening sequence

6.4 The crankshaft balancer can be held with a bar while the bolt is loosened or tightened

traces of old gasket material and carbon deposits.

10   Using a new gasket, install the manifold and hand tighten the fasteners. Following the recommended sequence **(see illustration)**, tighten the bolts/nuts to the torque listed in this Chapter's Specifications.

11   The remaining installation steps are the reverse of removal.

6   Crankshaft front oil seal - replacement

Refer to illustrations 6.4, 6.5, 6.6, 6.9 and 6.11

1   Detach the cable from the negative terminal of the battery. **Caution:** *If the vehicle is equipped with a Delco Loc II or Theftlock audio system, make sure you have the correct activation code before disconnecting the battery.*

2   Remove the drivebelt (see Chapter 1).

3   With the parking brake applied and the shifter in Park (automatic) or in gear (manual), raise the front of the vehicle and support it securely on jackstands.

4   Remove the bolt from the front of the crankshaft. A breaker bar will probably be necessary, since the bolt is very tight. Insert a bar through a hole in the balancer to prevent the crankshaft from turning **(see illustration)**.

5   Using a puller that bolts to the crankshaft hub, remove the crankshaft balancer from the crankshaft **(see illustration)**.

6   Pry the old oil seal out with a seal removal tool **(see illustration)** or a screwdriver. Be very careful not to nick or otherwise damage the crankshaft in the process and don't distort the timing chain cover.

7   Apply a thin coat of RTV-type sealant to the outer edge of the new seal. Lubricate the seal lip with multi-purpose grease or clean engine oil.

6.5 Use a puller that applies force to the crankshaft balancer hub - don't use a jaw-type puller that applies force to the outer edge or damage to the crankshaft balancer will occur

8   Place the seal squarely in position in the bore with the spring side facing in.

9   Carefully tap the seal into place with a large socket or section of pipe and a hammer **(see illustration)**. The outer diameter of the socket or pipe should be the same size as the seal outer diameter.

10   Apply a thin layer of clean multi-purpose grease or clean engine oil to the seal contact surface of the crankshaft balancer hub.

11   Position the crankshaft balancer on the crankshaft and slide it

6.6 Pry the old seal out with a seal removal tool (shown here) or a screwdriver

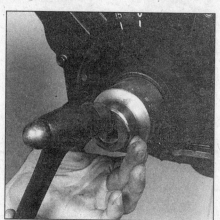

6.9 Install the new seal with a large socket or section of pipe

6.11 Align the keyway in the crankshaft balancer hub with the Woodruff key in the crankshaft (arrow)

7.3 Remove the three bolts and the engine mount bracket

7.5 Timing chain cover fastener locations (arrows)

7.8 Note how it's installed, then remove the oil slinger (arrow) from the crankshaft

7.10 Insert two 8 mm bolts (arrows) through the holes in the camshaft sprockets and into the holes in the timing chain housing - this locks the camshafts in the "timed" position

7.11 The mark on the crankshaft sprocket must align with the mark on the block (arrows)

through the seal until it bottoms against the crankshaft sprocket. Note that the slot (keyway) in the hub must be aligned with the Woodruff key in the end of the crankshaft (see illustration). The crankshaft bolt can also be used to press the crankshaft balancer into position.

12   Tighten the crankshaft bolt to the torque and angle of rotation listed in this Chapter's Specifications.

13   The remaining installation steps are the reverse of removal.

14   Start the engine and check for oil leaks at the seal.

7   Timing chain and sprockets - removal, inspection and installation

Note: Special tools are required for this procedure. Read through the entire procedure and acquire the necessary tools and equipment before beginning work.

Removal

Refer to illustrations 7.3, 7.5, 7.8, 7.10, 7.11, 7.12, 7.13, 7.15, 7.17 and 7.18

1   Detach the cable from the negative terminal of the battery. **Caution:** If the vehicle is equipped with a Delco Loc II or Theftlock audio system, make sure you have the correct activation code before disconnecting the battery.

2   Remove the coolant reservoir (see Chapter 3). Remove the drivebelt (see Chapter 1).

3   Support the engine from above, using an engine support fixture

(available at rental yards), or from below using a floor jack. Use a wood block between the floor jack and the engine to prevent damage. Remove the front engine mount (see Section 17). Remove the engine mount bracket and discard the bolts **(see illustration). Caution:** Replace the engine support bracket bolts with the manufacturers original type bolt anytime they are removed.

4   Remove the crankshaft balancer (see Section 6).

5   Working from above, remove the upper timing chain cover fasteners **(see illustration)**.

6   Working from below, remove the lower timing chain cover fasteners.

7   Detach the timing chain cover vent hose and remove the cover and gaskets from the housing.

8   Slide the oil slinger off the crankshaft **(see illustration)**.

9   Temporarily reinstall the crankshaft balancer bolt to use when turning the crankshaft.

10   Turn the crankshaft clockwise until the camshaft sprocket's timing pin holes align with the holes in the timing chain housing. Insert 8 mm pins or bolts into the holes to maintain alignment **(see illustration)**.

11   The mark on the crankshaft sprocket should align with the mark on the engine block **(see illustration)**. The crankshaft sprocket keyway should point up and align with the centerline of the cylinder bores.

12   Remove the three timing chain guides **(see illustration)**.

13   Make sure all the slack in the timing chain is above the tensioner assembly, then remove the chain tensioner assembly **(see illustration)**.

14   The timing chain must be disengaged from the wear grooves in the tensioner shoe in order to remove the tensioner assembly. Slide a

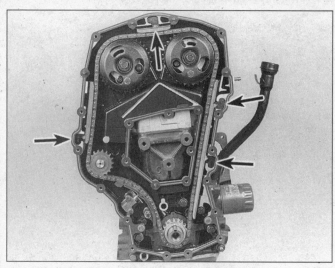

7.12 The three timing chain guides are wedged into the housing at four points (arrows) - just pull them out

7.13 The timing chain tensioner mounting bolt locations (1)

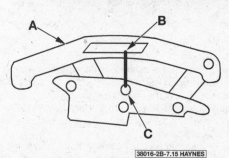

7.15 After retracting the tensioner shoe, insert a piece of wire (B) bent into a "U" shape between the tensioner shoe (A) and reset hole (C)

screwdriver blade under the timing chain while pulling the tensioner shoe out. **Note:** *If difficulty is encountered when removing the chain tensioner, proceed as follows:*

a) *Hold the intake camshaft sprocket with an appropriate tool and remove the sprocket bolt and washer.*

b) *Remove the washer from the bolt and thread the bolt back into the camshaft by hand.*

c) *Remove the intake camshaft sprocket, using a three-jaw puller in the three relief holes in the sprocket, if necessary.* **Caution:** *Don't try to pry the sprocket off the camshaft or damage to the sprocket could occur.*

15   Bend a piece of heavy wire into a "U", then apply light force to the tensioner shoe. While applying force, insert a small screwdriver into the reset hole **(see illustration)** and pry the ratchet pawl away from the ratchet teeth. When the shoe is fully retracted, insert the piece of bent wire into the hole and through the shoe to hold the shoe in place.

16   Remove the tensioner assembly retaining bolts and tensioner. **Warning:** *The tensioner plunger is spring loaded and could come out with great force, causing personal injury.* Remove the chain housing-to-block stud (timing chain tensioner shoe pivot).

17   Slip the timing chain off the sprockets **(see illustration)**.

18   To remove the camshaft sprockets, loosen the bolts while holding the sprockets with a screwdriver or punch inserted through one of the holes. Mark the sprockets for identification **(see illustration)**, remove the bolts, then pull on the sprockets by hand until they slip off the dowels. If necessary, use a small puller, with the legs inserted in the relief

holes, to pull the sprockets off.

19   The crankshaft sprocket should slip off the crankshaft by hand. If not, use a pulley.

20   The idler sprocket and bearing are pressed into place. If replacement is necessary, remove the timing chain housing (see Section 8) and take it to a dealer service department or automotive machine shop. Special tools are required and the bearing must be replaced each time it's pressed out.

Inspection

21   Visually inspect all parts for wear and damage. Look for loose pins, cracks, worn rollers and side plates. Check the sprockets for hook-shaped, chipped and broken teeth. **Note:** *Some scoring of the timing chain shoe and guides is normal. Replace the timing chain, sprockets, chain shoe and guides as a set if the engine has high mileage or fails the visual parts inspection.*

7.17 Begin removing the chain at the exhaust camshaft sprocket

7.18 Mark the sprockets *exhaust* and *intake* (arrows), then remove the bolts and pull the sprockets off

7.22 The camshaft sprocket dowel pin(s) should be near the top prior to sprocket installation

8.11 Position a new gasket over the dowel pins (arrows)

Installation

Refer to illustration 7.22

22   Make sure the camshafts are positioned with the dowel pins at the top **(see illustration)**. Install both camshaft sprockets (if removed). Apply thread locking compound to the camshaft sprocket bolt threads and make sure the washers are in place. Hold the camshaft from turning as described in Step 18 and tighten the bolts to the torque listed in this Chapter's Specifications.

23   Recheck the positions of the camshaft and crankshaft sprockets for correct valve timing **(see illustrations 7.10 and 7.11)**. **Note:** *If the camshafts are out of position and must be rotated more than 1/8-turn in order to install the alignment pins:*

a) *The crankshaft must be rotated 90-degrees clockwise past Top Dead Center to give the valves adequate clearance to open.*

b) *Once the camshafts are in position and the alignment pins installed, rotate the crankshaft counterclockwise back to Top Dead Center.* **Caution:** *Do not rotate the crankshaft clockwise to TDC (valve or piston damage could occur).*

24   Slip the timing chain over the exhaust camshaft sprocket, then around the idler and crankshaft sprockets.

25   Remove the alignment pin from the intake camshaft. Using an appropriate tool, rotate the intake camshaft sprocket counterclockwise enough to mesh the timing chain with it. The chain run between the two camshaft sprockets will tighten. If the valve timing is correct, the intake camshaft alignment pin should slide in easily. If it doesn't index, the camshafts aren't timed correctly; repeat the procedure.

26   Leave the alignment pins installed for now and check the timing marks. With slack removed from the timing chain between the intake camshaft sprocket and the crankshaft sprocket, the timing marks on the crankshaft sprocket and the engine block should be aligned. If the marks aren't aligned, move the chain one tooth forward or backward, remove the slack and recheck the marks.

27   Reload the timing chain tensioner assembly to its "zero" position as follows:

a) *Use the bent wire from the chain tensioner locking procedure in illustration 7.15 or reform another if necessary.*

b) *Apply light force on the tensioner shoe to compress the plunger.*

c) *Insert a small screwdriver into the reset access hole and depress the shoe.*

d) *Install the locking wire or keeper into the access hole and the shoe.*

28   Install the tensioner assembly in the chain housing. Tighten the bolts to the torque specified in this Chapter. **Note:** *Recheck the plunger assembly installation - it's correctly installed when the long end is toward the crankshaft.*

29   Remove bent piece of wire, squeeze the plunger into the tensioner body then let go to unload the plunger assembly.

30   Remove the camshaft sprocket alignment pins.

31   Slowly rotate the crankshaft clockwise two full turns (720-

degrees). Do not force it; if resistance is felt, recheck the installation procedure. Align the crankshaft timing mark with the mark on the engine block and temporarily reinstall the 8 mm alignment pins. The pins should slide in easily if the valve timing is correct. **Caution:** *If the valve timing is incorrect, severe engine damage could occur.*

32   Install the remaining components in the reverse order of removal, noting the following:

a) *Replace the engine mount bracket bolts with new bolts and tighten them to the torque listed in this Chapter's Specifications.*

b) *Check fluid levels, start the engine and check for proper operation and coolant/oil leaks.*

8   Timing chain housing - removal and installation

Refer to illustration 8.11

Warning: *Wait until the engine is completely cool before beginning this procedure.*

1   Detach the cable from the negative terminal of the battery. **Caution:** *If the vehicle is equipped with a Delco Loc II or Theftlock audio system, make sure you have the correct activation code before disconnecting the battery.*

2   Drain the cooling system (see Chapter 1), then remove the heater hose from the thermostat housing to drain the coolant from the engine block.

3   Remove the timing chain and sprockets (see Section 7). If you're installing a replacement timing chain housing, remove the water pump (see Chapter 3).

4   Remove the timing chain housing-to-belt tensioner bracket brace.

5   Remove the four oil pan-to-timing chain housing bolts.

6   Remove the timing chain housing-to-block lower fasteners.

7   Remove the lowest cover retaining stud from the timing chain housing.

8   Remove the eight chain housing-to-camshaft housing bolts.

9   Using an engine support fixture or floor jack, raise the engine slightly for additional clearance.

10   Remove the timing chain housing and gaskets. Thoroughly clean the mating surfaces to remove any traces of old sealant or gasket material.

11   Install the timing chain housing with new gaskets **(see illustration)**. Tighten the bolts to the torque listed in this Chapter's Specifications.

12   The remaining steps are the reverse of removal.

9   Camshafts, lifters and housings - removal, inspection and installation

Note: *Special tools are required for this procedure. Read through the entire procedure and acquire the necessary tools and equipment before beginning work.*

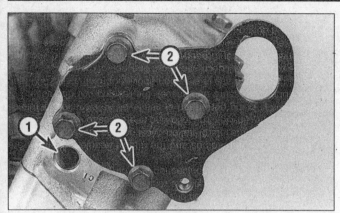

9.10 Oil pressure sending unit mounting hole (1) and engine lifting bracket bolts (2)

9.11 Gently lift the camshaft housing off the cylinder head and turn it over so the lifters don't fall out

9.14 Remove the oil seal (arrow) from the intake camshaft

Removal

Refer to illustrations 9.10, 9.11 and 9.14

1   Relieve the fuel system pressure (see Chapter 4).
2   Detach the cable from the negative terminal of the battery. **Caution:** *If the vehicle is equipped with a Delco Loc II or Theftlock audio system, make sure you have the correct activation code before disconnecting the battery.*
3   Position the engine at TDC on the compression stroke (see Section 3). Remove the timing chain and sprockets (see Section 7).
4   Remove the timing chain housing-to-camshaft housing bolts.
5   Remove the ignition coil and module assembly (see Chapter 5).
6   Remove the PCV oil/air separator from the side of the engine block (see Chapter 1). Remove the transaxle fluid level indicator and tube.
7   Without disconnecting the hoses, remove the power steering pump and secure it aside (see Chapter 10).
8   Disconnect the electrical connector from the camshaft position sensor.
9   Remove the fuel rail from the cylinder head and set it aside (see Chapter 4).
10   Disconnect the electrical connector from the oil pressure sending unit and unbolt the engine lifting bracket **(see illustration)**.
11   Loosen the camshaft housing-to-cylinder head bolts in 1/4-turn increments, in the reverse of the tightening sequence **(see illustration 9.28)**. Leave the two cover-to-housing bolts in place temporarily. Lift the housing off the cylinder head **(see illustration)**.
12   Remove the two camshaft cover-to-housing bolts. Push the cover off the housing by threading four of the housing-to-cylinder head bolts into the tapped holes in the cover. Carefully lift the camshaft out of the housing.
13   Remove all traces of old gasket material from the mating surfaces and clean them with lacquer thinner or acetone to remove any traces of oil.

14   Remove the oil seal from the intake camshaft **(see illustration)** and discard it.
15   Remove the lifters and store them in order so they can be reinstalled in their original locations. To minimize lifter bleed-down, store the lifters valve-side up, submerged in clean engine oil.

Inspection

Refer to illustrations 9.17a, 9.17b, 9.18 and 9.19

16   Refer to Chapter 2, Part C, for camshaft inspection procedures, but use this Chapter's Specifications. If the camshaft is damaged or worn beyond the specifications, replace the camshaft, do not attempt to salvage worn camshafts. Whenever a camshaft is replaced, replace all the lifters actuated by the camshaft as well.
17   Visually inspect the lifters for wear, galling, score marks and discoloration from overheating **(see illustrations)**.
18   Measure each lifter bore inside diameter and record the results **(see illustration)**.

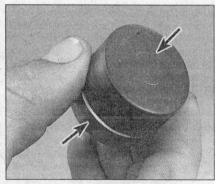

9.17a Check the camshaft lobe surfaces and the bore surfaces of the lifters for wear (arrows)

9.17b Check the valve-side of the lifters too, especially the valve stem contact area (arrow)

9.18 Use a telescoping gauge and micrometer to measure the lifter bores . . .

2B

9.19 . . . then measure the lifters with a micrometer - subtract each lift diameter from the corresponding bore diameter to obtain the lifter-to-bore clearances

9.24 The dowel pins (arrows) should be at the top (12 o'clock position)

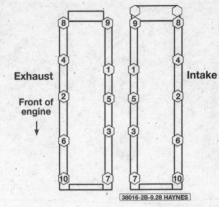

9.28 Camshaft housing-to-cylinder head bolt tightening sequence

19   Measure each lifter outside diameter and record the results **(see illustration)**.

20   Subtract the lifter outside diameter from the corresponding bore inside diameter to determine the clearance. Compare the results to this Chapter's specifications and replace parts as necessary.

Installation

Refer to illustrations 9.24 and 9.28

21   Using a new gasket, position the camshaft housing on the cylinder head and temporarily hold it in place with one bolt.

22   Coat the camshaft journals and lobes and the lifters with camshaft assembly lube and install them in their original locations.

23   On the intake camshaft only, lubricate the lip of the oil seal with clean engine oil, then position the seal on the camshaft journal with the spring side facing in.

24   Install the camshaft in the housing with the sprocket dowel pin UP (12 o'clock position) **(see illustration)**. Position the cover on the housing, holding it in place with the two bolts, as described previously.

25   Apply thread sealant to the threads of the camshaft housing and cover bolts.

26   Install new housing seals. **Note:** *Each housing seal is different shape and color. The intake seals are green with the inner seal configured differently than the outer seal. The exhaust seals are orange and they are also configured differently.*

27   Install the camshaft cover and bolts while positioning the oil seal (intake side only). Be sure the seal is installed to a precise 0.020 inch depth from the outer housing face.

28   Tighten the bolts in the sequence shown **(see illustration)** to the torque and angle of rotation listed in this Chapter's Specifications. Be sure to tighten the camshaft cover-to-housing bolts (rear two on the intake camshaft housing) at the prescribed lighter torque setting.

29   Install the power steering pump.

30   Install the remaining parts in the reverse order of removal.

31   Change the oil and filter (see Chapter 1). Add a can of engine oil supplement (GM part no. 1052367, or equivalent) if the camshafts and lifters have been replaced. **Note:** *If new lifters have been installed or the lifters bled down while the engine was disassembled, excessive lifter noise may be experienced after startup - this is normal. Use the following procedure to purge the lifters of air:*

   a) Start the engine and allow it to warm up at idle for five minutes.
   b) Increase engine speed to 2,000 rpm until the lifter noise is gone.
   c) Return the engine to idle for an additional five minutes.

32   Road test the vehicle and check for oil and coolant leaks.

10   Valve springs, retainers and seals - replacement

Refer to illustrations 10.5 and 10.18

Note: *Broken valve springs and defective valve stem seals can be replaced without removing the cylinder head. Two special tools and a compressed air source are normally required to perform this operation, so read through this Section carefully and rent or buy the tools before beginning the job. If compressed air isn't available, a length of nylon rope can be used to keep the valves from falling into the cylinder during this procedure.*

1   Remove the spark plug from the cylinder which has the defective part. Due to the design of this engine, the intake and exhaust camshaft housings can be removed separately to service their respective components. If all of the valve stem seals are being replaced, all of the spark plugs and both camshaft housings should be removed.

2   Refer to Chapter 5 and remove the ignition coil assembly.

3   Remove the camshaft(s), lifters and housing(s) as described in Section 9.

4   Turn the crankshaft until the piston in the affected cylinder is at top dead center on the compression stroke (refer to Chapter 2, Part C, for instructions). If you're replacing all of the valve stem seals, begin with cylinder number one and work on the valves for one cylinder at a time. Move from cylinder-to-cylinder following the firing order sequence (see this Chapter's Specifications).

5   Thread an adapter into the spark plug hole **(see illustration)** and connect an air hose from a compressed air source to it. Most auto parts stores can supply the air hose adapter. **Note:** *Many cylinder compression gauges utilize a screw-in fitting that may work with your air hose quick-disconnect fitting.*

6   Apply compressed air to the cylinder. **Warning:** *The piston may be forced down by compressed air, causing the crankshaft to turn suddenly. If the wrench used when positioning the number one piston at TDC is still attached to the bolt in the crankshaft nose, it could cause damage or injury when the crankshaft moves.*

7   The valves should be held in place by the air pressure.

8   If you don't have access to compressed air, an alternative method can be used. Position the piston at a point approximately 45-degrees before TDC on the compression stroke, then feed a long piece of nylon rope through the spark plug hole until it fills the combustion chamber. Be sure to leave the end of the rope hanging out of the engine so it can be removed easily. Use a large ratchet and socket to rotate the crankshaft in the normal direction of rotation (clockwise) until slight resistance is felt.

9   Stuff shop rags into the cylinder head holes adjacent to the valves to prevent parts and tools from falling into the engine, then use a valve spring compressor to compress the spring. Remove the keepers with small needle-nose pliers or a magnet.

10   Remove the retainer and valve spring, then remove the valve guide seal and rotator. **Note:** *If air pressure fails to hold the valve in the closed position during this operation, the valve face or seat is probably damaged. If so, the cylinder head will have to be removed for additional repair operations.*

11   Wrap a rubber band or tape around the top of the valve stem so the valve won't fall into the combustion chamber, then release the air pressure. **Note:** *If a rope was used instead of air pressure, turn the crankshaft slightly in a counterclockwise direction (opposite normal rotation).*

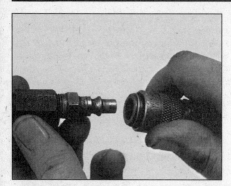

10.5 This is what the air hose adapter that threads into the spark plug hole looks like - they're commonly available from auto parts stores

10.18 Apply a small dab of grease to each keeper before installation to hold it in place on the valve stem until the spring is released

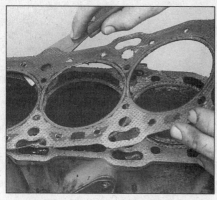

11.10 Remove the old gasket and clean the cylinder head thoroughly

2B

12   Inspect the valve stem for damage. Rotate the valve in the guide and check the end for eccentric movement, which would indicate the valve stem is bent.

13   Move the valve up-and-down in the guide and make sure it does not bind. If the valve stem binds, either the valve is bent or the guide is damaged. In either case, the cylinder head will have to be removed for repair.

14   Reapply air pressure to the cylinder to retain the valve in the closed position, then remove the tape or rubber band from the valve stem. If a rope was used instead of air pressure, rotate the crankshaft in the normal direction of rotation until slight resistance is felt.

15   Reinstall the valve rotator.

16   Lubricate the valve stem with engine oil and install a new guide seal.

17   Install the spring in position over the valve.

18   Install the valve spring retainer. Compress the valve spring and carefully install the keepers in the groove. Apply a small dab of grease to the inside of each keeper to hold it in place if necessary **(see illustration)**. Remove the pressure from the spring tool and make sure the keepers are seated.

19   Disconnect the air hose and remove the adapter from the spark plug hole. If a rope was used in place of air pressure, then turn the camshaft counterclockwise and pull it out of the cylinder.

20   Refer to Section 9 and install the camshaft(s), lifters and housing(s).

21   Install the spark plug(s) and the coil assembly.

22   Start and run the engine, then check for oil leaks and unusual sounds coming from the camshaft housings.

11   Cylinder head - removal and installation

Warning: *Wait until the engine is completely cool before beginning this procedure.*

Removal

1   Detach the cable from the negative terminal of the battery. **Caution:** *If the vehicle is equipped with a Delco Loc II or Theftlock audio system, make sure you have the correct activation code before disconnecting the battery.*

2   Drain the cooling system (see Chapter 1), then remove the heater hose from the thermostat housing to drain the coolant from the engine block.

3   Refer to Section 4 and remove the intake manifold. Refer to Section 5 and detach the exhaust manifold.

4   Remove the timing chain, camshafts and housings as described in Sections 7, 8 and 9.

5   Using the new cylinder head gasket, outline the cylinders and bolt pattern on a piece of cardboard to make a holder for the cylinder head bolts. Be sure to indicate the front of the engine for reference. Punch holes at the bolt locations **(see illustration 9.13a in Part A)**.

6   Loosen the cylinder head bolts in 1/4-turn increments until they can be removed by hand. Work from bolt-to-bolt in a pattern that's the reverse of the tightening sequence **(see illustration 11.15)**. Store the bolts in the cardboard holder as they're removed - this will ensure they are reinstalled in their original locations.

7   Lift the cylinder head off the engine. If resistance is felt, don't pry between the cylinder head and block as damage to the mating surfaces will result. To dislodge the cylinder head, place a wood block against the end of it and strike the wood block with a hammer. Store the cylinder head on wood blocks to prevent damage to the gasket sealing surfaces.

8   Cylinder head disassembly and inspection procedures are covered in detail in Chapter 2, Part C.

Installation

Refer to illustrations 11.10 and 11.15

9   The mating surfaces of the cylinder head and block must be perfectly clean when the cylinder head is installed.

10   Use a gasket scraper to remove all traces of carbon and old gasket material **(see illustration)**, then clean the mating surfaces with lacquer thinner or acetone. If there's oil on the mating surfaces when the cylinder head is installed, the gasket may not seal correctly and leaks could develop. **Note:** *Since the cylinder head is made of aluminum, aggressive scraping can cause damage. Be extra careful not to nick or gouge the mating surface with the scraper. Use a vacuum cleaner to remove debris that falls into the cylinders.* **Caution:** *Do not use a wire brush or an abrasive pad to clean the cylinder head mating surface.*

11   Check the block and cylinder head mating surfaces for nicks, deep scratches and other damage. If damage is slight, it can be removed with a flat mill file; if it's excessive, machining may be the only alternative.

12   Use a nylon bristle brush to clean the threads in the cylinder head bolt holes. Mount each bolt in a vise and run a die down the threads to remove corrosion and restore the threads. Dirt, corrosion, sealant and damaged threads will affect torque readings.

13   Position the new gasket over the dowel pins in the block.

14   Carefully position the cylinder head on the block without disturbing the gasket.

15   Install the bolts in their original locations and tighten them finger

11.15 Cylinder head bolt TIGHTENING sequence

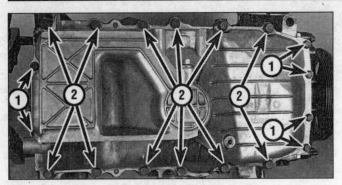

12.13 Oil pan bolt locations

1   *16 mm bolts*   2   *8 mm bolts*

tight. Following the recommended sequence **(see illustration)**, tighten the bolts in several steps to the torque and angle of rotation listed in this Chapter's Specifications.
16   The remaining installation steps are the reverse of removal.
17   Refill the cooling system and change the oil and filter (see Chapter 1, if necessary).
18   Run the engine and check for leaks and proper operation.

12   Oil pan - removal and installation

Warning: *Wait until the engine is completely cool before beginning this procedure.*
Note: *The following procedure is based on the assumption the engine is installed in the vehicle. If it has been removed, simply unbolt the oil pan and detach it from the block.*

Removal

Refer to illustration 12.13
1   Detach the cable from the negative terminal of the battery. **Caution:** *If the vehicle is equipped with a Delco Loc II or Theftlock audio system, make sure you have the correct activation code before disconnecting the battery.*
2   Remove the right front wheel and the lower splash shield from the fenderwell.
3   Drain the engine oil and the coolant (see Chapter 1).
4   Remove the drivebelt (see Chapter 1).
5   Remove the air conditioning compressor from the bracket (see Chapter 3).
6   Remove the engine mount strut bracket (brace) from under the engine compartment (see Section 17).
7   Detach the radiator outlet pipes from the oil pan brackets.
8   Unbolt the exhaust manifold brace (see Section 5).
9   Remove the flywheel/driveplate inspection cover.
10   Remove the radiator outlet pipe from the lower radiator hose (see Chapter 3).
11   Disconnect and remove the oil level sensor from the oil pan.
12   Remove the oil pan-to-transaxle bolts.
13   Remove the oil pan mounting bolts **(see illustration)**.
14   Carefully separate the pan from the block. Don't pry between the block and pan or damage to the sealing surfaces may result and oil leaks may develop. **Note:** *The crankshaft may have to be rotated to gain clearance for oil pan removal.*

Installation

15   Clean the sealing surfaces with lacquer thinner or acetone. Make sure the bolt holes in the block are clean.
16   The gasket should be checked carefully and replaced with a new one if damage is noted. Minor imperfections can be repaired with RTV sealant. **Caution:** *Use only enough sealant to restore the gasket to its original size and shape. Excess sealant may cause part misalignment and oil leaks.*
17   Carefully install the pan gasket and hold the pan against the block

and install the bolts finger tight.
18   Tighten the bolts in three steps to the torque specified in this Chapter **(see illustration 12.13)**. Start at the center of the pan and work out toward the ends in a spiral pattern. Note that the bolts are not all tightened to the same torque figure.
19   The remaining steps are the reverse of removal. **Caution:** *Don't forget to refill the engine with oil and coolant before starting it (see Chapter 1).*
20   Start the engine and check carefully for oil leaks at the oil pan.

13   Oil pump - removal, inspection and installation

Removal

1   Remove the oil pan as described in Section 12.
2   Remove the balance shaft chain cover and tensioner (see Section 14).
3   Remove the oil pump mounting bolts and separate the oil pump from the balance shaft assembly.
4   Remove the oil pump cover and the gears from the oil pump housing.

Inspection

5   Clean all parts thoroughly.
6   Visually inspect all parts for wear, cracks and other damage. Replace the pump if it's defective, if the engine has high mileage or if the engine is being rebuilt.

Installation

6   Position the oil pump onto the balance shaft assembly. Tighten the mounting bolts to the torque listed in this Chapter's Specifications.
7   Install the balance shaft chain tensioner and cover (see Section 14).
8   Install the oil pan.
9   Add oil and run the engine. Check for oil pressure and leaks.

14   Balance shaft assembly - removal, inspection and installation

Note: *Special tools are normally required to perform this operation. Read through the entire Section carefully and acquire the necessary tools before beginning this procedure.*

Removal

1   Remove the oil pan (see Section 12). Remove the balance shaft chain cover.
2   Remove the oil pump (see Section 13).
3   Loosen, but don't remove, the balance shaft chain guide.
4   Remove the balance shaft driven sprocket. **Caution:** *The bolt is a left handed thread and must be loosened in a clockwise direction. Before removal of the driven sprocket, if it is to be reused, mark the face of the sprocket so it can be installed the same way it came off. The balance shaft may try to rotate as the bolt is loosened. Wedge a screwdriver in the flywheel/driveplate ring gear teeth to hold the crankshaft still (which will also prevent the balance shafts from turning).*
5   Just break loose the bolts holding the upper and lower housing halves together. DO NOT loosen or remove at this time. The bolts that retain only the balance shaft housing and do not extend into the engine block must be left alone at this time.
6   Remove the balance shaft assembly-to-block bolts. **Warning:** *Support the assembly securely before removal of the bolts.*
7   Remove the balance shaft assembly and place it on a workbench for disassembly and inspection.

Inspection

8   Remove the bolts and separate the upper and lower housing.
9   Pry out the oil pump pick-up screen. Clean or replace before reassembly.

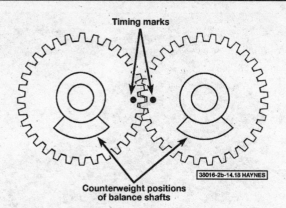

14.18 Be sure the timing marks are set correctly on the balance shaft gears

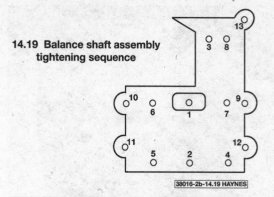

14.19 Balance shaft assembly tightening sequence

10 Remove the thrust plate bolts and plate.
11 Inspect the thrust plate for gouges or burrs.
12 Remove each balance shaft and gear assembly.
13 Inspect all parts. Look for damage such as nicks, cracks, scored bearing bores, damaged threaded holes, broken or worn guides, etc. and replace any necessary parts. **Caution:** *If the housing is damaged in any way, replace the entire assembly.*
14 Remove the bearings from the housings and inspect the bearings for scoring, overheating, etc. in both the upper and lower housings. Replace if necessary. **Caution:** *Balance shafts must be replaced together. Any time balance shafts are replaced, the bearings must also be replaced.*
15 Inspect the chain for damaged links. **Caution:** *DO NOT replace individual links in the chain. The entire chain must be replaced if any damage is found. If the chain is to be replaced, the sprockets must also be replaced. Chain replacement requires removal of the crankshaft.*

Installation

Refer to illustrations 14.18 and 14.19

16 Assemble the thrust plate and tighten the bolts to the torque listed in this Chapter's Specifications.
17 Install the bearing halves in the upper and lower housings and lubricate the bearing faces with engine assembly lube.
18 Install the balance shafts in the housings. Align the timing marks **(see illustration). Caution:** *The engine will make noise or vibrate if the marks are not properly aligned.*
19 Assemble the upper and lower housings and tighten the bolts to 44 inch-lbs following the correct sequence **(see illustration)**. Final tightening will be done after the balance shaft assembly is installed on the block.
20 Place the number 1 piston at TDC (Top Dead Center), see Section 3.
21 Rotate the crankshaft, clockwise, 90-degrees.
22 Bolt the balance shaft assembly to the engine block. **Note:** *Use Locktite 242, or equivalent, thread locking compound on the housing-to-block bolts.*
23 Tighten the housing-to-block bolts, in sequence, to 44 inch-lbs **(see illustration 14.19)**. Make sure the balance shafts spin freely.
24 Tighten all bolts, in sequence **(see illustration 14.19)** to the torque listed in this Chapter's Specifications.
25 Install the oil pick-up screen. **Caution:** *The screen must not be installed until all bolts have be tightened to the final specification.*
26 Assemble the driveshaft sprocket and chain and bolt the sprocket to the balance shaft. **Caution:** *If reusing the old sprocket, be sure the mark, made on disassembly, shows.*
27 Immobilize the crankshaft as described in Step 4 and tighten the bolt to torque listed in this Chapter's Specifications. The balance shafts must not turn while the driven sprocket is being tightened. Remember, the balance shaft sprocket bolt is reverse threaded, so turn it counter-clockwise to tighten.
28 Loosely install the chain tensioner and bolts.

29 Adjust the chain tension by inserting a 0.040-inch brass feeler gauge between the chain and chain guide. Apply light pressure (about 3 lbs) to the chain guide and tighten the chain guide bolt to the torque listed in this Chapter's Specifications. **Caution:** *A brass feeler gauge is necessary to measure chain clearance. A steel gauge will not bend and will give a incorrect chain-to-guide clearance.*
30 The remainder of the installation is the reverse of the removal procedure.
31 Add oil and a new filter, run the engine and check for leaks.

15 Flywheel/driveplate - removal and installation

This procedure is essentially the same for all engines. Refer to Chapter 2A and follow the procedure outlined there. Be sure to use the bolt torque listed in this Chapter's Specifications.

16 Rear main oil seal - replacement

Refer to illustrations 16.5, 16.6, 16.7 and 16.8

1 Remove the transaxle (see Chapter 7). Support the engine from above using an engine support fixture (available at rental yards). If the special support fixture is unavailable, position a jack under the engine oil pan. Place a large wood block between the jack head and the oil pan, then carefully raise the engine just enough to support the weight. **Warning:** *DO NOT place any part of your body under the engine when it's supported only by a jack!*
2 If equipped with a manual transaxle, remove the pressure plate and clutch disc (see Chapter 8).
3 Remove the flywheel or driveplate (see Section 15).
4 Remove the seal housing-to-oil pan bolts.
5 Remove the seal housing-to-engine block bolts **(see illustration)**.

16.5 Remove the seal housing bolts (arrows)

16.6 After removing the housing from the engine, support it on wood blocks and drive out the old seal with a punch and hammer

16.7 Drive the new seal into the housing with a wood block - be careful not to cock the seal in the housing bore

Detach the seal housing and remove the old gasket material.

6   Support the seal housing between two wood blocks on a workbench and drive the old seal out from the back side with a punch and hammer **(see illustration)**.

7   Drive the new seal into the housing with a wood block **(see illustration)**.

8   Lubricate the crankshaft seal journal and the lip of the new seal with multi-purpose grease or clean engine oil. Position a new gasket on the engine block **(see illustration)**.

9   Inspect the oil pan gasket. The gasket should be checked carefully and replaced with a new one if damage is noted. Minor imperfections can be repaired with RTV sealant. **Caution:** *Use only enough sealant to restore the gasket to its original size and shape. Excess sealant may cause part misalignment and oil leaks.*

10   Slowly and carefully push the new seal onto the crankshaft. The seal lip is stiff, so work it onto the crankshaft with a smooth object such as the end of an extension as you push the housing against the block.

11   Install and tighten the housing bolts and the oil pan bolts to the torque listed in this Chapter's specifications.

12   Install the flywheel and clutch components.

13   Reinstall the transaxle.

17 Engine mounts - check and replacement

1   Engine mounts seldom require attention, but broken or deteriorated mounts should be replaced immediately or the added strain placed on the driveline components may cause damage or wear.

Check

2   During the check, the engine must be raised slightly to remove the weight from the mounts.

3   Raise the vehicle and support it securely on jackstands. Support the engine from above using an engine support fixture (available at rental yards). If the special support fixture is unavailable, position a jack under the engine oil pan. Place a large wood block between the jack head and the oil pan, then carefully raise the engine just enough to take the weight off the mounts. **Warning:** *DO NOT place any part of your body under the engine when it's supported only by a jack!*

4   Check the mounts to see if the rubber is cracked, hardened or separated from the metal plates. Sometimes the rubber will split right down the center.

5   Check for relative movement between the mount plates and the engine or frame (use a large screwdriver or pry bar to attempt to move the mounts). If movement is noted, lower the engine and tighten the mount fasteners.

Replacement

Note: *Rubber preservative should be applied to the mounts to slow deterioration.*

16.8 Position a new gasket over the dowel pins (arrows)

6   Detach the cable from the negative terminal of the battery. **Caution:** *If the vehicle is equipped with a Delco Loc II or Theftlock audio system, make sure you have the correct activation code before disconnecting the battery.*

Front engine mount

7   Remove the coolant reservoir (see Chapter 3).

8   Raise the engine slightly to take the weight off the mount.

9   Remove the mounting nuts and bolts and remove the engine mount.

10   Place the new mount in position and install the nuts. Gently lower the engine and tighten the nuts to the torque listed in this Chapter's Specifications.

Engine mount strut

11   Raise the vehicle and support it securely on jackstands. Remove the right lower splash shield.

12   Working under the mount, remove the nut from the through-bolt.

13   Raise the engine slightly to take the weight off the mount and separate the strut from the bracket.

14   Installation is the reverse of removal. Tighten the fasteners securely.

Engine mount strut bracket

15   Raise the vehicle and support it securely on jackstands. Remove the right lower splash shield.

16   Remove the engine mount strut.

17   Remove the engine mount strut bracket bolts and remove the bracket.

18   Installation is the reverse of removal. Tighten the bolts to the torque listed in this Chapter's Specifications.

Chapter 2 Part C
General engine overhaul procedures

Contents

Specifications

2.2 liter OHV four-cylinder engine

General
VIN engine code	4
Displacement	134 cubic inches
Cylinder compression pressure	
Minimum	100 psi
Maximum variation between cylinders	30-percent
Oil pressure (minimum)	15 psi at 900 rpm
Oil pressure (maximum)	56 psi at 3,000 rpm

Cylinder head
Warpage limit	0.005 inch (0.127 mm)

Valves and related components
Valve face angle	45-degrees
Valve seat angle	46-degrees
Margin width	1/32 inch (minimum)
Valve stem-to-guide clearance	
1995 through 1997	
Intake	0.0010 to 0.0027 inch (0.025 to 0.069 mm)
Exhaust	0.0014 to 0.0031 inch (0.035 to 0.081 mm)
1998 and later	
Intake	0.0007 to 0.0020 inch (0.020 to 0.053 mm)
Exhaust	0.0014 to 0.0029 inch (0.035 to 0.076 mm)

2.2 liter OHV four-cylinder engine (continued)

Valves and related components (continued)

Valve spring
  Free length
    1995 through 1997 .... 1.95 inch (49.5 mm)
    1998 and later .... 1.91 inch (48.7 mm)
  Installed height
    1995 through 1997 .... 1.71 inch (43.4 mm)
    1998 and later .... 1.60 inch (40.6 mm)

Crankshaft and connecting rods

Crankshaft endplay .... 0.002 to 0.007 inch (0.05 to 0.18 mm)
Connecting rod side clearance (endplay) .... 0.004 to 0.015 inch (0.10 to 0.38 mm)
Main bearing journal
  Diameter .... 2.4945 to 2.4954 inches (63.360 to 63.385 mm)
  Taper/out-of-round limits .... 0.0002 inch (0.005 mm)
Main bearing oil clearance .... 0.0006 to 0.0019 inch (0.015 to 0.047 mm)
Connecting rod journal
  Diameter .... 1.9983 to 1.9994 inches (50.758 to 50.784 mm)
  Taper/out-of-round limits .... 0.0002 inch (0.005 mm)
Connecting rod bearing oil clearance .... 0.001 to 0.0031 inch (0.025 to 0.079 mm)

Engine block

Cylinder bore
  Diameter .... 3.5036 to 3.5067 inches (88.991 to 89.00 mm)
  Out-of-round limit .... 0.0005 inch (0.013 mm)
Taper limit (thrust side) .... 0.0005 inch (0.013 mm)
Block deck warpage limit .... If more than 0.010 inch (0.25 mm) must
be removed, replace the block

Pistons and rings

Piston-to-bore clearance .... 0.0007 to 0.0017 inch (0.015 to 0.045 mm)
Piston ring side clearance
  Compression rings .... 0.0019 to 0.0011 inch (0.05 to 0.027 mm)
  Oil control ring .... 0.0019 to 0.082 inch (0.05 to 0.21 mm)
Piston ring end gap
  Compression rings .... 0.010 to 0.020 inch (0.25 to 0.50 mm)
  Oil control ring .... 0.010 to 0.050 inch (0.25 to 1.27 mm)

Camshaft

Lobe lift (intake and exhaust)
  1995 through 1997 .... 0.288 inch (7.31 mm)
  1998 and later .... 0.263 inch (6.69 mm)
Bearing journal diameter .... 1.868 to 1.869 inches (47.45 to 47.48 mm)
Bearing oil clearance .... 0.001 to 0.0039 inch (0.026 to 0.101 mm)

Torque specifications*

Ft-lbs (unless otherwise indicated)
Main bearing cap bolts 70
Connecting rod cap nuts 38
Camshaft thrust plate-to-block bolts .... 108 in-lbs
*Note: *Refer to Chapter 2A for additional torque specifications.*

2.3 and 2.4 liter OHC four-cylinder engines

General

VIN engine code
  1995 .... D
  1996 and later .... T
Displacement
  1995 .... 138 cubic inches
  1996 and later .... 146 cubic inches
Cylinder compression pressure
  Minimum .... 100 psi
  Maximum variation between cylinders 30-percent
Oil pressure
  1995 through 1997
    At 900 rpm .... 15 psi minimum
    At 2000 rpm .... 30 psi minimum
  1998 and later
    At 900 rpm .... 10 psi minimum
    At 3000 rpm .... 30 psi minimum

Cylinder head
Warpage limit.. 0.008 inch (0.203 mm)

Valves and related components
Valve face angle
  1995 (intake and exhaust) 44-degrees
  1996 and later
    Intake 46-degrees
    Exhaust 45.5-degrees
Valve seat angle (intake and exhaust) 45-degrees
Valve margin width .... 1/32 inch (minimum)
Valve stem-to-guide clearance
  1995
    Intake .... 0.0010 to 0.0020 inch (0.025 to 0.069 mm)
    Exhaust .... 0.0015 to 0.0032 inch (0.038 to 0.081 mm)
  1996 and later
    Intake .... 0.0009 to 0.0025 inch (0.025 to 0.069 mm)
    Exhaust .... 0.0016 to 0.0032 inch (0.038 to 0.081 mm)
Valve spring
  Free length .... Not available
  Installed height*
    1995 .... 0.9840 to 1.0040 inches (25.00 to 25.50 mm)
    1996 and later .... 0.9787 to 1.0024 inches (24.86 to 25.62 mm)
*Measured from tip of stem to top of camshaft housing mounting surface

Crankshaft and connecting rods
Crankshaft
  Endplay .... 0.0034 to 0.0095 inch (0.087 to 0.243 mm)
  Runout
    At center main journal.. 0.00098 inch (0.025 mm)
    At flywheel flange .... 0.00098 inch (0.025 mm)
Main bearing journal
  Diameter
    1995 .... 2.0470 to 2.0480 inches (51.996 to 52.020 mm)
    1996 and 1997 .... 2.3634 to 2.3626 inches (60.031 to 60.010 mm)
    1998 and later .... 2.3622 to 2.3631 inches (60.000 to 60.024 mm)
  Out-of-round limit... 0.0002 inch (0.005 mm)
  Taper limit.. 0.0003 inch (0.007 mm)
Main bearing oil clearance
  1995 .... 0.0005 to 0.0023 inch (0.013 to 0.058 mm)
  1996 and 1997 .... 0.0005 to 0.0030 inch (0.013 to 0.075 mm)
  1998 and later .... 0.0004 to 0.0023 inch (0.010 to 0.060 mm)
Connecting rod bearing journal
  Diameter .... 1.8887 to 1.8897 inches (47.975 to 48.00 mm)
  Out-of-round .... 0.0002 inch (0.005 mm)
  Taper limit.. 0.0002 inch (0.005 mm)
Connecting rod bearing oil clearance
  1995 through 1997 .... 0.0005 to 0.0020 inch (0.013 to 0.053 mm)
  1998 and later .... 0.0004 to 0.0026 inch (0.010 to 0.068 mm)
Connecting rod side clearance (endplay) .... 0.0059 to 0.0177 inch (0.150 to 0.450 mm)
Seal journal
  Diameter .... 3.2210 to 3.2299 inches (81.96 to 82.04 mm)
  Runout limit .... 0.002 inch (0.05 mm)

Engine block
Cylinder bore
  Diameter
    1995 .... 3.6217 to 3.6223 inches (91.992 to 92.008 mm)
    1996 .... 3.5431 to 3.5435 inches (89.994 to 90.006 mm)
    1997 and later .... 3.5110 to 3.5435 inches (89.994 to 90.006 mm)
  Out-of-round limit... 0.0004 inch (0.010 mm)
  Taper limit.. 0.0003 inch (0.008 mm)
Block deck warpage limit .... If more than 0.010 inch (0.25 mm) must
    be removed, replace the block

Pistons and rings
Piston diameter
  1995 and 1996 .... 3.6203 to 3.6210 inches (91.957 to 91.973 mm)
  1997 and later .... 3.5420 to 3.5427 inches (89.968 to 89.984 mm)

Pistons and rings (continued)

Piston-to-bore clearance
    1995 and 1996 .... 0.0007 to 0.0020 inch (0.019 to 0.051 mm)
    1997 and later .... 0.0006 to 0.0018 inch (0.010 to 0.042 mm)

Piston ring end gap
    1995
        Top compression ring .... 0.0138 to 0.0236 inch (0.35 to 0.60 mm)
        Second compression ring .... 0.0157 to 0.0256 inch (0.40 to 0.65 mm)
        Oil control ring .... 0.0157 to 0.0551 inch (0.40 to 1.40 mm)
    1996 and 1997
        Top compression ring .... 0.0060 to 0.0120 inch (0.15 to 0.30 mm)
        Second compression ring .... 0.0119 to 0.0161 inch (0.30 to 0.41 mm)
        Oil control ring .... 0.0098 to 0.0256 inch (0.03 to 0.65 mm)
    1998 and later
        Top compression ring .... 0.0060 to 0.0120 inch (0.15 to 0.30 mm)
        Second compression ring .... 0.0098 to 0.0160 inch (0.25 to 0.40 mm)
        Oil control ring .... 0.0098 to 0.0299 inch (0.25 to 0.76 mm)

Piston ring side clearance
    1995
        Top compression ring .... 0.0020 to 0.0039 inch (0.050 to 0.100 mm)
        Second compression ring .... 0.00157 to 0.00315 inch (0.040 to 0.080 mm)
    1996 and later
        Top compression ring .... 0.0016 to 0.0031 inch (0.040 to 0.080 mm)
        Second compression ring .... 0.0012 to 0.0028 inch (0.030 to 0.070 mm)

Camshaft

Lobe lift
    Intake
        1995 .... 0.375 inch (9.525 mm)
        1996 and later .... 0.354 inch (9.00 mm)
    Exhaust
        1995 .... 0.375 inch (9.525 mm)
        1996 and later .... 0.346 inch (8.80 mm)
Journal diameter
    Number 1 .... 1.5720 to 1.5728 inches (39.93 to 39.95 mm)
    Numbers 2 through 5 .... 1.3751 to 1.3760 inches (34.93 to 34.95 mm)
Endplay .... 0.0009 to 0.0088 inch (0.025 to 0.225 mm)

Torque specifications**

Ft-lbs

Main bearing cap bolts
    Step 1 15
    Step 2 .... Tighten an additional 90-degrees
Connecting rod cap nuts
    Step 1 18
    Step 2 .... Tighten an additional 80-degrees

Note: *Refer to Chapter 2B for additional torque specifications.*

1   General information

Included in this portion of Chapter 2 are the general overhaul procedures for the cylinder head and internal engine components.

The information ranges from advice concerning preparation for an overhaul and the purchase of replacement parts to detailed, step-by-step procedures covering Removal and installation of internal engine components and the inspection of parts.

The following Sections have been written based on the assumption that the engine has been removed from the vehicle. For information concerning in-vehicle engine repair, as well as Removal and Installation of the external components necessary for the overhaul, see Chapter 2A (2.2L OHV engine) or 2B (2.3 and 2.4L OHC engines) and Section 8 of this Chapter.

The Specifications included in this Chapter are only those necessary for the inspection and overhaul procedures which follow. Refer to Chapter 2A or 2B for additional Specifications.

It's not always easy to determine when, or if, an engine should be completely overhauled, as a number of factors must be considered.

High mileage is not necessarily an indication that an overhaul is needed, while low mileage doesn't preclude the need for an overhaul.

Frequency of servicing is probably the most important consideration. An engine that's had regular and frequent oil and filter changes, as well as other required maintenance, will most likely give many thousands of miles of reliable service. Conversely, a neglected engine may require an overhaul very early in its life.

Excessive oil consumption is an indication that piston rings, valve seals and/or valve guides are in need of attention. Make sure that oil leaks aren't responsible before deciding that the rings and/or guides are bad. Perform a cylinder compression check to determine the extent of the work required (see Section 4). Also check the vacuum readings under various conditions (see Section 3).

Loss of power, rough running, knocking or metallic engine noises, excessive valve train noise and high fuel consumption rates may also point to the need for an overhaul, especially if they're all present at the same time. If a complete tune-up doesn't remedy the situation, major mechanical work is the only solution.

An engine overhaul involves restoring the internal parts to the specifications of a new engine. During an overhaul, the piston rings are replaced and the cylinder walls are reconditioned (re-bored and/or honed). If a re-bore is done by an automotive machine shop, new over-size pistons will also be installed. The main bearings, connecting rod bearings and camshaft bearings are generally replaced with new ones and, if necessary, the crankshaft may be reground to restore the jour-

2.2a The oil pressure sending unit (arrow) on 2.2L engines is located on the right (firewall) side of the engine block near the oil filter

2.2b The oil pressure sending unit (arrow) on 2.3L and 2.4L engines is located at the rear of the camshaft housing on the driver's side of the vehicle

3.6 A compression gauge with a threaded fitting for the spark plug hole is preferred over the type that requires hand pressure to maintain the seal - be sure to open the throttle valve as far as possible during the compression check

nals. Generally, the valves are serviced as well, since they're usually in less-than-perfect condition at this point. While the engine is being overhauled, other components, such as the distributor, starter and alternator, can be rebuilt as well. The end result should be a like new engine that will give many trouble free miles. **Note:** *Critical cooling system components such as the hoses, drivebelts, thermostat and water pump should be replaced with new parts when an engine is overhauled. The radiator should be checked carefully to ensure that it isn't clogged or leaking (see Chapter 3). If you purchase a rebuilt engine or short block, some rebuilders will not warranty their engines unless the radiator has been professionally flushed. Also, we don't recommend overhauling the oil pump - always install a new one when an engine is rebuilt.*

Before beginning the engine overhaul, read through the entire procedure to familiarize yourself with the scope and requirements of the job. Overhauling an engine isn't difficult, but it is time-consuming. Plan on the vehicle being tied up for a minimum of two weeks, especially if parts must be taken to an automotive machine shop for repair or reconditioning. Check on availability of parts and make sure that any necessary special tools and equipment are obtained in advance. Most work can be done with typical hand tools, although a number of precision measuring tools are required for inspecting parts to determine if they must be replaced. Often an automotive machine shop will handle the inspection of parts and offer advice concerning reconditioning and replacement. **Note:** *Always wait until the engine has been completely disassembled and all components, especially the engine block, have been inspected before deciding what service and repair operations must be performed by an automotive machine shop.* Since the block's condition will be the major factor to consider when determining whether to overhaul the original engine or buy a rebuilt one, never purchase parts or have machine work done on other components until the block has been thoroughly inspected. As a general rule, time is the primary cost of an overhaul, so it doesn't pay to install worn or substandard parts.

As a final note, to ensure maximum life and minimum trouble from a rebuilt engine, everything must be assembled with care in a spotlessly-clean environment.

2   Oil Pressure check

Refer to illustrations 2.2a and 2.2b

1   Low engine oil pressure can be a sign of an engine in need of rebuilding. A "low oil pressure" indicator (often called an "idiot light") is not a test of the oiling system. Such indicators only come on when the oil pressure is dangerously low. Even a factory oil pressure gauge in the

instrument panel is only a relative indication, although much better for driver information than a warning light. A better test is with a mechanical (not electrical) oil pressure gauge. When used in conjunction with an accurate tachometer, an engine's oil pressure performance can be compared to factory Specifications for that year and model.
2   Locate the oil pressure indicator sending unit **(see illustrations)**.
3   Remove the oil pressure sending unit and install a fitting which will allow you to directly connect your hand-held, mechanical oil pressure gauge. Use Teflon tape or sealant on the threads of the adapter and the fitting on the end of your gauge's hose.
4   Connect an accurate tachometer to the engine, according to the tachometer manufacturer's instructions.
5   Check the oil pressure with the engine running (full operating temperature) at the specified engine speed, and compare it to this Chapter's Specifications. If it's extremely low, the bearings and/or oil pump are probably worn out.

3   Cylinder compression check

Refer to illustration 3.6

1   A compression check will tell you what mechanical condition the upper end (pistons, rings, valves, head gaskets) of the engine is in. Specifically, it can tell you if the compression is down due to leakage caused by worn piston rings, defective valves and seats or a blown head gasket. **Note:** *The engine must be at normal operating temperature and the battery must be fully charged for this check.*
2   Begin by cleaning the area around the spark plugs before you remove them. Compressed air should be used, if available, otherwise a small brush or even a bicycle tire pump will work. The idea is to prevent dirt from getting into the cylinders as the compression check is being done.
3   Remove all of the spark plugs from the engine (see Chapter 1).
4   Block the throttle wide open.
5   Relieve the fuel system pressure (see Chapter 4), then disable the fuel and ignition systems by removing the PCM IGN fuse from the instrument panel fuse block and the IGNITION fuse from the underhood fuse block (1995 models), or the PCM BATT fuse from the instrument panel fuse block and the IGNITION fuse from the underhood fuse block (1996 and later models).
6   Install the compression gauge in the number one spark plug hole **(see illustration)**.
7   Crank the engine over at least seven compression strokes and watch the gauge. The compression should build up quickly in a healthy engine. Low compression on the first stroke, followed by gradually increasing pressure on successive strokes, indicates worn piston

0279H

0282H

0280H

4.1 Low, steady reading

4.2 Low, fluctuating needle

4.3 Regular drops

0281H

0278H

4.4 Irregular drops

4.5 Rapid vibration

rings. A low compression reading on the first stroke, which doesn't build up during successive strokes, indicates leaking valves or a blown head gasket (a cracked head could also be the cause). Deposits on the undersides of the valve heads can also cause low compression. Record the highest gauge reading obtained.

8   Repeat the procedure for the remaining cylinders, turning the engine over for the same length of time for each cylinder, and compare the results to this Chapter's Specifications.

9   If the readings are below normal, add some engine oil (about three squirts from a plunger-type oil can) to each cylinder, through the spark plug hole, and repeat the test.

10   If the compression increases after the oil is added, the piston rings are definitely worn. If the compression doesn't increase significantly, the leakage is occurring at the valves or head gasket. Leakage past the valves may be caused by burned valve seats and/or faces or warped, cracked or bent valves.

11   If two adjacent cylinders have equally low compression, there's a strong possibility the head gasket between them is blown. The appearance of coolant in the combustion chambers or the crankcase would verify this condition.

12   If one cylinder is about 20-percent lower than the others, and the engine has a slightly rough idle, a worn exhaust lobe on the camshaft could be the cause.

13   If the compression is unusually high, the combustion chambers are probably coated with carbon deposits. If that's the case, the cylinder heads should be removed and decarbonized.

14   If compression is way down or varies greatly between cylinders, it would be a good idea to have a leak-down test performed by an automotive repair shop. This test will pinpoint exactly where the leakage is occurring and how severe it is.

15   Install the fuses and drive the vehicle to restore the "block learn" memory.

4   Vacuum gauge diagnostic checks

Refer to illustrations 4.1, 4.2, 4.3, 4.4, 4.5, 4.7 and 4.9

A vacuum gauge provides valuable information about what is going on in the engine at a low cost. You can check for worn rings or

cylinder walls, leaking head or intake manifold gaskets, incorrect carburetor adjustments, restricted exhaust, stuck or burned valves, weak valve springs, improper ignition or valve timing and ignition problems.

Unfortunately, vacuum gauge readings are easy to misinterpret, so they should be used in conjunction with other tests to confirm the diagnosis.

Both the gauge readings and the rate of needle movement are important for accurate interpretation. Most gauges measure vacuum in inches of mercury (in-Hg). The following references to vacuum assume the diagnosis is being performed at sea level. As elevation increases (or atmospheric pressure decreases), the reading will decrease. For every 1,000-foot increase in elevation above approximately 2000 feet, the gauge readings will decrease about one inch of mercury.

Connect the vacuum gauge directly to intake manifold vacuum, not to ported (before throttle plate) vacuum. Be sure no hoses are left disconnected during the test or false readings will result.

Before you begin the test, allow the engine to warm up completely. Block the wheels and set the parking brake. With the transmission in Park, start the engine and allow it to run at normal idle speed.

Read the vacuum gauge; an average, healthy engine should normally produce about 17 to 22 inches of vacuum with a fairly steady needle. Refer to the following vacuum gauge readings and what they indicate about the engine's condition:

1   A low, steady reading usually indicates a leaking gasket between the intake manifold and throttle body, a leaky vacuum hose, late ignition timing or incorrect camshaft timing **(see illustration)**. Eliminate all other possible causes, utilizing the tests provided in this Chapter before you remove the timing chain cover to check the timing marks.

2   If the reading is three to eight inches below normal and it fluctuates at that low reading, suspect an intake manifold gasket leak at an intake port **(see illustration)**.

3   If the needle has regular drops of about two to four inches at a steady rate, the valves are probably leaking. Perform a compression or leak-down test to confirm this **(see illustration)**.

4   An irregular drop or down-flick of the needle can be caused by a sticking valve or an ignition misfire. Perform a compression or leak-down test and read the spark plugs **(see illustration)**.

5   A rapid vibration of about four inches-Hg vibration at idle combined with exhaust smoke indicates worn valve guides **(see illustration)**. Perform a leak-down test to confirm this. If the rapid vibration

0284H

4.7 Large fluctuation

0283H

4.9 Slow return after revving

occurs with an increase in engine speed, check for a leaking intake manifold gasket or head gasket, weak valve springs, burned valves or ignition misfire.

6   A slight fluctuation, say one inch up and down, may mean ignition problems. Check all the usual tune-up items and, if necessary, run the engine on an ignition analyzer.

7   If there is a large fluctuation, perform a compression or leak-down test to look for a weak or dead cylinder or a blown head gasket **(see illustration)**.

8   If the needle moves slowly through a wide range, check for a clogged PCV system, incorrect idle fuel mixture, throttle body or intake manifold gasket leaks.

9   Check for a slow return after revving the engine by quickly snapping the throttle open until the engine reaches about 2,500 rpm and let it shut **(see illustration)**. Normally the reading should drop to near zero, rise above normal idle reading (about 5 in-Hg over) and then return to the previous idle reading. If the vacuum returns slowly and doesn't peak when the throttle is snapped shut, the rings may be worn. If there is a long delay, look for a restricted exhaust system (often the muffler or catalytic converter). An easy way to check this is to temporarily disconnect the exhaust ahead of the suspected part and re-test.

5   Engine removal - methods and precautions

If you've decided the engine must be removed for overhaul or major repair work, several preliminary steps should be taken.

Locating a suitable place to work is extremely important. Adequate work space, along with storage space for the vehicle, will be needed. If a shop or garage isn't available, at the very least a flat, level, clean work surface made of concrete or asphalt is required.

Cleaning the engine compartment and engine before beginning the removal procedure will help keep tools clean and organized.

An engine hoist or A-frame will also be necessary **(see illustration)**. Make sure the equipment is rated in excess of the combined weight of the engine and transaxle. Safety is of primary importance, considering the potential hazards involved in lifting the engine out of the vehicle.

If the engine is being removed by a novice, a helper should be available. Advice and aid from someone more experienced would also be helpful. There are many instances when one person cannot simultaneously perform all of the operations required when lifting the engine out of the vehicle.

Plan the operation ahead of time. Arrange for or obtain all of the tools and equipment you'll need prior to beginning the job. Some of the equipment necessary to perform engine removal and installation safely and with relative ease are (in addition to an engine hoist) a heavy duty floor jack, complete sets of wrenches and sockets as described in the front of this manual, wooden blocks and plenty of rags and cleaning solvent for mopping up spilled oil, coolant and gasoline. If the hoist must be rented, be sure to arrange for it in advance and perform all of

the operations possible without it beforehand. This will save you money and time.

Plan for the vehicle to be out of use for quite a while. A machine shop will be required to perform some of the work which the do-it-yourselfer can't accomplish without special equipment. These shops often have a busy schedule, so it would be a good idea to consult them before removing the engine in order to accurately estimate the amount of time required to rebuild or repair components that may need work.

Always be extremely careful when removing and installing the engine. Serious injury can result from careless actions. Plan ahead, take your time and a job of this nature, although major, can be accomplished successfully.

6   Engine - removal and installation

Refer to illustrations 6.5, 6.12 and 6.28

Warning 1: *Gasoline is extremely flammable, so take extra precautions when disconnecting any part of the fuel system. Don't smoke or allow open flames or bare light bulbs in or near the work area and don't work in a garage where a natural gas appliance (such as a clothes dryer or water heater) is installed. If you spill gasoline on your skin, rinse it off immediately. Have a fire extinguisher rated for gasoline fires handy and know how to use it! Also, the air conditioning system is under high pressure - have a dealer service department or service station discharge the system before disconnecting any of the hoses or fittings.*

Warning 2: *These models are equipped with airbags. Impact sensors for the airbag system are located just above the radiator grille on the right side. The airbag(s) could accidentally deploy if these sensors are disturbed, so be extremely careful when working in this area. Airbag system components are also located in the steering wheel, steering column and base of the steering column, so be extremely careful in these areas and don't disturb any airbag system component or wiring. You could easily be injured if an airbag accidentally deploys, and the airbag might not deploy correctly in a collision if any components or wiring in the system have been disturbed. Refer to Chapters 10 and 12 for additional information on disarming the airbag and removal procedures.*

Removal

Note: *Read through the following steps carefully and familiarize yourself with the procedure before beginning work.*

1   On air-conditioned models, have the air conditioning system discharged by a dealer service department or service station (see Chapter 3 for additional information).

2   Refer to Chapter 4 and relieve the fuel system pressure, then disconnect the negative cable from the battery. **Caution:** *If the vehicle is equipped with a Delco Loc II or Theftlock audio system, make sure you have the correct activation code before disconnecting the battery.*

3   Cover the fenders and cowl and remove the hood (see Chapter 11). Special pads are available to protect the fenders, but an old bedspread or blanket will also work.

6.5 Label each wire before unplugging the connector

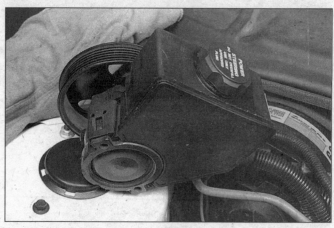

6.12 Unbolt the power steering pump and move it aside, then use wire or rope to hold it in place

4   Remove the air cleaner assembly and the intake ducts (see Chapter 4).

5   Remove the plastic firewall covers. Label the vacuum lines, emissions system hoses, wiring connectors, ground straps and fuel lines to ensure correct reinstallation, then detach them. The relay panel and bracket can be detached as an assembly. Pieces of masking tape with numbers or letters written on them work well for marking wires and hoses **(see illustration)**. If there's any possibility of confusion, make a sketch of the engine compartment and clearly label the lines, hoses and wires.

6   Raise the vehicle and support it securely on jackstands. Drain the cooling system (see Chapter 1).

7   Label and detach all coolant hoses from the engine.

8   Remove the coolant reservoir, cooling fan, shroud and radiator (see Chapter 3).

9   Remove the drivebelt and idler, if equipped (see Chapter 1). Remove the crankshaft pulley/balancer (see Chapter 2A).

10   Disconnect the fuel lines running from the engine to the chassis (see Chapter 4). Plug or cap all open fittings/lines.

11   Disconnect the throttle linkage (and TV linkage/speed control cable, if equipped) from the engine (see Chapters 4 and 7).

12   Unbolt the power steering pump and set it aside (see Chapter 10). Leave the lines/hoses attached and make sure the pump is kept in an upright position in the engine compartment **(see illustration)**.

13   Unbolt the air conditioning compressor (see Chapter 3) and set it aside. Do not disconnect the hoses, if possible.

14   Drain the engine oil and remove the filter (see Chapter 1).

15   Remove the starter and the alternator (see Chapter 5).

16   On OHC models, remove the oil/air separator (Chapter 6).

17   Detach the brake master cylinder from the power booster to allow clearance for the transaxle. Position the master cylinder and ABS unit to the side as far as possible or remove the unit completely, if necessary (see Chapter 9).

18   Disconnect the exhaust system from the engine (see Chapter 4).

19   Disconnect the components from the transaxle, including the driveaxles, cables, wiring, linkage, etc. (see Chapter 7).

20   Attach an engine sling or a length of chain to the lifting brackets on the engine.

21   Roll the hoist into position and connect the sling to it. Raise the engine assembly slightly to take the weight off the mounts. **Warning:** *DO NOT place any part of your body under the engine when it's supported only by a hoist or other lifting device.*

23   Remove the engine mount strut and the strut bracket from the engine.

24   Remove the front engine mount-to-chassis bolts/nuts and remove the engine mount.

25   If equipped with an automatic transaxle, remove the transaxle mount-to-body bolts. Remove the transaxle mount support bolts from the transaxle and remove the mount and support assembly.

26   If equipped with a manual transaxle, remove the transaxle-to-body mount through-bolts and pry the mount out of the frame bracket. Remove the mounts from the transaxle.

27   Recheck to be sure nothing is still connecting the engine to the vehicle (or transaxle, where applicable). Disconnect anything still remaining.

28   Slowly raise the engine and transaxle as an assembly out of the vehicle. Check carefully to make sure nothing is hanging up as the hoist is raised **(see illustration)**.

29   Once the engine/transaxle assembly is out of the vehicle, remove the transaxle-to-engine block bolts. Carefully separate the engine from the transaxle. If you're working on a vehicle with an automatic transaxle, be sure the torque converter stays in place (clamp a pair of vise-grips to the housing to keep the converter from sliding out). If you're working on a vehicle with a manual transaxle, the input shaft must be completely disengaged from the clutch.

30   Remove the clutch and flywheel or driveplate and mount the engine on an engine stand.

Installation

31   Check the engine and transaxle mounts. If they're worn or damaged, replace them.

32   If you're working on a manual transaxle equipped vehicle, install the clutch and pressure plate (see Chapter 7). Now is a good time to install a new clutch. Apply a dab of high-temperature grease to the input shaft.

33   **Caution:** *DO NOT use the transaxle-to-engine bolts to force the transaxle and engine together.* If you're working on an automatic transaxle equipped vehicle, take great care when installing the torque converter, following the procedure outlined in Chapter 7.

34   Carefully lower the engine into the engine compartment - make sure the mounts line up. Reinstall the remaining components in the reverse order of removal. Double-check to make sure everything is connected correctly.

35   Add coolant, oil, power steering and transmission fluid as needed. If the brake master cylinder was removed, bleed the brakes (see Chapter 9). Recheck the fluid level and test the brakes.

36   Run the engine and check for leaks and proper operation of all accessories, then install the hood and test drive the vehicle.

37   If the air conditioning system was discharged, have it evacuated, recharged and leak tested by the shop that discharged it.

---

7   Engine rebuilding alternatives

The home mechanic is faced with a number of options when performing an engine overhaul. The decision to replace the engine block, piston/connecting rod assemblies and crankshaft depends on a num-

6.28 Lift the engine/transaxle assembly from the engine compartment and guide it carefully around any obstacles as an assistant raises the hoist until it clears the front of the vehicle

ber of factors, with the number one consideration being the condition of the block. Other considerations are cost, access to machine shop facilities, parts availability, time required to complete the project and the extent of prior mechanical experience.

Some of the rebuilding alternatives include:

Individual parts - If the inspection procedures reveal the engine block and most engine components are in reusable condition, purchasing individual parts may be the most economical alternative. The block, crankshaft and piston/connecting rod assemblies should all be inspected carefully. Even if the block shows little wear, the cylinder bores should be surface honed.

Short block - A short block consists of an engine block with a crankshaft and piston/connecting rod assemblies already installed. All new bearings are incorporated and all clearances will be correct. The existing camshaft, valve train components, cylinder head(s) and external parts can be bolted to the short block with little or no machine shop work necessary.

Long block - A long block consists of a short block plus an oil pump, oil pan, cylinder head(s), rocker arm cover(s), camshaft and valve train components, timing sprockets and chain and timing chain cover. All components are installed with new bearings, seals and gaskets incorporated throughout. The installation of manifolds and external parts is all that's necessary.

Give careful thought to which alternative is best for you and discuss the situation with local automotive machine shops, auto parts dealers and experienced rebuilders before ordering or purchasing replacement parts.

8   Engine overhaul - disassembly sequence

1   It's much easier to disassemble and work on the engine if it's mounted on a portable engine stand. A stand can often be rented quite cheaply from an equipment rental yard. Before it's mounted on a stand, the flywheel/driveplate should be removed from the engine.

2   If a stand isn't available, it's possible to disassemble the engine with it blocked up on the floor. Be extra careful not to tip or drop the engine when working without a stand.

3   If you're going to obtain a rebuilt engine, all external components must come off first, to be transferred to the replacement engine, just as they will if you're doing a complete engine overhaul yourself. These include:

Alternator and brackets
Emissions control components
Ignition coil/module assembly, spark plug wires and spark plugs
Thermostat and housing cover
Water pump

EFI components
Intake/exhaust manifolds
Oil filter
Engine mounts
Clutch and flywheel/driveplate

Note: *When removing the external components from the engine, pay close attention to details that may be helpful or important during installation. Note the installed position of gaskets, seals, spacers, pins, brackets, washers, bolts and other small items.*

4   If you're obtaining a short block, which consists of the engine block, crankshaft, pistons and connecting rods all assembled, then the cylinder head(s), oil pan and oil pump will have to be removed as well. See Engine rebuilding alternatives for additional information regarding the different possibilities to be considered.

5   If you're planning a complete overhaul, the engine must be disassembled and the internal components removed in the following general order:

2.2 liter OHV engines

Rocker arm cover
Intake and exhaust manifolds
Rocker arms and pushrods
Valve lifters
Cylinder head
Timing chain cover
Timing chain and sprockets
Camshaft
Oil pan
Oil pump
Piston/connecting rod assemblies
Crankshaft and main bearings

2.3 and 2.4 liter (OHC) engines

Timing chain and sprockets
Timing chain housing
Cylinder head and camshafts
Oil pan
Oil pump
Piston/connecting rod assemblies
Rear main oil seal housing
Crankshaft and main bearings

6   Before beginning the disassembly and overhaul procedures, make sure the following items are available. Also, refer to Engine overhaul - reassembly sequence for a list of tools and materials needed for engine reassembly.

Common hand tools
Small cardboard boxes or plastic bags for storing parts
Gasket scraper
Ridge reamer
Vibration damper puller
Micrometers
Telescoping gauges
Dial indicator set
Valve spring compressor
Cylinder surfacing hone
Piston ring groove cleaning tool
Electric drill motor
Tap and die set
Wire brushes
Oil gallery brushes
Cleaning solvent

9   Cylinder head - disassembly

Refer to illustrations 9.2, 9.3 and 9.4
Note: *New and rebuilt cylinder heads are commonly available for most engines at dealerships and auto parts stores. Due to the fact that some specialized tools are necessary for the disassembly and inspection pro-*

2C

9.2 A small plastic bag, with an appropriate label, can be used to store the valve train components so they can be kept together and reinstalled in the original positions

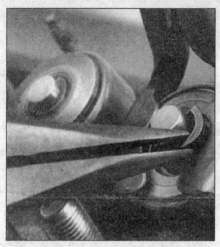

9.3 Use a valve spring compressor to compress the spring, then remove the keepers from the valve stem

9.4 If the valve won't pull through the guide, deburr the edge of the stem end and the area around the top of the keeper groove with a file or whetstone

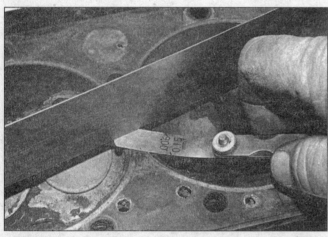

10.12 Check the cylinder head gasket surface for warpage by trying to slip a feeler gauge under the straightedge (see this Chapter's Specifications for the maximum warpage allowed and use a feeler gauge of that thickness)

cedures, and replacement parts aren't always readily available, it may be more practical and economical for the home mechanic to purchase replacement head(s) rather than taking the time to disassemble, inspect and recondition the original(s).

1   Cylinder head disassembly involves removal of the intake and exhaust valves and related components. If you're working on a OHC engine, the camshafts and housings must be removed before beginning the cylinder head disassembly procedure (see Chapter 2B). If you're working on a 2.2 liter OHV engine, remove the rocker arm nuts, pivot balls and rocker arms from the cylinder head studs. Label the parts or store them separately so they can be reinstalled in their original locations.

2   Before the valves are removed, arrange to label and store them, along with their related components, so they can be kept separate and reinstalled in their original locations **(see illustration)**.

3   Compress the springs on the first valve with a spring compressor and remove the keepers **(see illustration)**. Carefully release the valve spring compressor and remove the retainer, the spring and the spring seat (if used). Note that valve rotators are installed under the springs on the OHC engine.

4   Pull the valve out of the head, then remove the oil seal from the guide. If the valve binds in the guide (won't pull through), push it back into the head and deburr the area around the keeper groove with a fine

file or whetstone **(see illustration)**.

5   Repeat the procedure for the remaining valves. Remember to keep all the parts for each valve together so they can be reinstalled in the same locations.

6   Once the valves and related components have been removed and stored in an organized manner, the head should be thoroughly cleaned and inspected. If a complete engine overhaul is being done, finish the engine disassembly procedures before beginning the cylinder head cleaning and inspection process.

---

10 Cylinder head - cleaning and inspection

1   Thorough cleaning of the cylinder head(s) and related valve train components, followed by a detailed inspection, will enable you to decide how much valve service work must be done during the engine overhaul. **Note:** *If the engine was severely overheated, the cylinder head is probably warped (see Step 12).*

Cleaning

2   Scrape all traces of old gasket material and sealant off the head gasket, intake manifold and exhaust manifold mating surfaces. Be very careful not to gouge the cylinder head. Special gasket removal solvents that soften gaskets and make removal much easier are available at auto parts stores.

3   Remove all built up scale from the coolant passages.

4   Run a stiff wire brush through the various holes to remove deposits that may have formed in them.

5   Run an appropriate size tap into each of the threaded holes to remove corrosion and thread sealant that may be present. If compressed air is available, use it to clear the holes of debris produced by this operation. **Warning:** *Wear eye protection when using compressed air!*

6   Clean the rocker arm pivot stud threads with a wire brush (except OHC engine).

7   Clean the cylinder head with solvent and dry it thoroughly. Compressed air will speed the drying process and ensure that all holes and recessed areas are clean. **Note:** *Decarbonizing chemicals are available and may prove very useful when cleaning cylinder heads and valve train components. They're very caustic and should be used with caution. Be sure to follow the instructions on the container.*

8   Clean the rocker arms, pivot balls, nuts and pushrods (except OHC engine) with solvent and dry them thoroughly (don't mix them up during the cleaning process). Compressed air will speed the drying process and can be used to clean out the oil passages.

9   Clean all the valve springs, spring seats, rotators (where applica-

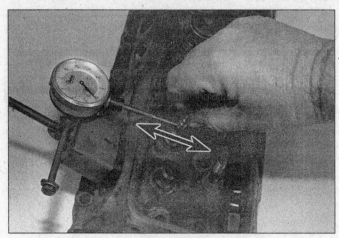

10.14 A dial indicator can be used to determine the valve stem-to-guide clearance (move the valve stem as indicated by the arrows)

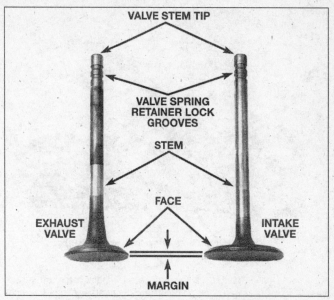

10.15 Check for valve wear at the points shown here

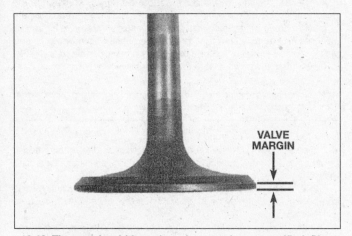

10.16 The margin width on the valve must be as specified (if no margin exists, the valve cannot be re-used)

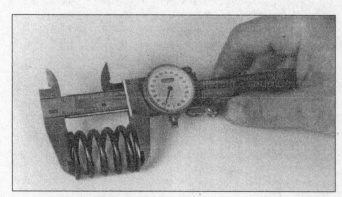

10.17 Measure the free length of each valve spring with a dial or vernier caliper

2C

ble), keepers and retainers with solvent and dry them thoroughly. Do the components from one valve at a time to avoid mixing up the parts.

10 Scrape off any heavy deposits that may have formed on the valves, then use a motorized wire brush to remove deposits from the valve heads and stems. Again, make sure the valves don't get mixed up.

Inspection

Refer to illustrations 10.12, 10.14, 10.15, 10.16, 10.17 and 10.18
Note: *Be sure to perform all of the following inspection procedures before concluding machine shop work is required. Make a list of the items that need attention.*

Cylinder head

11 Inspect the head very carefully for cracks, evidence of coolant leakage and other damage. If cracks are found, check with an automotive machine shop concerning repair. If repair isn't possible, a new cylinder head should be obtained.

12 Using a straightedge and feeler gauge, check the head gasket mating surface for warpage **(see illustration)**. If the warpage exceeds the limit in this Chapter's Specifications, it can be resurfaced at an automotive machine shop.

13 Examine the valve seats in each of the combustion chambers. If they're pitted, cracked or burned, the head will require valve service that's beyond the scope of the home mechanic.

14 Check the valve stem-to-guide clearance by measuring the lateral movement of the valve stem with a dial indicator attached securely to the head **(see illustration)**. The valve must be in the guide and approximately 1/16-inch off the seat. The total valve stem movement indicated by the gauge needle must be divided by two to obtain the actual clearance. After this is done, if there's still some doubt regarding the condition of the valve guides, they should be checked by an automotive machine shop (the cost should be minimal).

Valves

15 Carefully inspect each valve face for uneven wear, deformation, cracks, pits and burned areas. Check the valve stem for scuffing and galling and the neck for cracks. Rotate the valve and check for any obvious indication that it's bent. Look for pits and excessive wear on the end of the stem. The presence of any of these conditions **(see illustration)** indicates the need for valve service by an automotive machine shop.

16 Measure the margin width on each valve **(see illustration)**. Any valve with a margin narrower than specified in this Chapter will have to be replaced with a new one.

Valve components

17 Check each valve spring for wear (on the ends) and pits. Measure the free length and compare it to this Chapter's Specifications **(see illustration)**. Any springs that are shorter than specified have sagged and shouldn't be reused. The tension of all springs should be checked with a special fixture before deciding they're suitable for use in a rebuilt engine (take the springs to an automotive machine shop for this check).

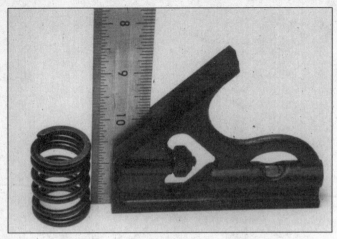

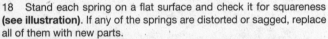

10.18 Check each valve spring for squareness

12.4 Gently tap the new seal onto the guide using a deep socket

18   Stand each spring on a flat surface and check it for squareness **(see illustration)**. If any of the springs are distorted or sagged, replace all of them with new parts.

19   Check the spring retainers and keepers for obvious wear and cracks. Any questionable parts should be replaced with new ones, as extensive damage will occur if they fail during engine operation. Make sure the rotators (OHC engine only) operate smoothly with no binding or excessive play.

20   On OHV models, check the rocker arm faces (the areas that contact the pushrod ends and valve stems) for pits, wear, galling, score marks and rough spots. Check the rocker arm pivot contact areas and pivot balls as well. Look for cracks in each rocker arm and nut.

21   Inspect the pushrod ends for scuffing and excessive wear. Roll each pushrod on a flat surface, like a piece of plate glass, to determine if it's bent.

22   Check the rocker arm studs in the cylinder heads for damaged threads and secure installation.

23   Any damaged or excessively worn parts must be replaced with new ones.

24   On OHC models, refer to Chapter 2B for the inspection procedures for the camshafts, lifters and housings.

25   If the inspection process indicates the valve components are in generally poor condition and worn beyond the limits specified, which is usually the case in an engine that's being overhauled, reassemble the valves in the cylinder head and refer to Section 11 for valve servicing recommendations.

11   Valves - servicing

1   Because of the complex nature of the job and the special tools and equipment needed, servicing of the valves, the valve seats and the valve guides, commonly known as a valve job, should be done by a professional.

2   The home mechanic can remove and disassemble the head, do the initial cleaning and inspection, then reassemble and deliver it to a dealer service department or an automotive machine shop for the actual service work. Doing the inspection will enable you to see what condition the head and valvetrain components are in and will ensure that you know what work and new parts are required when dealing with an automotive machine shop.

3   The dealer service department, or automotive machine shop, will remove the valves and springs, recondition or replace the valves and valve seats, recondition the valve guides, check and replace the valve springs, rotators, spring retainers and keepers (as necessary), replace the valve seals with new ones, reassemble the valve components and make sure the installed spring height is correct. The cylinder head gasket surface will also be resurfaced if it's warped.

4   After the valve job has been performed by a professional, the head will be in like new condition. When the head is returned, be sure to clean it again before installation on the engine to remove any metal particles and abrasive grit that may still be present from the valve service or head resurfacing operations. Use compressed air, if available, to blow out all the oil holes and passages.

12   Cylinder head - reassembly

Refer to illustrations 12.4 and 12.6

1   Regardless of whether or not the head was sent to an automotive repair shop for valve servicing, make sure it's clean before beginning reassembly.

2   If the head was sent out for valve servicing, the valves and related components will already be in place. Begin the reassembly procedure with Step 8.

3   Install the spring seats or valve rotators (if equipped) before the valve seals.

4   Install new seals on each of the valve guides. Using a hammer and a deep socket or seal installation tool, gently tap each seal into place until it's completely seated on the guide **(see illustration)**. Don't twist or cock the seals during installation or they won't seal properly on the valve stems.

5   Beginning at one end of the head, lubricate and install the first valve. Apply moly-base grease or clean engine oil to the valve stem.

6   Position the valve springs (and shims, if used) over the valves. Compress the springs with a valve spring compressor and carefully install the keepers in the groove, then slowly release the compressor and make sure the keepers seat properly. Apply a small dab of grease to each keeper to hold it in place if necessary **(see illustration)**.

7   Repeat the procedure for the remaining valves. Be sure to return the components to their original locations - don't mix them up!

8   Check the installed valve spring height with a ruler graduated in 1/32-inch increments or a dial caliper. If the head was sent out for service work, the installed height should be correct (but don't automatically assume it is). On OHV models, the measurement is taken from the top of each spring seat, rotator or top shim to the bottom of the retainer. On OHC models, measure from the tip of the valve stem to the top of the camshaft housing mounting surface. If the height is greater than specified in this Chapter, shims can be added under the springs to correct it. **Caution:** *Do not, under any circumstances, shim the springs to the point where the installed height is less than specified.*

9   On OHV models, apply moly-base grease to the rocker arm faces and the pivot balls, then install the rocker arms and pivots on the cylinder head studs (see Chapter 2A).

10   If you're working on a OHC engine, refer to Chapter 2B and install the camshafts, lifters and housings on the head.

12.6 Apply a small dab of grease to each keeper as shown here before installation - it'll hold them in place on the valve stem as the spring is released

13.3 The dial indicator plunger must be positioned directly above and in line with the pushrod (use a short length of vacuum hose to hold the plunger over the pushrod end if you encounter difficulty keeping the plunger on the pushrod)

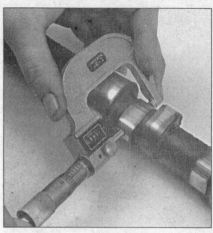

13.9 Measure the lobe heights on each camshaft - if any lobe height is less than the specified allowable minimum, replace that camshaft

2C

13.12a Remove the oil pump drive (2.2L OHV engine)

13.12b Be sure to install a new O-ring

13   Camshaft and bearings (2.2L OHV engine) - removal, inspection and installation

Note: *This procedure applies to the 2.2 liter OHV engine. Since there isn't enough room to remove the camshaft with the engine in the vehicle, the engine must be out of the vehicle and mounted on a stand for this procedure.*

Camshaft lobe lift check

With cylinder head installed

Refer to illustration 13.3

1   In order to determine the extent of cam lobe wear, the lobe lift should be checked prior to camshaft removal. Refer to Chapter 2A and remove the valve cover.

2   Position the number one piston at TDC on the compression stroke (see Section 4).

3   Beginning with the number one cylinder valves, loosen the rocker arm nuts and pivot the rocker arms sideways. Mount a dial indicator on the engine and position the plunger against the top of the first pushrod. The plunger should be directly in line with the pushrod **(see illustration)**.

4   Zero the dial indicator, then very slowly turn the crankshaft in the normal direction of rotation (clockwise) until the indicator needle stops and begins to move in the opposite direction. The point at which it stops indicates maximum cam lobe lift.

5   Record this figure for future reference, then reposition the piston

at TDC on the compression stroke.

6   Move the dial indicator to the remaining number one cylinder pushrod and repeat the check. Be sure to record the results for each valve.

7   Repeat the check for the remaining valves. Since each piston must be at TDC on the compression stroke for this procedure, work from cylinder-to-cylinder following the firing order sequence.

8   After the check is complete, compare the results to this Chapter's Specifications. If camshaft lobe lift is less than specified, cam lobe wear has occurred and a new camshaft should be installed.

With cylinder head removed

Refer to illustration 13.9

9   If the cylinder head(s) have already been removed, an alternate method of lobe measurement can be used. Remove the camshaft as described below. Using a micrometer, measure the lobe at its highest point. Then measure the base circle perpendicular (90-degrees) to the lobe **(see illustration)**. Do this for each lobe and record the results.

10   Subtract the base circle measurement from the lobe height. The difference is the lobe lift. See Step 8 above.

Removal

Refer to illustrations 13.12a, 13.12b, 13.14 and 13.15

11   Refer to the appropriate Sections in Chapter 2A or 2C and remove the timing chain and sprockets, lifters and pushrods.

12   Remove the oil pump drive. If it's stuck, remove the oil pump and push the drive out with a long socket extension. Be sure to replace the O-ring before reassembly **(see illustrations)**.

13.14 Thread long bolts into the sprocket bolt holes to use as a handle when removing and installing the camshaft

13.15 Support the camshaft near the block to avoid damaging the bearings

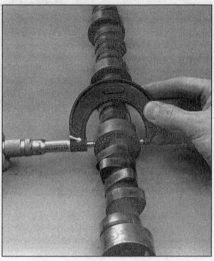

13.17 Measure the camshaft journal diameter

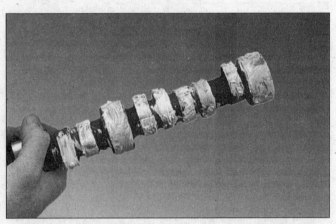

13.22 Be sure to apply moly-base grease or engine assembly lube to the cam lobes and bearing journals before installing the camshaft

13   Remove the bolts and detach the camshaft thrust plate from the engine block.
14   Thread long bolts into the camshaft sprocket bolt holes to use as a handle when removing the camshaft from the block **(see illustration)**.
15   Carefully pull the camshaft out. Support the cam near the block so the lobes don't nick or gouge the bearings as it's withdrawn **(see illustration)**.

Inspection

Refer to illustration 13.17

16   After the camshaft has been removed from the engine, cleaned with solvent and dried, inspect the bearing journals for uneven wear, pitting and evidence of seizure. If the journals are damaged, the bearing inserts in the block are probably damaged as well. Both the camshaft and bearings will have to be replaced.
17   Measure the bearing journals with a micrometer **(see illustration)** to determine if they're excessively worn or out-of-round.
18   Check the camshaft lobes and oil pump drive and driven gears for heat discoloration, score marks, chipped areas, pitting and uneven wear. If the lobes and gears are in good condition and if the lobe lift measurements are as specified, the components can be reused.
19   Check the bearings in the block for wear and damage. Look for galling, pitting and discolored areas.
20   The inside diameter of each bearing can be determined with a

small hole gauge and outside micrometer or an inside micrometer. Subtract the camshaft bearing journal diameter(s) from the corresponding bearing inside diameter(s) to obtain the bearing oil clearance. If it's excessive, new bearings will be required regardless of the condition of the originals.
21   Camshaft bearing replacement requires special tools and expertise that place it outside the scope of the home mechanic. Take the block to an automotive machine shop to ensure the job is done correctly.

Installation

Refer to illustration 13.22

22   Lubricate the camshaft bearing journals and cam lobes with moly-base grease or engine assembly lube **(see illustration)**.
23   Slide the camshaft into the engine. Support the cam near the block and be careful not to scrape or nick the bearings. Install the thrust plate and tighten the bolts.
24   Refer to Chapter 2A to complete the installation of the lifters, timing chain and sprockets.

14   Pistons and connecting rods - removal

Refer to illustrations 14.1, 14.3 and 14.6

Note: *Prior to removing the piston/connecting rod assemblies, remove the cylinder head(s), the oil pan and the oil pump by referring to the appropriate Sections in Parts A or B of Chapter 2.*
1   Use your fingernail to feel if a ridge has formed at the upper limit of ring travel (about 1/4-inch down from the top of each cylinder). If carbon deposits or cylinder wear have produced ridges, they must be completely removed with a special tool **(see illustration)**. Follow the manufacturer's instructions provided with the tool. Failure to remove the ridges before attempting to remove the piston/connecting rod assemblies may result in piston breakage.
2   After the cylinder ridges have been removed, turn the engine upside-down so the crankshaft is facing up.
3   Before the connecting rods are removed, check the endplay with feeler gauges. Slide them between the first connecting rod and the crankshaft throw until the play is removed **(see illustration)**. The endplay is equal to the thickness of the feeler gauge(s). If the endplay exceeds the service limit, new connecting rods will be required. If new rods (or a new crankshaft) are installed, the endplay may fall under the minimum specified in this Chapter (if it does, the rods will have to be machined to restore it - consult an automotive machine shop for advice if necessary). Repeat the procedure for the remaining connecting rods.

14.1 A ridge reamer is required to remove the ridge from the top of each cylinder - do this before removing the pistons!

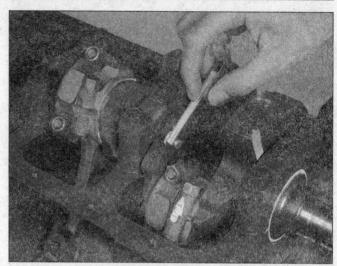

14.3 Check the connecting rod side clearance with a feeler gauge as shown

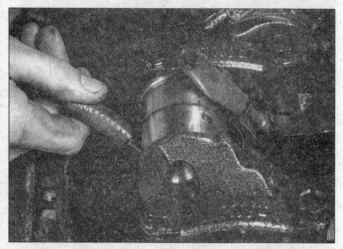

14.6 To prevent damage to the crankshaft journals and cylinder walls, slip sections of rubber or plastic hose over the rod bolts before removing the pistons/rods

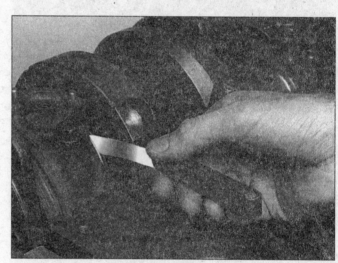

15.3 Checking crankshaft endplay with a feeler gauge

2C

4   Check the connecting rods and caps for identification marks. If they aren't plainly marked, use a small center punch to make the appropriate number of indentations on each rod and cap (1, 2, 3, etc., depending on the engine type and cylinder they're associated with).

5   Loosen each of the connecting rod cap nuts 1/2-turn at a time until they can be removed by hand. Remove the number one connecting rod cap and bearing insert. Don't drop the bearing insert out of the cap.

6   Slip a short length of plastic or rubber hose over each connecting rod cap bolt to protect the crankshaft journal and cylinder wall as the piston is removed (see illustration).

7   Remove the bearing insert and push the connecting rod/piston assembly out through the top of the engine. Use a wooden or plastic hammer handle to push on the upper bearing surface in the connecting rod. If resistance is felt, double-check to make sure all of the ridge was removed from the cylinder.

8   Repeat the procedure for the remaining cylinders.

9   After removal, reassemble the connecting rod caps and bearing inserts in their respective connecting rods and install the cap nuts finger tight. Leaving the old bearing inserts in place until reassembly will help prevent the connecting rod bearing surfaces from being accidentally nicked or gouged.

10   Don't separate the pistons from the connecting rods (see Section 19 for additional information).

15 Crankshaft - removal

Refer to illustrations 15.3, 15.4a and 15.4b

Note: *The crankshaft can be removed only after the engine has been removed from the vehicle. It's assumed the flywheel or driveplate, crankshaft balancer, timing chain, oil pan, oil pump and piston/connecting rod assemblies have already been removed. The rear main oil seal housing (OHC only) must be unbolted and separated from the block before proceeding with crankshaft removal.*

1   Before the crankshaft is removed, check the endplay. Mount a dial indicator with the stem in line with the crankshaft and just touching one of the crank throws.

2   Push the crankshaft all the way to the rear and zero the dial indicator. Next, pry the crankshaft to the front as far as possible and check the reading on the dial indicator. The distance it moves is the endplay. If it's greater than specified in this Chapter, check the crankshaft thrust surfaces for wear. If no wear is evident, new main bearings should correct the endplay.

3   If a dial indicator isn't available, feeler gauges can be used. Gently pry or push the crankshaft all the way to the front of the engine. Slip feeler gauges between the crankshaft and the front face of the thrust main bearing to determine the clearance (see illustration).

4   Check the main bearing caps to see if they're marked to indicate

15.4a Use a center-punch or number stamping dies to mark the main bearing caps to ensure installation in their original locations on the block (make the punch marks near one of the bolt heads)

15.4b The arrow on the main bearing cap indicates the front of the engine

16.4a A hammer and large punch can be used to knock the core plugs sideways in their bores

16.4b Pull the core plugs from the block with pliers

their locations. They should be numbered consecutively from the front of the engine to the rear. If they aren't, mark them with number stamping dies or a center punch **(see illustration)**. Main bearing caps generally have a cast-in arrow, which points to the front of the engine **(see illustration)**. Loosen the main bearing cap bolts 1/4-turn at a time each, until they can be removed by hand. Note if any stud bolts are used and make sure they're returned to their original locations when the crankshaft is reinstalled.

5   On OHC models, remove the balance shaft assembly. Be sure to note the exact location of the oil pump drive and drive chain before disassembly to ensure the correct timing of the balance shafts to the crankshaft rotation. Refer to Chapter 2B for additional information on the balance shaft assembly removal and installation procedure.

6   Gently tap the caps with a soft-face hammer, then separate them from the engine block. If necessary, use the bolts as levers to remove the caps. Try not to drop the bearing inserts if they come out with the caps.

7   Carefully lift the crankshaft out of the engine. It may be a good idea to have an assistant available, since the crankshaft is quite heavy. With the bearing inserts in place in the engine block and main bearing caps, return the caps to their respective locations on the engine block and tighten the bolts finger tight.

16 Engine block - cleaning

Refer to illustrations 16.4a, 16.4b, 16.8 and 16.10

1   Remove the main bearing caps and separate the bearing inserts from the caps and the engine block. Tag the bearings, indicating which cylinder they were removed from and whether they

were in the cap or the block, then set them aside.

2   Using a gasket scraper, remove all traces of gasket material from the engine block. Be very careful not to nick or gouge the gasket sealing surfaces.

3   Remove all of the covers and threaded oil gallery plugs from the block. The plugs are usually very tight - they may have to be drilled out and the holes retapped. Use new plugs when the engine is reassembled.

4   Remove the core plugs from the engine block. To do this, Knock one side of the core plug into the block with a hammer and punch, then grasp them with a pair of large adjustable pliers and lever them out **(see illustrations)**. **Caution:** *The core plugs (also known as freeze or soft plugs) may be difficult or impossible to retrieve if they're driven into the block coolant passages.*

5   If the engine is extremely dirty, it should be taken to an automotive machine shop to be steam cleaned or hot tanked.

6   After the block is returned, clean all oil holes and oil galleries one more time. Brushes specifically designed for this purpose are available at most auto parts stores. Flush the passages with warm water until the water runs clear, dry the block thoroughly and wipe all machined surfaces with a light, rust preventive oil. If you have access to compressed air, use it to speed the drying process and blow out all the oil holes and galleries. **Warning:** *Wear eye protection when using compressed air!*

7   If the block isn't extremely dirty or sludged up, you can do an adequate cleaning job with hot soapy water and a stiff brush. Take plenty of time and do a thorough job. Regardless of the cleaning method used, be sure to clean all oil holes and galleries very thoroughly, dry the block completely and coat all machined surfaces with light oil.

8   The threaded holes in the block must be clean to ensure accurate torque readings during reassembly. Run the proper size tap into each of

16.8 All bolt holes in the block - particularly the main bearing cap and head bolt holes - should be cleaned and restored with a tap (be sure to remove debris from the holes after this is done)

16.10 A large socket on an extension can be used to drive the new core plugs into the bores

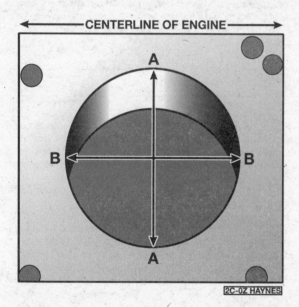

17.4a Measure the diameter of each cylinder at a right angle to engine centerline (A), and parallel to engine centerline (B) - out-of-round is the difference between A and B; taper is the difference between the diameter at the top of the cylinder and the diameter at the bottom of the cylinder

2C

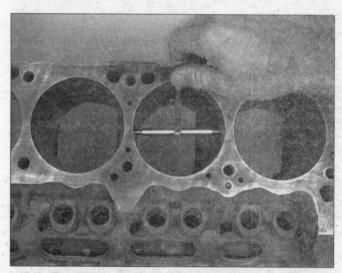

17.4b The ability to "feel" when the telescoping gauge is at the correct point will be developed over time, so work slowly and repeat the check until you're satisfied the bore measurement is accurate

the holes to remove rust, corrosion, thread sealant or sludge and restore damaged threads **(see illustration)**. If possible, use compressed air to clear the holes of debris produced by this operation. Now is a good time to clean the threads on the head bolts and the main bearing cap bolts as well.

9   Reinstall the main bearing caps and tighten the bolts finger tight.

10   After coating the sealing surfaces of the new core plugs with Permatex no. 2 sealant, or equivalent, install them in the engine block. Make sure they're driven in straight and seated properly or leakage could result. Special tools are available for this purpose, but a large socket, with an outside diameter that will just slip into the core plug, a 1/2-inch drive extension and a hammer will work just as well **(see illustration)**.

11   Apply non-hardening sealant (such as Permatex no. 2 or Teflon pipe sealant) to the new oil gallery plugs and thread them into the holes in the block. Make sure they're tightened securely.

12   If the engine isn't going to be reassembled right away, cover it with a large plastic trash bag to keep it clean.

17 Engine block - inspection

Refer to illustrations 17.4a, 17.4b and 17.4c

Note: *The manufacturer recommends checking the block deck and transaxle mounting bolt hole bosses for warpage and the main bearing bore concentricity and alignment. Since special measuring tools are needed, the checks should be done by an automotive machine shop. Also, if you're working on a OHC engine, it may be a good idea to verify the condition of the oil flow check valve located in the oil passage in the right front corner of the block deck.*

1   Before the block is inspected, it should be cleaned as described in Section 16.

2   Visually check the block for cracks, rust and corrosion. Look for stripped threads in the threaded holes. It's also a good idea to have the block checked for hidden cracks by an automotive machine shop that has the special equipment to do this type of work. If defects are found, have the block repaired, if possible, or replaced.

3   Check the cylinder bores for scuffing and scoring.

4   Check the cylinders for taper and out-of-round conditions as follows **(see illustrations):**

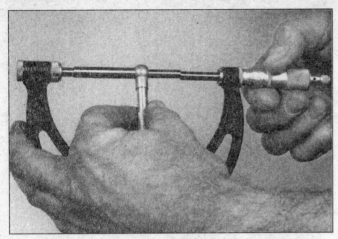

17.4c The gauge is then measured with a micrometer to determine the bore size

18.3a A "bottle brush" hone will produce better results if you've never honed cylinders before

11   If the cylinders are in reasonably good condition and not worn to the outside of the limits, and if the piston-to-cylinder clearances can be maintained properly, they don't have to be rebored. Honing is all that's necessary (see Section 18).

18   Cylinder honing

Refer to illustrations 18.3a and 18.3b

1   Prior to engine reassembly, the cylinder bores must be honed so the new piston rings will seat correctly and provide the best possible combustion chamber seal. **Note:** *If you don't have the tools or don't want to tackle the honing operation, most automotive machine shops will do it for a reasonable fee.*

2   Before honing the cylinders, install the main bearing caps and tighten the bolts to the specified torque.

3   Two types of cylinder hones are commonly available - the flex hone or "bottle brush" type and the more traditional surfacing hone with spring-loaded stones. Both will do the job, but for the less experienced mechanic the "bottle brush" hone will probably be easier to use. You'll also need some honing oil (kerosene will work if honing oil isn't available), rags and an electric drill motor. Proceed as follows:

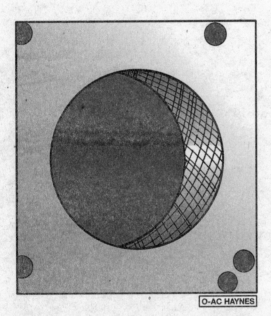

18.3b The cylinder hone should leave a smooth, crosshatch pattern with the lines intersecting at approximately a 60-degree angle

a) *Mount the hone in the drill motor, compress the stones and slip it into the first cylinder (see illustration). Be sure to wear safety goggles or a face shield!*

b) *Lubricate the cylinder with plenty of honing oil, turn on the drill and move the hone up-and-down in the cylinder at a pace that will produce a fine crosshatch pattern on the cylinder walls. Ideally, the crosshatch lines should intersect at approximately a 60-degree angle (see illustration). Be sure to use plenty of lubricant and don't take off any more material than is absolutely necessary to produce the desired finish. Note: Piston ring manufacturers may specify a smaller crosshatch angle than the traditional 60-degrees - read and follow any instructions included with the new rings.*

c) *Don't withdraw the hone from the cylinder while it's running. Instead, shut off the drill and continue moving the hone up-and-down in the cylinder until it comes to a complete stop, then compress the stones and withdraw the hone. If you're using a "bottle brush" type hone, stop the drill motor, then turn the chuck in the normal direction of rotation while withdrawing the hone from the cylinder.*

d) *Wipe the oil out of the cylinder and repeat the procedure for the remaining cylinders.*

5   Measure the diameter of each cylinder at the top (just under the ridge area), center and bottom of the cylinder bore, parallel to the crankshaft axis.

6   Next, measure each cylinder's diameter at the same three locations perpendicular to the crankshaft axis.

7   The taper of each cylinder is the difference between the bore diameter at the top of the cylinder and the diameter at the bottom. The out-of-round specification of the cylinder bore is the difference between the parallel and perpendicular readings. Compare your results to this Chapter's Specifications.

8   If the cylinder walls are badly scuffed or scored, or if they're out-of-round or tapered beyond the limits given in this Chapter's Specifications, have the engine block rebored and honed at an automotive machine shop.

9   If a rebore is done, oversize pistons and rings will be required.

10   Using a precision straightedge and feeler gauge, check the block deck (the surface the cylinder heads mate with) for distortion as you did with the cylinder heads (see Section 10). If it's distorted beyond the specified limit, the block decks can be resurfaced by an automotive machine shop.

4   After the honing job is complete, chamfer the top edges of the cylinder bores with a small file so the rings won't catch when the pistons are installed. Be very careful not to nick the cylinder walls with the end of the file.

9.4a The piston ring grooves can be cleaned with a special tool, as shown here . . .

19.4b . . . or a section of broken ring

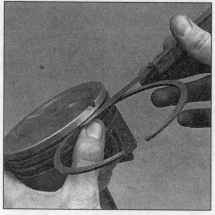

19.10 Check the ring side clearance with a feeler gauge at several points around the groove

5   The entire engine block must be washed again very thoroughly with warm, soapy water to remove all traces of the abrasive grit produced during the honing operation. **Note:** *The bores can be considered clean when a lint-free white cloth - dampened with clean engine oil - used to wipe them out doesn't pick up any more honing residue, which will show up as gray areas on the cloth.* Be sure to run a brush through all oil holes and galleries and flush them with running water.

6   After rinsing, dry the block and apply a coat of light rust preventive oil to all machined surfaces. Wrap the block in a plastic trash bag to keep it clean and set it aside until reassembly.

19   Pistons and connecting rods - inspection

Refer to illustrations 19.4a, 19.4b, 19.10 and 19.11

1   Before the inspection process can be carried out, the piston/connecting rod assemblies must be cleaned and the original piston rings removed from the pistons. **Note:** *Always use new piston rings when the engine is reassembled.*

2   Using a piston ring installation tool, carefully remove the rings from the pistons. Be careful not to nick or gouge the pistons in the process.

3   Scrape all traces of carbon from the top of the piston. A hand held wire brush or a piece of fine emery cloth can be used once the majority of the deposits have been scraped away. Do not, under any circumstances, use a wire brush mounted in a drill motor to remove deposits from the pistons. The piston material is soft and may be eroded away by the wire brush.

4   Use a piston ring groove cleaning tool to remove carbon deposits from the ring grooves. If a tool isn't available, a piece broken off the old ring will do the job. Be very careful to remove only the carbon deposits - don't remove any metal and do not nick or scratch the sides of the ring grooves **(see illustrations)**.

5   Once the deposits have been removed, clean the piston/rod assemblies with solvent and dry them with compressed air (if available). **Warning:** *Wear eye protection. Make sure the oil return holes in the back sides of the ring grooves are clear.*

6   If the pistons and cylinder walls aren't damaged or worn excessively, and if the engine block isn't rebored, new pistons won't be necessary. Normal piston wear appears as even vertical wear on the piston thrust surfaces and slight looseness of the top ring in its groove. New piston rings, however, should always be used when an engine is rebuilt.

7   Carefully inspect each piston for cracks around the skirt, at the pin bosses and at the ring lands.

8   Look for scoring and scuffing on the thrust faces of the skirt, holes in the piston crown and burned areas at the edge of the crown. If the skirt is scored or scuffed, the engine may have been suffering from

19.11 Measure the piston diameter at a 90-degree angle to the piston pin and in line with it

overheating and/or abnormal combustion, which caused excessively high operating temperatures. The cooling and lubrication systems should be checked thoroughly. A hole in the piston crown is an indication that abnormal combustion (preignition) was occurring. Burned areas at the edge of the piston crown are usually evidence of spark knock (detonation). If any of the above problems exist, the causes must be corrected or the damage will occur again. The causes may include intake air leaks, incorrect fuel/air mixture, low octane fuel, ignition timing and EGR system malfunctions.

9   Corrosion of the piston, in the form of small pits, indicates coolant is leaking into the combustion chamber and/or the crankcase. Again, the cause must be corrected or the problem may persist in the rebuilt engine.

10   Measure the piston ring side clearance by laying a new piston ring in each ring groove and slipping a feeler gauge in beside it **(see illustration)**. Check the clearance at three or four locations around each groove. Be sure to use the correct ring for each groove - they are different. If the side clearance is greater than specified in this Chapter, new pistons will have to be used.

11   Check the piston-to-bore clearance by measuring the bore (see Section 17) and the piston diameter. Make sure the pistons and bores are correctly matched. Measure the piston across the skirt, at a 90-degree angle to the piston pin **(see illustration)**. The measurement must be taken at a specific point, depending on the engine type, to be accurate.

a)   *The piston diameter on 2.2 liter OHV engines is measured directly in line with the piston pin centerline.*

2C

20.1 The oil holes should be chamfered so sharp edges don't gouge or scratch the new bearings

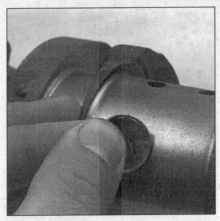

20.3 Rubbing a penny lengthwise on each journal will reveal its condition - if copper rubs off and is embedded in the crankshaft, the journals should be reground

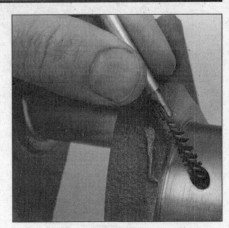

20.4 Use a wire or stiff plastic bristle brush to clean the oil passages in the crankshaft

20.6 Measure the diameter of each crankshaft journal at several points to detect taper and out-of-round conditions

20.8 If the seals have worn grooves in the crankshaft journals, or if the seal contact surfaces are nicked or scratched, the new seals will leak

b) *If you're working on a OHC engine, measure the piston approximately 1/2-inch (12.0 mm) up from the lower edge of the skirt.*

12 Subtract the piston diameter from the bore diameter to obtain the clearance. If it's greater than specified, the block will have to be rebored and new pistons and rings installed.

13 Check the piston-to-rod clearance by twisting the piston and rod in opposite directions. Any noticeable play indicates excessive wear, which must be corrected. The piston/connecting rod assemblies should be taken to an automotive machine shop to have the pistons and rods resized and new pins installed.

14 If the pistons must be removed from the connecting rods for any reason, they should be taken to an automotive machine shop. While they are there have the connecting rods checked for bend and twist, since automotive machine shops have special equipment for this purpose. **Note:** *Unless new pistons and/or connecting rods must be installed, do not disassemble the pistons and connecting rods.*

15 Check the connecting rods for cracks and other damage. Temporarily remove the rod caps, lift out the old bearing inserts, wipe the rod and cap bearing surfaces clean and inspect them for nicks, gouges and scratches. After checking the rods, replace the old bearings, slip the caps into place and tighten the nuts finger tight. **Note:** *If the engine is being rebuilt because of a connecting rod knock, be sure to install new rods.*

20 Crankshaft - inspection

Refer to illustrations 20.1, 20.3, 20.4, 20.6 and 20.8

1 Remove all burrs from the crankshaft oil holes with a stone, file or scraper **(see illustration)**.

2 Check the main and connecting rod bearing journals for uneven wear, scoring, pits and cracks.

3 Rub a penny across each journal several times **(see illustration)**. If a journal picks up copper from the penny, it's too rough and must be reground.

4 Clean the crankshaft with solvent and dry it with compressed air (if available). **Warning:** *Wear eye protection when using compressed air.* Be sure to clean the oil holes with a stiff brush **(see illustration)** and flush them with solvent.

5 Check the rest of the crankshaft for cracks and other damage. It should be magnafluxed to reveal hidden cracks - an automotive machine shop will handle the procedure.

6 Using a micrometer, measure the diameter of the main and connecting rod journals and compare the results to this Chapter's Specifications **(see illustration)**. By measuring the diameter at a number of points around each journal's circumference, you'll be able to determine

whether or not the journal is out-of-round. Take the measurement at each end of the journal, near the crank throws, to determine if the journal is tapered.

7   If the crankshaft journals are damaged, tapered, out-of-round or worn beyond the limits given in the Specifications, have the crankshaft reground by an automotive machine shop. Be sure to use the correct size bearing inserts if the crankshaft is reconditioned.

8   Check the oil seal journals at each end of the crankshaft for wear and damage. If the seal has worn a groove in the journal, or if it's nicked or scratched **(see illustration)**, the new seal may leak when the engine is reassembled. In some cases, an automotive machine shop may be able to repair the journal by pressing on a thin sleeve. If repair isn't feasible, a new or different crankshaft should be installed.

9   Refer to Section 21 and examine the main and rod bearing inserts.

21   Main and connecting rod bearings - inspection

Refer to illustration 21.1

1   Even though the main and connecting rod bearings should be replaced with new ones during the engine overhaul, the old bearings should be retained for close examination, as they may reveal valuable information about the condition of the engine **(see illustration)**.

2   Bearing failure occurs because of lack of lubrication, the presence of dirt or other foreign particles, overloading the engine and corrosion. Regardless of the cause of bearing failure, it must be corrected before the engine is reassembled to prevent it from happening again.

3   When examining the bearings, remove them from the engine block, the main bearing caps, the connecting rods and the rod caps and lay them out on a clean surface in the same general position as their location in the engine. This will enable you to match any bearing problems with the corresponding crankshaft journal.

4   Dirt and other foreign particles get into the engine in a variety of ways. It may be left in the engine during assembly, or it may pass through filters or the PCV system. It may get into the oil, and from there into the bearings. Metal chips from machining operations and normal engine wear are often present. Abrasives are sometimes left in engine components after reconditioning, especially when parts aren't thoroughly cleaned using the proper cleaning methods. Whatever the source, these foreign objects often end up embedded in the soft bearing material and are easily recognized. Large particles won't embed in the bearing and will score or gouge the bearing and journal. The best prevention for this cause of bearing failure is to clean all parts thoroughly and keep everything spotlessly clean during engine assembly. Frequent and regular engine oil and filter changes are also recommended.

5   Lack of lubrication (or lubrication breakdown) has a number of interrelated causes. Excessive heat (which thins the oil), overloading (which squeezes the oil from the bearing face) and oil leakage or throw off (from excessive bearing clearances, worn oil pump or high engine speeds) all contribute to lubrication breakdown. Blocked oil passages, which usually are the result of misaligned oil holes in a bearing shell, will also oil starve a bearing and destroy it. When lack of lubrication is the cause of bearing failure, the bearing material is wiped or extruded from the steel backing of the bearing. Temperatures may increase to the point where the steel backing turns blue from overheating.

6   Driving habits can have a definite effect on bearing life. Full throttle, low speed operation (lugging the engine) puts very high loads on bearings, which tends to squeeze out the oil film. These loads cause the bearings to flex, which produces fine cracks in the bearing face (fatigue failure). Eventually the bearing material will loosen in pieces and tear away from the steel backing. Short trip driving leads to corrosion of bearings because insufficient engine heat is produced to drive off the condensed water and corrosive gases. These products collect in the engine oil, forming acid and sludge. As the oil is carried to the engine bearings, the acid attacks and corrodes the bearing material.

7   Incorrect bearing installation during engine assembly will lead to bearing failure as well. Tight fitting bearings leave insufficient oil clear-

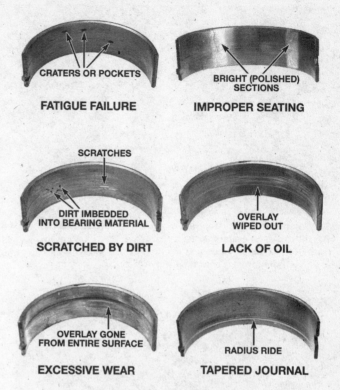

21.1 Typical bearing failures

ance and will result in oil starvation. Dirt or foreign particles trapped behind a bearing insert result in high spots on the bearing which lead to failure.

22   Engine overhaul - reassembly sequence

1   Before beginning engine reassembly, make sure you have all the necessary new parts, gaskets and seals as well as the following items on hand:

> *Common hand tools*
> *Torque wrench (1/2-inch drive)*
> *Piston ring installation tool*
> *Piston ring compressor*
> *Vibration damper installation tool*
> *Short lengths of rubber or plastic hose to fit over connecting rod bolts*
> *Plastigage*
> *Feeler gauges*
> *Fine-tooth file*
> *New engine oil*
> *Engine assembly lube or moly-base grease*
> *Gasket sealant*
> *Thread locking compound*

2   In order to save time and avoid problems, engine reassembly must be done in the following general order:

2.2 liter OHV engine

> *New camshaft bearings (must be done by automotive machine shop)*
> *Crankshaft and main bearings*
> *Piston/connecting rod assemblies*
> *Oil pump*
> *Camshaft and lifters*
> *Timing chain and sprockets*
> *Timing chain cover*

2C

23.3 When checking piston ring end gap, the ring must be square in the cylinder bore (this is done by pushing the ring down with the top of a piston as shown)

23.4 With the ring square in the cylinder, measure the end gap with a feeler gauge

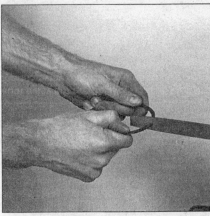

23.5 If the end gap is too small, clamp a file in a vise and file the ring ends (from the outside in only) to enlarge the gap slightly

23.9a Installing the spacer/expander in the oil control ring groove

23.9b DO NOT use a piston ring installation tool when installing the oil ring side rails

Oil pan
Cylinder head, pushrods and rocker arms
Intake and exhaust manifolds
Rocker arm cover
Engine rear plate
Flywheel/driveplate

2.3 and 2.4 liter (OHC) engines

Crankshaft and main bearings
Rear main oil seal housing
Piston/connecting rod assemblies
Oil pump
Oil pan
Cylinder head and camshafts
Timing chain housing
Timing chain and sprockets

23   Piston rings - installation

Refer to illustrations 23.3, 23.4, 23.5, 23.9a, 23.9b and 23.12

1   Before installing the new piston rings, the ring end gaps must be checked. It's assumed the piston ring side clearance has been checked and verified correct (see Section 19).

2   Lay out the piston/connecting rod assemblies and the new ring sets so the ring sets will be matched with the same piston and cylinder during the end gap measurement and engine assembly.

3   Insert the top (number one) ring into the first cylinder and square it up with the cylinder walls by pushing it in with the top of the piston **(see illustration)**. The ring should be near the bottom of the cylinder, at the lower limit of ring travel.

4   To measure the end gap, slip feeler gauges between the ends of the ring until a gauge equal to the gap width is found **(see illustration)**. The feeler gauge should slide between the ring ends with a slight amount of drag. Compare the measurement to this Chapter's Specifications. If the gap is larger or smaller than specified, double-check to make sure you have the correct rings before proceeding.

5   If the gap is too small, it must be enlarged or the ring ends may come in contact with each other during engine operation, which can cause serious engine damage. The end gap can be increased by filing the ring ends very carefully with a fine file. Mount the file in a vise equipped with soft jaws, slip the ring over the file with the ends contacting the file teeth and slowly move the ring to remove material from the ends. When performing this operation, file only from the outside in **(see illustration)**.

6   Excess end gap isn't critical unless it's greater than 0.040-inch. Again, double-check to make sure you have the correct rings for the engine.

7   Repeat the procedure for each ring that will be installed in the first cylinder and for each ring in the remaining cylinders. Remember to keep rings, pistons and cylinders matched up.

8   Once the ring end gaps have been checked/corrected, the rings can be installed on the pistons.

23.12 Installing the compression rings with a ring expander - the mark (arrow) must face up

24.11 Lay the Plastigage strips on the main bearing journals, parallel to the crankshaft centerline

24.15 Measuring the width of the crushed Plastigage to determine the main bearing oil clearance (be sure to use the correct scale - standard and metric ones are included)

2C

9   The oil control ring (lowest one on the piston) is usually installed first. It's composed of three separate components. Slip the spacer/expander into the groove (see illustration). If an anti-rotation tang is used, make sure it's inserted into the drilled hole in the ring groove. Next, install the lower side rail. Don't use a piston ring installation tool on the oil ring side rails, as they may be damaged. Instead, place one end of the side rail into the groove between the spacer/expander and the ring land, hold it firmly in place and slide a finger around the piston while pushing the rail into the groove (see illustration). Next, install the upper side rail in the same manner.

10   After the three oil ring components have been installed, check to make sure both the upper and lower side rails can be turned smoothly in the ring groove.

11   The number two (middle) ring is installed next. It's usually stamped with a mark, which must face up, toward the top of the piston. Note: *Always follow the instructions printed on the ring package or box - different manufacturers may require different approaches. Don't mix up the top and middle rings, as they have different cross sections.*

12   Use a piston ring installation tool and make sure the identification mark is facing the top of the piston, then slip the ring into the middle groove on the piston (see illustration). Don't expand the ring any more than necessary to slide it over the piston.

13   Install the number one (top) ring in the same manner. Make sure the mark is facing up. Be careful not to confuse the number one and number two rings.

14   Repeat the procedure for the remaining pistons and rings.

24   Crankshaft - installation and main bearing oil clearance check

Refer to illustrations 24.11 and 24.15

1   Crankshaft installation is the first step in engine reassembly. It's assumed at this point that the engine block and crankshaft have been cleaned, inspected and repaired or reconditioned.

2   Position the engine with the bottom facing up.

3   Remove the main bearing cap bolts and lift out the caps. Lay them out in the proper order to ensure correct installation.

4   If they're still in place, remove the original bearing inserts from the block and the main bearing caps. Wipe the bearing surfaces of the block and caps with a clean, lint-free cloth. They must be kept spotlessly clean.

Main bearing oil clearance check

Note: *Don't touch the faces of the new bearing inserts with your fingers. Oil and acids from your skin can etch the bearings.*

5   Clean the back sides of the new main bearing inserts and lay one in each main bearing saddle in the block. If one of the bearing inserts from each set has a large groove in it, make sure the grooved insert is installed in the block. Lay the other bearing from each set in the corresponding main bearing cap. Make sure the tab on the bearing insert fits into the recess in the block or cap. Caution: *The oil holes in the block must line up with the oil holes in the bearing inserts. Do not hammer the bearing into place and don't nick or gouge the bearing faces. No lubrication should be used at this time.*

6   On 2.2 liter OHV engines, the flanged thrust bearing must be installed in the number four cap and saddle (counting from the front of the engine). On OHC engines, the thrust bearing must be installed in the number three (center) cap and saddle.

7   Clean the faces of the bearings in the block and the crankshaft main bearing journals with a clean, lint-free cloth.

8   Check or clean the oil holes in the crankshaft, as any dirt here can go only one way - straight through the new bearings.

9   Once you're certain the crankshaft is clean, carefully lay it in position in the main bearings.

10   Before the crankshaft can be permanently installed, the main bearing oil clearance must be checked.

11   Cut several pieces of the appropriate size Plastigage (they should be slightly shorter than the width of the main bearings) and place one piece on each crankshaft main bearing journal, parallel with the journal axis (see illustration).

12   Clean the faces of the bearings in the caps and install the caps in their original locations (don't mix them up) with the arrows pointing toward the front of the engine. Don't disturb the Plastigage.

13   Starting with the center main and working out toward the ends, tighten the main bearing cap bolts, in three steps, to the torque figure listed in this Chapter's Specifications. Don't rotate the crankshaft at any time during this operation.

14   Remove the bolts and carefully lift off the main bearing caps. Keep them in order. Don't disturb the Plastigage or rotate the crankshaft. If any of the main bearing caps are difficult to remove, tap them gently from side-to-side with a soft-face hammer to loosen them.

15   Compare the width of the crushed Plastigage on each journal to the scale printed on the Plastigage envelope to obtain the main bearing oil clearance (see illustration). Check the Specifications to make sure it's correct.

16   If the clearance is not as specified, the bearing inserts may be the wrong size (which means different ones will be required). Before deciding different inserts are needed, make sure no dirt or oil was between the bearing inserts and the caps or block when the clearance was measured. If the Plastigage was wider at one end than the other, the journal may be tapered (refer to Section 20).

17   Carefully scrape all traces of the Plastigage material off the main

25.3 Tap around the outer edge of the new oil seal with a hammer and blunt punch to seat it squarely in the bore

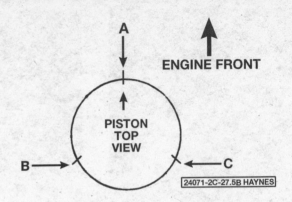

26.5 Piston ring gap positions

A   *Top compression ring gap*
B   *Second compression ring gap*
C   *Oil ring assembly gap*

bearing journals and/or the bearing faces. Use your fingernail or the edge of a credit card - don't nick or scratch the bearing faces.

Final crankshaft installation

18   Carefully lift the crankshaft out of the engine.
19   Clean the bearing faces in the block, then apply a thin, uniform layer of moly-base grease or engine assembly lube to each of the bearing surfaces. Be sure to coat the thrust faces as well as the journal face of the thrust bearing.
20   Make sure the crankshaft journals are clean, then lay the crankshaft back in place in the block. **Note:** *On OHC engines, make sure the balance shaft chain is properly installed and does not interfere with the engine block when installing the crankshaft. Refer to Chapter 2B for additional information on the balance shaft removal and installation procedure.*
21   Clean the faces of the bearings in the caps, then apply lubricant to them.
22   Install the caps in their original locations with the arrows pointing toward the front of the engine.
23   Install the bolts.
24   Tighten all except the thrust bearing cap bolts to the specified torque (work from the center out and approach the final torque in three steps).
25   Tighten the thrust bearing cap bolts to 10-to-12 ft-lbs.
26   Tap the ends of the crankshaft forward and backward with a lead or brass hammer to line up the main bearing and crankshaft thrust surfaces.
27   Retighten all main bearing cap bolts to the torque specified in this Chapter, starting with the center main and working out toward the ends.
28   Rotate the crankshaft a number of times by hand to check for any obvious binding.
29   The final step is to check the crankshaft endplay with feeler gauges or a dial indicator as described in Section 15. The endplay should be correct if the crankshaft thrust faces aren't worn or damaged and new bearings have been installed.
30   Refer to Section 26 and install the new rear main oil seal.

25   Rear main oil seal - installation

2.2 liter OHV engines

Refer to illustration 25.3

1   Clean the bore in the block/cap and the seal journal on the crankshaft. Check the crankshaft journal for scratches and nicks that could damage the new seal lip and cause oil leaks. If the crankshaft is

damaged, the only alternative is a new or different crankshaft.
2   Apply a light coat of engine oil or multi-purpose grease to the outer edge of the new seal. Lubricate the seal lip with moly-base grease or engine assembly lube.
3   Press the new seal into place with the special tool, if available (**see Chapter 2A**). The seal lip must face toward the front of the engine. If the special tool isn't available, carefully work the seal lip over the end of the crankshaft and tap the seal in with a hammer and blunt punch until it's seated in the bore (**see illustration**).

2.3 and 2.4 liter (OHC) engine

4   This engine is equipped with a one-piece seal that fits into a housing attached to the block. The crankshaft must be installed first and the main bearing caps bolted in place, then the new seal should be installed in the housing and the housing bolted to the block (see Chapter 2B).
5   Before installing the crankshaft, check the seal journal very carefully for scratches and nicks that could damage the new seal lip and cause oil leaks. If the crankshaft is damaged, the only alternative is a new or different crankshaft.
6   The old seal can be removed from the housing with a hammer and punch by driving it out from the back side (see Section 15 in Chapter 2B). Be sure to note how far it's recessed into the housing bore before removing it; the new seal will have to be recessed an equal amount. Be very careful not to scratch or otherwise damage the bore in the housing or oil leaks could develop.
7   Make sure the housing is clean, then apply a thin coat of engine oil to the outer edge of the new seal. The seal must be pressed squarely into the housing bore (see Section 15 in Chapter 2B). Work slowly and make sure the seal enters the bore squarely.
8   The seal lips must be lubricated with moly-base grease or engine assembly lube before the seal/housing is slipped over the crankshaft and bolted to the block. Use a new gasket - no sealant is required - and make sure the dowel pins are in place before installing the housing.
9   Tighten the bolts a little at a time until they're all at the torque listed in the Chapter 2B Specifications.

26   Pistons and connecting rods - installation and rod bearing oil clearance check

Refer to illustrations 26.5, 26.9, 26.11, 26.13 and 26.17

1   Before installing the piston/connecting rod assemblies, the cylinder walls must be perfectly clean, the top edge of each cylinder must be chamfered, and the crankshaft must be in place.
2   Remove the cap from the end of the number one connecting rod

26.9 The notch or arrow on each piston must face the front (timing chain) end of the engine as the pistons are installed

26.11 Drive the piston into the cylinder bore with the end of a wooden or plastic hammer handle

26.13 Lay the Plastigage strips on each rod bearing journal, parallel to the crankshaft centerline

(check the marks made during removal). Remove the original bearing inserts and wipe the bearing surfaces of the connecting rod and cap with a clean, lint-free cloth. They must be kept spotlessly clean.

Connecting rod bearing oil clearance check

Note: *Don't touch the faces of the new bearing inserts with your fingers. Oil and acids from your skin can etch the bearings.*

3   Clean the back side of the new upper bearing insert, then lay it in place in the connecting rod. Make sure the tab on the bearing fits into the recess in the rod. Don't hammer the bearing insert into place and be very careful not to nick or gouge the bearing face. Don't lubricate the bearing at this time.

4   Clean the back side of the other bearing insert and install it in the rod cap. Again, make sure the tab on the bearing fits into the recess in the cap, and don't apply any lubricant. It's critically important that the mating surfaces of the bearing and connecting rod are perfectly clean and oil free when they're assembled.

5   Position the piston ring gaps at 120-degree intervals around the piston **(see illustration)**.

6   Slip a section of plastic or rubber hose over each connecting rod cap bolt.

7   Lubricate the piston and rings with clean engine oil and attach a piston ring compressor to the piston. Leave the skirt protruding about 1/4-inch to guide the piston into the cylinder. The rings must be compressed until they're flush with the piston.

8   Rotate the crankshaft until the number one connecting rod journal is at BDC (bottom dead center) and apply a coat of engine oil to the cylinder walls.

9   With the mark or notch on top of the piston **(see illustration)** facing the front of the engine, gently insert the piston/connecting rod assembly into the number one cylinder bore and rest the bottom edge of the ring compressor on the engine block. If you're working on a OHC engine, make sure the oil hole in the lower end of the connecting rod is facing the right (exhaust manifold) side of the engine.

10   Tap the top edge of the ring compressor to make sure it's contacting the block around its entire circumference.

11   Gently tap on the top of the piston with the end of a wooden or plastic hammer handle **(see illustration)** while guiding the end of the connecting rod into place on the crankshaft journal. The piston rings may try to pop out of the ring compressor just before entering the cylinder bore, so keep some downward pressure on the ring compressor. Work slowly, and if any resistance is felt as the piston enters the cylinder, stop immediately. Find out what's hanging up and fix it before proceeding. Do not, for any reason, force the piston into the cylinder - you might break a ring and/or the piston.

12   Once the piston/connecting rod assembly is installed, the connecting rod bearing oil clearance must be checked before the rod cap is permanently bolted in place.

26.17 Measuring the width of the crushed Plastigage to determine the rod bearing oil clearance (be sure to use the correct scale - standard and metric ones are included)

13   Cut a piece of the appropriate size Plastigage slightly shorter than the width of the connecting rod bearing and lay it in place on the number one connecting rod journal, parallel with the journal axis **(see illustration)**.

14   Clean the connecting rod cap bearing face, remove the protective hoses from the connecting rod bolts and install the rod cap. Make sure the mating mark on the cap is on the same side as the mark on the connecting rod.

15   Install the nuts and tighten them to the torque listed in this Chapter's Specifications. Work up to it in three steps. **Note:** *Use a thin-wall socket to avoid erroneous torque readings that can result if the socket is wedged between the rod cap and nut. If the socket tends to wedge itself between the nut and the cap, lift up on it slightly until it no longer contacts the cap. Do not rotate the crankshaft at any time during this operation.*

16   Remove the nuts and detach the rod cap, being very careful not to disturb the Plastigage.

17   Compare the width of the crushed Plastigage to the scale printed on the Plastigage envelope to obtain the oil clearance **(see illustration)**. Compare it to this Chapter's Specifications to make sure the clearance is correct.

18   If the clearance is not as specified, the bearing inserts may be the wrong size (which means different ones will be required). Before deciding different inserts are needed, make sure no dirt or oil was between the bearing inserts and the connecting rod or cap when the clearance

was measured. Also, recheck the journal diameter. If the Plastigage was wider at one end than the other, the journal may be tapered (refer to Section 20).

Final connecting rod installation

19   Carefully scrape all traces of the Plastigage material off the rod journal and/or bearing face. Be very careful not to scratch the bearing - use your fingernail or the edge of a credit card.

20   Make sure the bearing faces are perfectly clean, then apply a uniform layer of clean moly-base grease or engine assembly lube to both of them. You'll have to push the piston into the cylinder to expose the face of the bearing insert in the connecting rod - be sure to slip the protective hoses over the rod bolts first.

21   Slide the connecting rod back into place on the journal, remove the protective hoses from the rod cap bolts, install the rod cap and tighten the nuts to the torque specified in this Chapter. Again, work up to the torque in three steps.

22   Repeat the entire procedure for the remaining pistons/connecting rods.

23   The important points to remember are:
  a) Keep the back sides of the bearing inserts and the insides of the connecting rods and caps perfectly clean when assembling them.
  b) Make sure you have the correct piston/rod assembly for each cylinder.
  c) The arrow or mark on the piston must face the front (timing chain end) of the engine.
  d) Lubricate the cylinder walls with clean oil.
  e) Lubricate the bearing faces when installing the rod caps after the oil clearance has been checked.

24   After all the piston/connecting rod assemblies have been properly installed, rotate the crankshaft a number of times by hand to check for any obvious binding.

25   As a final step, the connecting rod endplay must be checked. Refer to Section 14 for this procedure.

26   Compare the measured endplay to this Chapter's Specifications to make sure it's correct. If it was correct before disassembly and the original crankshaft and rods were reinstalled, it should still be right. If new rods or a new crankshaft were installed, the endplay may be inadequate. If so, the rods will have to be removed and taken to an automotive machine shop for resizing.

27   Initial start-up and break-in after overhaul

Warning: *Have a fire extinguisher handy when starting the engine for the first time.*

1   Once the engine has been installed in the vehicle, double-check the oil and coolant levels.

2   With the spark plugs out of the engine and the ECM fuse removed, crank the engine until oil pressure registers on the gauge or the light goes out.

3   Install the spark plugs, hook up the plug wires (except OHC engine) and install the ECM fuse.

4   Start the engine. It may take a few moments for the fuel system to build up pressure, but the engine should start without a great deal of effort. **Note:** *If the engine keeps backfiring, recheck the valve timing (and spark plug wires, where applicable).*

5   After the engine starts, it should be allowed to warm up to normal operating temperature. While the engine is warming up, make a thorough check for fuel, oil and coolant leaks.

6   Shut the engine off and recheck the engine oil and coolant levels.

7   Drive the vehicle to an area with minimum traffic, accelerate from 30 to 50 mph, then allow the vehicle to slow to 30 mph with the throttle closed. Repeat the procedure 10 or 12 times. This will load the piston rings and cause them to seat properly against the cylinder walls. Check again for oil and coolant leaks.

8   Drive the vehicle gently for the first 500 miles (no sustained high speeds) and keep a constant check on the oil level. It's not unusual for an engine to use oil during the break-in period.

9   At approximately 500 to 600 miles, change the oil and filter.

10   For the next few hundred miles, drive the vehicle normally. Don't pamper it or abuse it.

11   After 2000 miles, change the oil and filter again and consider the engine broken in.

Chapter 3
Cooling, heating and air conditioning systems

Contents

Specifications

General

Coolant capacity	See Chapter 1
Coolant reservoir pressure cap rating	15 psi
Thermostat opening temperature	
2.2L OHV engine	195-degrees F
2.3L and 2.4L OHC engines	180-degrees F
Refrigerant type	R-134a
Refrigerant capacity	1.5 pounds

Torque specifications

	Ft-lbs (unless otherwise indicated)
Transaxle oil cooler lines	27
Thermostat housing nuts/bolts	
2.2L OHV engine	
1995	89 in-lbs
1996 and later	124 in-lbs
2.3 and 2.4L OHC engines	124 in-lbs
Thermostat outlet pipe-to-oil pan bolt (2.3 and 2.4L OHC engines)	19
Water pump attaching bolts	
2.2L OHV engine	18
2.3L and 2.4L OHC engines	
Water pump-to-timing chain housing nuts	19
Water pump cover-to-water pump housing bolts	124 in-lbs.
Water pump cover-to-engine block bolts	19
Radiator outlet pipe-to-water pump housing bolts	124 in-lbs.
Water pump pulley bolts (2.2L OHV engine)	18

1 General information

All vehicles covered by this manual employ a pressurized engine cooling system with thermostatically controlled coolant circulation. Coolant is drawn from the radiator by an impeller-type water pump mounted at the front of the block. The coolant is then circulated through the engine block, intake manifold and the cylinder heads before it's redirected back into the radiator.

A wax pellet type thermostat is located in the thermostat housing near the front of the engine. Locations differ slightly depending upon the engine and year of the vehicle. On 1995 2.2L OHV models, the thermostat is located in a housing at the end of the cylinder head over the transaxle. On 1996 and later 2.2L OHV models, the thermostat is located in a housing up front near the radiator. 2.3L and 2.4L OHC models are equipped with a special designed thermostat housing that

mounts the thermostat in the metal outlet line on the right (firewall) side of the engine block. During warm up, the closed thermostat prevents coolant from circulating through the radiator. When the engine reaches normal operating temperature, the thermostat opens and allows hot coolant to travel through the radiator, where it is cooled before returning to the engine.

The cooling system is pressurized by a spring-loaded coolant reservoir cap, which, by maintaining pressure, increases the boiling point of the coolant. If the coolant temperature goes above this increased boiling point, the extra pressure in the system forces the reservoir cap valve off its seat and allows the coolant to escape through the overflow tube into the coolant reservoir. When the system cools, the excess coolant is automatically drawn from the reservoir tank back into the radiator.

The coolant reservoir serves as both the point at which fresh coolant is added to the cooling system to maintain the proper fluid

2.4 An inexpensive hydrometer can be used to test the condition of your coolant

level and as a holding tank for overheated coolant.

The heating system works by directing air through the heater core mounted in the dash and then to the interior of the vehicle by a system of ducts. Temperature is controlled by mixing heated air with fresh air, using a system of doors in the ducts, and a blower motor.

Air conditioning is an optional accessory, consisting of an evaporator core located under the dash, a condenser in front of the radiator, an accumulator in the engine compartment and a belt-driven compressor mounted at the front of the engine.

2 Antifreeze - general information

Refer to illustration 2.4

Warning: *Do not allow antifreeze to come in contact with your skin or painted surfaces of the vehicle. Rinse off spills immediately with plenty of water. Antifreeze is highly toxic if ingested. Never leave antifreeze lying around in an open container or in puddles on the floor; children and pets are attracted by it's sweet smell and may drink it. Check with local authorities about disposing of used antifreeze. Many communities have collection centers which will see that antifreeze is disposed of safely. Never dump used anti-freeze on the ground or pour it into drains.*

Caution: *The manufacturer recommends using only DEX-COOL coolant for these systems. DEX-COOL is a long-lasting coolant designed for 100,000 miles or 5 years. Never mix green-colored ethylene glycol antifreeze and orange-colored "DEX-COOL" silicate-free coolant because doing so will destroy the efficiency of the "DEX-COOL".*

The cooling system should be filled with a water/ethylene glycol based antifreeze solution which will prevent freezing down to at least -20-degrees F (even lower in cold climates). It also provides protection against corrosion and increases the coolant boiling point.

The cooling system should be drained, flushed and refilled at least every other year (see Chapter 1). The use of antifreeze solutions for periods of longer than two years is likely to cause damage and encourage the formation of rust and scale in the system. However, these models are filled with a new, long-life "Dex-Cool coolant," which the manufacturer claims is good for five years.

Before adding antifreeze to the system, check all hose connections. Antifreeze can leak through very minute openings.

The exact mixture of antifreeze to water which you should use depends on the relative weather conditions. The mixture should contain at least 50-percent antifreeze, but should never contain more than 70-percent antifreeze. Consult the mixture ratio chart on the antifreeze container before adding coolant. Hydrometers are available at most auto parts stores to test the coolant **(see illustration)**. Use antifreeze which meets the vehicle manufacturer's specifications. Double-check with the hydrometer manufacturer concerning the specific gravity ratios between ethylene glycol and the modern DEX COOL antifreeze solutions when measuring coolant strength.

3 Thermostat - check and replacement

Warning: *The engine must be completely cool when this procedure is performed.*
Note: *Don't drive the vehicle without a thermostat! The computer may stay in open loop mode and emissions and fuel economy will suffer.*

Check

1 Before condemning the thermostat, check the coolant level, drivebelt tension and temperature gauge (or light) operation.
2 If the engine takes a long time to warm up, the thermostat is probably stuck open. Replace the thermostat.
3 If the engine runs hot, check the temperature of the upper radiator hose. If the hose isn't hot, the thermostat is probably stuck shut. Replace the thermostat.
4 If the upper radiator hose is hot, it means the coolant is circulating and the thermostat is open. Refer to the *Troubleshooting* at the front of this manual for the cause of overheating.
5 If an engine has been overheated, you may find damage such as leaking head gaskets, scuffed pistons and warped or cracked cylinder heads.

Replacement
2.2L OHV engine
Refer to illustrations 3.8, 3.9 and 3.12
6 Drain coolant (about 1 gallon) from the radiator, until the coolant level is below the thermostat housing (see Chapter 1).
7 Disconnect the radiator hose from the thermostat cover.
8 Remove the bolts and lift the cover off **(see illustration)**. It may be necessary to tap the cover with a soft-face hammer to break the gasket seal.

3.8 Thermostat cover bolt locations (arrows) - 1996 2.2L OHV engine shown

3.9 Note how it's installed, then remove the thermostat (the spring end points toward the engine)

3.12 Install a new rubber seal around the thermostat

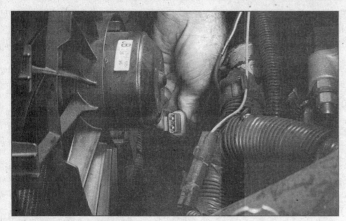

4.1 Disconnect the electrical connector from the fan and apply fused battery power and ground to test the fan

4.2 The cooling fan relay and fuse (arrows) are easily identifiable by viewing the decal on the inside of the underhood fuse/relay box cover

9   Note how it's installed, then remove the thermostat **(see illustration)**. Be sure to use a replacement thermostat with the correct opening temperature (see this Chapter's Specifications).
10   If a gasket was used, use a scraper or putty knife to remove all traces of old gasket material and sealant from the mating surfaces. **Caution:** *Be careful not to gouge or damage the gasket surfaces, because a leak could develop after assembly. Make sure no gasket material falls into the coolant passage; it's a good idea to stuff a rag in the passage. Wipe the mating surfaces with a rag saturated with lacquer thinner or acetone.*
11   Install the thermostat and make sure the correct end faces out - the spring is directed toward the engine.
12   Most models will not have a traditional gasket, but rather a rubber ring around the thermostat. If so, replace this ring and install the thermostat cover without gasket sealant **(see illustration)**. *Note: If a gasket was used, apply a thin coat of RTV sealant to both sides of the new gasket and position it on the engine side, over the thermostat, and make sure the gasket holes line up with the bolt holes in the housing.*
13   Carefully position the cover and install the bolts. Tighten them to the torque listed in this Chapter's Specifications - do not over-tighten the bolts or the cover may crack or become distorted.
14   Reattach the radiator hose to the cover and tighten the clamp - now may be a good time to check and replace the hoses and clamps (see Chapter 1).
15   Refer to Chapter 1 and refill the system, then run the engine and check carefully for leaks.
16   Repeat steps 1 through 5 to be sure the repairs corrected the previous problem(s).

2.3L and 2.4L OHC engines

17   Drain the coolant from the radiator (see Chapter 1).
18   Remove the exhaust manifold heat shield (see Chapter 2B).
19   Remove the cover that protects the outlet pipe on the thermostat housing by accessing cover bolts through the exhaust manifold runners.
20   Raise the vehicle and secure it on jackstands.
21   Disconnect the radiator hose from the metal outlet pipe
22   Remove the outlet pipe bolt and carefully separate the outlet pipe from the thermostat housing and the engine block.
23   Note how it's installed, then remove the thermostat from the outlet pipe. Be sure to use a replacement thermostat with the correct opening temperature (see this Chapter's Specifications).
24   If a gasket was used, use a scraper or putty knife to remove all traces of old gasket material and sealant from the mating surfaces. **Caution:** *Be careful not to gouge or damage the gasket surfaces, because a leak could develop after assembly. Make sure no gasket material falls into the coolant passage; it's a good idea to stuff a rag in the passage. Wipe the mating surfaces with a rag saturated with lacquer thinner or acetone.*
25   Install the thermostat and make sure the correct end faces out - the spring is directed toward the engine.

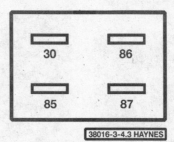

4.3 Test for continuity between relay terminals 30 and 87 - there should be no continuity until positive battery voltage is applied to terminal 85 and ground to terminal 86

38016-3-4.3 HAYNES

26   Most of these models do not have a traditional gasket, but rather a rubber ring around the thermostat. If so, replace this ring and install the thermostat cover without gasket sealant. *Note: If a gasket was used, apply a thin coat of RTV sealant to both sides of the new gasket and position it on the engine side, over the thermostat, and make sure the gasket holes line up with the bolt holes in the housing.*
27   Reattach the metal outlet pipe to the thermostat housing and tighten the bolts to the torque listed in this Chapter's Specifications. Now may be a good time to check and replace the hoses and clamps (see Chapter 1).
28   Refer to Chapter 1 and refill the system, then run the engine and check carefully for leaks.
29   The remaining steps are the reverse of the removal procedure.
30   Repeat steps 1 through 5 to be sure the repairs corrected the previous problem(s).

4   Engine cooling fan and circuit - check and component replacement

Warning: *Keep hands, tools and clothing away from the fan. To avoid injury or damage DO NOT operate the engine with a damaged fan. Do not attempt to repair fan blades - replace a damaged fan with a new one.*

Check

Refer to illustrations 4.1, 4.2 and 4.3

1   To test a fan motor, unplug the electrical connector at the motor and use fused jumper wires to connect battery power and ground directly to the fan **(see illustration)**. If the fan still doesn't work, replace the motor.
2   If the motor tests OK, check the cooling fan relay, located in the underhood fuse/relay panel **(see illustration)**.
3   Remove the cooling fan relay and test for continuity between terminals 30 and 87 **(see illustration)**. There should be no continuity.
4   Apply positive battery power to terminal 85, and ground terminal 86, and test for continuity again between 30 and 87. There should now be

3

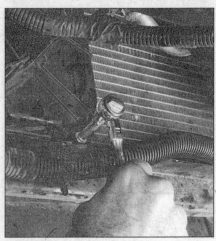

4.10 Detach the bolts securing the fan shroud to the radiator support . . .

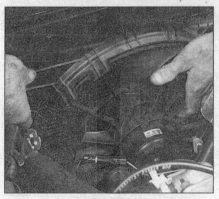

4.11 . . . then remove the fan assembly by lifting it to release the tabs that lock the top portion onto the air conditioning condenser

4.12 Remove the nut securing the fan assembly to the fan motor

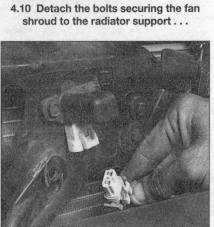

5.3 Disconnect the airbag sensor harness connector

5.6 Remove the upper and lower (arrow) transaxle cooler lines using an open end wrench. Use a back-up wrench to avoid damaging the radiator housing by twisting the cooler line if the fitting is tight or corroded

5.7a Remove the hood latch bolts (arrows)

continuity; if not, replace the relay.

5   Test the number 30 terminal (on the fuse/relay panel). There should be power at all times, probing with a grounded test light. The number 85 terminal should have power only with the key ON.

6   If the relay and the fan motor are good, check the wiring from the relays to the PCM (computer) for open or short circuits. Refer to the wiring schematics at the end of Chapter 12.

7   If the circuit checks OK but the fan still don't come on, check the engine coolant temperature sensor (see Chapter 6).

Replacement

Refer to illustrations 4.10, 4.11 and 4.12

Warning: *Keep hands, tools and clothing away from the fan. To avoid injury or damage DO NOT operate the engine with a damaged fan. Do not attempt to repair fan blades - replace a damaged fan with a new one.*

8   Disconnect the cable from the negative terminal of the battery. **Caution:** *On models equipped with a Delco Loc II or Theftlock audio system, be sure the lockout feature is turned off before performing any procedure which requires disconnecting the battery.*

9   Disconnect the electrical connector from the fan motor.

10   Remove the bolts and detach the fan shroud from the radiator **(see illustration)**.

11   Pull the fan assembly up slightly to dislodge its tabs from the radiator, then guide the fan assembly out from the engine compartment **(see illustration)**, making sure that all wiring clips are disconnected. Be careful not to contact the radiator cooling fins.

12   Detach the fan retaining bolt from the fan assembly **(see illustration)**. **Note:** *It is extremely important to replace the original fan motor with the correct part because different models have different wattage ratings and insufficient cooling could occur if the original type fan motor is not installed back into the system.*

13   Installation is the reverse of removal.

---

5   Radiator and coolant reservoir - removal and installation

---

Warning 1: *The models covered by this manual are equipped with airbags. Always disable the airbag system before working in the vicinity of the impact sensors, steering column or instrument panel to avoid the possibility of accidental deployment of the airbag(s), which could cause personal injury (see Chapter 12).*

Warning 2: *The engine must be completely cool when this procedure is performed.*

Radiator

Refer to illustrations 5.3, 5.6, 5.7a, 5.7b, 5.9, 5.10a, 5.10b, 5.10c and 5.11

1   Disconnect the cable from the negative terminal of the battery. **Caution:** *On models equipped with a Delco Loc II or Theftlock audio system, be sure the lockout feature is turned off before performing any procedure which requires disconnecting the battery.*

5.7b Use a flat bladed screwdriver to lift the fan assembly tabs that lock over the condenser and separate the radiator from the condenser

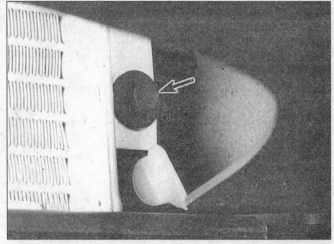

5.9 Location of one of the condenser mounting bolts (arrow)

2   Drain the cooling system as described in Chapter 1. Refer to the coolant **Warning** in Section 2.

3   Disable the airbag system (see Chapter 12). Disconnect the forward discriminating sensor harness from the front fenderwell **(see illustration)**. It will be necessary to remove the hood latch mechanism in conjunction with the airbag sensor and secure them together and out of the way during the radiator removal procedure.

4   Remove the cooling fan assembly (see Section 4).

5   Disconnect the upper and lower radiator hoses from the radiator, then disconnect the coolant recovery hoses at the radiator neck.

6   Remove the automatic transaxle cooler lines (if equipped) from the left tank of the radiator **(see illustration)**. Be careful not to damage the lines or fittings. Plug the ends of the disconnected lines to prevent leakage and stop dirt from entering the system. Have a drip pan ready to catch any spills.

7   Remove the hood latch support from the upper tie bar on the front of the engine compartment (see Chapter 11) **(see illustrations)**.

8   Remove the right and the left headlight assemblies from the vehicle (see Chapter 12).

9   Remove the condenser mounting bolts from the radiator assembly **(see illustration)**.

10   Remove the tie bar (crossbeam) from the engine compartment **(see illustrations)**.

11   The radiator can be removed from the top with some wiggling to clear various components surrounding the radiator **(see illustration)**.

12   Prior to installation of the radiator, replace any damaged hose clamps, radiator hoses and radiator mounts. If leaks have been noticed or there have been cooling problems, have the radiator cleaned and

5.10a Remove the tie bar mounting bolts (arrows)

tested at a radiator shop.

13   Radiator installation is the reverse of removal. Be sure to seat the radiator securely in the lower mounts before installing the upper support panel or upper fan shroud.

14   After installation, fill the system with the proper mixture of antifreeze, bleed the air from the cooling system as described in Chapter 1, and check the automatic transaxle fluid level. Also check the engine oil level, if equipped with an oil cooler (see Chapter 1).

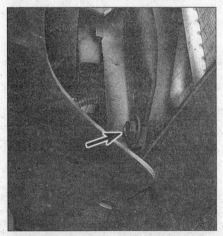

5.10b Remove the strut support (arrow)

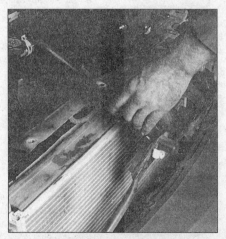

5.10c Lift the support from the engine compartment

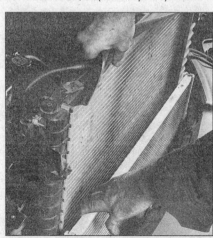

5.11 Carefully lift the radiator from the engine compartment

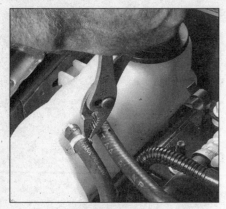

5.17a Remove the hose clamps and separate the upper coolant reservoir hoses

5.17b Disconnect the lower coolant reservoir hose clamp and hose

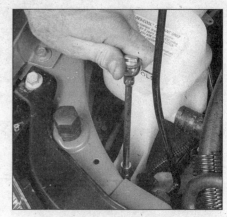

5.18 Remove the coolant reservoir mounting bolt

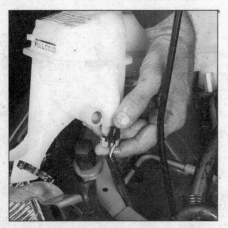

5.19 Lift the coolant reservoir from the engine compartment and disconnect the coolant level sensor connector

6.2 The weep hole (arrow) is located on the top of the water pump (2.2L OHV engine shown)

6.4 Check the pump for loose or rough bearings (this can be done with the belt removed and the pulley in place)

Coolant reservoir

Refer to illustration 5.17a, 5.17b, 5.18 and 5.19

15   Disconnect the cable from the negative terminal of the battery. **Caution:** *On models equipped with a Delco Loc II or Theftlock audio system, be sure the lockout feature is turned off before performing any procedure which requires disconnecting the battery.*

16   Drain the cooling system as described in Chapter 1 until the coolant reservoir is empty. Refer to the coolant **Warning** in Section 2.

17   Remove the coolant recovery hoses from the coolant reservoir **(see illustrations)**.

18   Detach the reservoir mounting bolts and remove it from the vehicle **(see illustration)**.

19   Disconnect the connector from the low coolant warning sensor **(see illustration)**.

20   Prior to installation make sure the reservoir is clean and free of debris which could be drawn into the radiator (wash it with soapy water and a brush if necessary, then rinse thoroughly).

21   Installation is the reverse of removal.

6   Water pump - check

Refer to illustrations 6.2 and 6.4

1   Water pump failure can cause overheating and serious damage to the engine. There are three ways to check the operation of the water pump while it is installed on the engine. If any one of the following quick-checks indicates water pump problems, it should be replaced immediately.

2   A seal protects the water pump impeller shaft bearing from contamination by engine coolant. If this seal fails, a weep hole in the water pump snout will leak coolant **(see illustration)** (an inspection mirror can be used to look at the underside of the pump if the hole isn't on top). If the weep hole is leaking, shaft bearing failure will follow. Replace the water pump immediately.

3   The water pump impeller shaft bearing can also prematurely wear out. When the bearing wears out, it emits a high-pitched squealing sound. If such a noise is coming from the water pump during engine operation, the shaft bearing has failed - replace the water pump immediately. **Note:** *Do not confuse belt noise with bearing noise.*

4   To identify excessive bearing wear, grasp the water pump pulley (2.2L OHV models only) and try to force it up-and-down or from side-to-side. If the pulley can be moved either horizontally or vertically, the bearing is nearing the end of its service life **(see illustration)**. Replace the water pump. Don't mistake drivebelt slippage, which causes a squealing sound, for water pump bearing failure.

5   It is possible for a water pump to be bad, even if it doesn't howl or leak water. Sometimes the fins on the back of the impeller can corrode away until the pump is no longer effective. The only way to check for this is to remove the pump for examination.

7   Water pump - removal and installation

Warning: *Wait until the engine is completely cool before starting this procedure.*

7.4 Use a large flat-bladed screwdriver to lock the pulley in place and loosen the water pump pulley bolts

7.6 Remove the water pump retaining bolts (arrow indicates one of four)

7.25 Lubricate the splines of the water pump drive with grease

Removal

1   Disconnect the cable from the negative terminal of the battery. **Caution:** *On models equipped with a Delco Loc II or Theftlock audio system, be sure the lockout feature is turned off before performing any procedure which requires disconnecting the battery.*
2   Drain the coolant (see Chapter 1).
3   Remove the cooling fan (see Section 4) then remove the serpentine drivebelt (see Chapter 1).

2.2L OHV engine

Refer to illustrations 7.4 and 7.6
4   Remove the water pump pulley from the front of the water pump **(see illustration)**.
5   Remove the coolant reservoir hoses (see Section 5).
6   Detach the water pump retaining bolts and remove the pump from the vehicle **(see illustration)**. If necessary, strike the pump with a soft-face hammer or a block of wood or wooden hammer handle to break the gasket seal. Do not pry between the pump and the block.

2.3L and 2.4L OHC engines

7   Disconnect the oxygen sensor harness connector (see Chapter 6).
8   Raise the vehicle and support it securely on jackstands.
9   Remove the exhaust manifold brace and loosen the exhaust manifold bolts (see Chapter 2B).
10   Remove the radiator outlet pipe assembly to the water pump housing bolts and separate the assembly from the engine. **Note:** *Do not rotate the flex coupling more than 4 degrees or damage to the flex coupling may occur.*
11   Remove the exhaust manifold pipe from the exhaust manifold.
12   Remove the radiator outlet pipe from the oil pan. **Note:** *On manual transaxle vehicles, remove the exhaust manifold brace and keep the radiator inlet hose attached to the assembly. Pull down on the radiator outlet pipe in order to disengage the pipe from the water pump while leaving the radiator outlet pipe hanging.*
13   Lower the vehicle. Remove the brake booster vacuum hose from the camshaft housing. Also, remove the exhaust manifold from the cylinder head (see Chapter 2B).
14   Remove the front timing chain cover from the engine (see Chapter 2B).
15   Remove the timing chain tensioner (see Chapter 2B). **Caution:** *The timing chain tensioner must be removed to unload chain tension before removing the water pump from the housing otherwise the water pump will become jammed in the timing chain housing.*
16   Remove the water pump cover bolts from the engine block and the timing chain housing.
17   Remove the water pump bolts and separate the assembly from the housing.

Installation

2.2L OHV engine

18   Clean the sealing surfaces of all gasket material on both the water pump and block. Wipe the mating surfaces with a rag saturated with lacquer thinner or acetone.
19   Apply a thin layer of RTV sealant to both sides of the new gasket and install the gasket on the water pump.
20   Place the water pump in position and install the bolts finger tight. Use caution to ensure that the gasket doesn't slip out of position. Remember to replace any mounting brackets secured by the water pump mounting bolts/studs. Tighten the bolts to the torque listed in this Chapter's Specifications. **Note:** *Coat the water pump bolts with Teflon thread sealant to prevent any seepage.*
21   The remainder of the installation procedure is the reverse of removal.
22   Add coolant to the specified level (see Chapter 1). Start the engine and check for the proper coolant level and the water pump and hoses for leaks. Bleed the cooling system of air as described in Chapter 1.

2.3L and 2.4L OHC engines

Refer to illustration 7.25
Note: *Read the entire procedure before installing the water pump into the housing to avoid an incorrect tightening sequence and leaky gasket mating surfaces.*
23   Clean the sealing surfaces of all gasket material on both the water pump and the housing. Wipe the mating surfaces with a rag saturated with lacquer thinner or acetone.
24   First, install the water pump cover and gasket to the water pump on the bench but leave the bolts finger tight. Apply a thin layer of RTV sealant to both sides of the new gasket and install the gasket on the water pump. This entire procedure must be performed in a timely manner before the RTV sealer sets up and dries.
25   Lubricate the splines of the water pump drive with grease and position the assembly into the timing chain housing and install the nuts finger tight **(see illustration)**. Use caution to ensure that the gasket doesn't slip out of position. Leave the nuts loose to allow repositioning when the block bolts are installed in the next step.
26   Install the water pump cover to engine block bolts and tighten them by hand also.
27   Lubricate the O-ring on the metal outlet pipe with coolant and install it into the water pump housing. Hand tighten the bolts.
28   Now that the entire water pump housing assembly is installed and is fitted against the engine block, follow the tightening sequence to insure proper seating.
   a)   Tighten the water pump-to-timing chain housing nuts to the torque listed in this Chapter's Specifications.

3

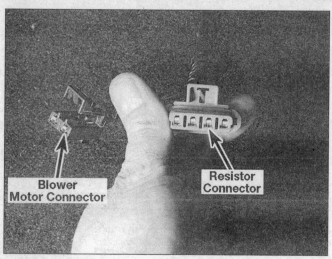

9.4 Connect a voltmeter to the heater blower motor connector by backprobing the connector using pins, and check the running voltage at each blower switch position

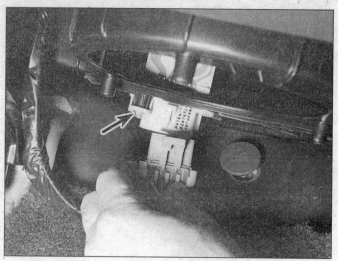

9.8a The blower motor resistor (arrow) is located just to the rear of the blower motor

b) *Tighten the water pump cover-to-water pump housing bolts to the torque listed in this Chapter's Specifications.*
c) *Tighten the water pump cover-to-engine block bolts to the torque listed in this Chapter's Specifications. Tighten the bottom bolts first then the top bolt.*
d) *Tighten the radiator outlet pipe-to-water pump housing to the torque listed in this Chapter's Specifications.*

29   Install the timing chain tensioner and cover (see Chapter 2B).
30   Install the exhaust manifold (see Chapter 2B).
31   Install the exhaust pipe to the exhaust manifold.
32   The remainder of the installation procedure is the reverse of removal. Add coolant to the specified level (see Chapter 1). Start the engine and check for the proper coolant level and the water pump and hoses for leaks. Bleed the cooling system of air as described in Chapter 1.

8 Coolant temperature gauge sending unit - check and replacement

Note: *Some 1995 models are equipped with a temperature light. This light activates when the engine has overheated. Have the temperature sending unit checked by a dealer service department or other qualified automotive repair facility.*

Check

1   The coolant temperature indicator system is composed of a temperature gauge or warning light mounted in the dash and a coolant temperature sensor mounted on the engine. This coolant temperature sensor doubles as an information sensor for the fuel and emissions systems (see Chapter 6) and as a sending unit for the temperature gauge.
2   If an overheating indication occurs, check the coolant level in the system and then make sure the wiring between the gauge and the sending unit is secure and all fuses are intact.
3   Check the operation of the coolant temperature sensor (see Chapter 6). If the sensor is defective, replace it with a new part of the same specification.
4   If the coolant temperature sensor is good, have the temperature gauge checked by a dealer service department. This test will require a SCAN tool to access the information as it is processed by the On Board computer.

Replacement

5   Refer to Chapter 6 for the coolant temperature sensor replacement procedure.

9 Blower motor and circuit - check

Refer to illustrations 9.4, 9.8a, 9.8b and 9.8c
Warning: *The models covered by this manual are equipped with airbags. Always disable the airbag system before working in the vicinity of the impact sensors, steering column or instrument panel to avoid the possibility of accidental deployment of the airbag(s), which could cause personal injury (see Chapter 12). The yellow wires and connectors routed through the instrument panel are for this system. Do not use electrical test equipment on these yellow wires or tamper with them in any way while working under the instrument panel.*
1   Check the fuse (marked HVAC) and all connections in the circuit for looseness and corrosion. Make sure the battery is fully charged. **Note:** *The heater/blower relay is located in the fuse/relay center in the engine compartment and the HVAC fuse is located in the fuse panel under the dash.*
2   With the transaxle in Park, the parking brake securely set, turn the ignition switch to the Run position. It isn't necessary to start the vehicle.
3   Remove the lower right dash insulator panel (below the glove box) for access to the blower motor.
4   Backprobe the blower motor electrical connector with two straight pins and connect a voltmeter to the blower motor connector and ground **(see illustration)**.
5   Move the blower switch through each of its positions and note the voltage readings. Changes in voltage indicate that the motor speeds will also vary as the switch is moved to the different positions.
6   If there is voltage present, but the blower motor does not operate, the blower motor is probably faulty. Disconnect the blower motor connector, then hook one side of the blower motor terminals to a chassis ground and the other to a fused source of battery voltage. If the blower doesn't operate, it is faulty.
7   If there was no voltage present at the blower motor at one or more speeds, and the motor itself tested OK, check the blower motor resistor.
8   Disconnect the electrical connector from the blower motor resistor and remove the resistor from the heater unit **(see illustrations)**. Inspect the resistor windings for damage. Using an ohmmeter, check for continuity between the terminals of the blower motor resistor. Continuity should exist between each terminal with varying resistance. Replace the resistor if defective.
9   Test the blower motor relay for battery voltage and for correct relay operation. Also, check the blower fuse (BLO). The blower motor relay is identical to the cooling fan relay - follow the relay checks in Section 4, Steps 2 through 6. The blower fuse and relay are located in the fuse/relay center in the engine compartment.

9.8b If the blower resistor is suspected faulty, it should first be visually inspected for an open circuit in the resistor windings . . .

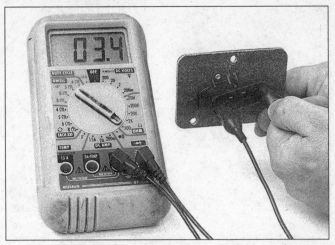

9.8c . . . then it should be checked for continuity between the terminals

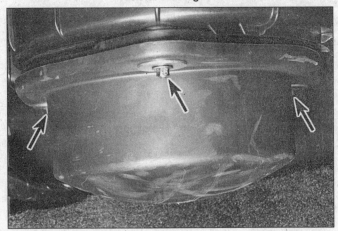

10.2 Remove the screws (arrows) retaining the blower motor to the housing

10.3 Lower the blower motor and fan assembly straight down to remove it from the vehicle

10 Check the ground terminal for contact and corrosion against the chassis metal. The blower motor ground is located on the left hand, lower side of the instrument panel. It will be necessary to partially dismantle the dash components to access this ground bolt. Refer to Chapter 11 for additional information:

10 Blower motor - removal and installation

Refer to illustrations 10.2, 10.3 and 10.4

Warning: *The models covered by this manual are equipped with airbags. Always disable the airbag system before working in the vicinity of the impact sensors, steering column or instrument panel to avoid the possibility of accidental deployment of the airbag(s), which could cause personal injury (see Chapter 12). The yellow wires and connectors routed through the instrument panel are for this system. Do not use electrical test equipment on these yellow wires or tamper with them in any way while working under the instrument panel.*

1 Remove the lower right dash insulator panel (below the glove box) for access to the blower motor.

2 Disconnect the electrical connector from the blower motor and remove the three screws from the blower housing **(see illustration)**.

3 Pull the blower motor and fan straight down **(see illustration)**.

4 To remove the fan from the blower motor, use a narrow punch and tap the shaft out of the fan assembly **(see illustration)**.

5 Install the fan onto the motor and install the blower motor into the heater housing.

11 Heater core - removal and installation

Warning 1: *The models covered by this manual are equipped with airbags. Always disable the airbag system before working in the vicinity of the impact sensors, steering column or instrument panel to avoid the possibility of accidental deployment of the airbag(s), which could cause personal injury (see Chapter 12). The yellow wires and connectors routed through the instrument panel are for this system. Do not use electrical test equipment on these yellow wires or tamper with them in any way while working under the instrument panel.*
Warning 2: *The air conditioning system is under high pressure. DO NOT loosen any fittings or remove any components until after the system has been discharged. Air conditioning refrigerant should be properly discharged into an EPA-approved container at a dealership service department or an automotive air conditioning facility. Always wear eye protection when disconnecting air conditioning system fittings.*

Removal

Refer to illustrations 11.3, 11.5, 11.7a, 11.7b, 11.7c, 11.7d, 11.8a and 11.8b

1 Disconnect the battery cable at the negative battery terminal.
Caution: *On models equipped with a Delco Loc II or Theftlock audio system, be sure the lockout feature is turned off before performing any procedure which requires disconnecting the battery.*

2 Drain the cooling system (see Chapter 1).

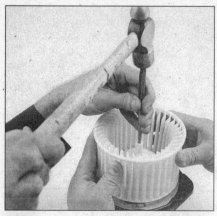

10.4 Have an assistant hold the blower assembly while tapping the motor shaft with a narrow punch and carefully drive the shaft out of the blower fan

11.3 Disconnect the heater core hoses (arrows) at the engine compartment firewall

11.5 Disconnect the ducts from the heater core cover

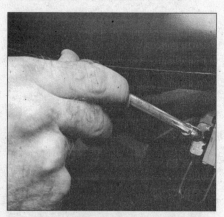

11.7a Remove the heater core cover bolts from the face of the assembly

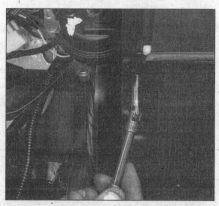

11.7b Remove the bolt from under the ledge area of the cover. Don't forget the recessed mounting bolt near the middle, rear section of the heater core cover

11.7c Separate the lower panel from the heater core housing

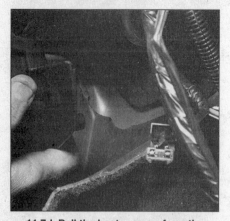

11.7d Pull the heater cover from the bottom to dislodge the condenser drain hoses

11.8a Remove the heater core straps (arrows) from the assembly

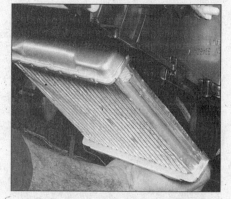

11.8b Lower the forward section of the heater core and angle the assembly toward the rear of the vehicle to separate it from the heater core housing

3   Disconnect the heater hoses at the heater core inlet and outlet on the engine side of the firewall (passenger side) **(see illustration)** and plug the open fittings. If the hoses are stuck to the pipes, cut them off.

4   From the inside of the car, remove the lower right dash insulator panel (below the glove box) and the glove box (see Chapter 11).

5   disconnect the ducts from the heater cover **(see illustration)**.

6   Remove the instrument panel from the passenger compartment (see Chapter 11).

7   Remove the heater core cover **(see illustrations)**.

8   Remove the heater core clamp bolts and clamps **(see illustration)**, then slide the heater core out carefully **(see illustration)**.

Installation

9   Installation is the reverse of removal. **Note:** *When reinstalling the heater core, make sure any original insulating/sealing materials are in place around the heater core pipes and around the core.*

10   Refill the cooling system (see Chapter 1).

11   Start the engine and check for proper operation.

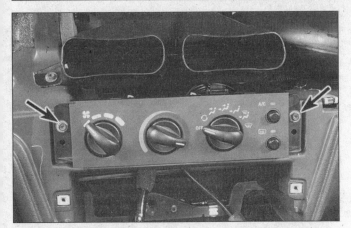

12.3a Remove the screws (arrows) retaining the heater/air conditioning control assembly to the instrument panel

12.3b Disconnect the blower switch electrical connector

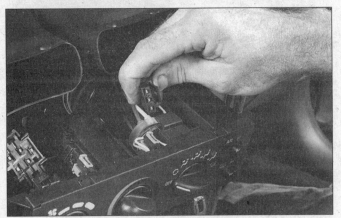

12.3c Disconnect the air conditioning and defogger switch electrical connector

12.3d Disconnect the vacuum harness

12 Heater and air conditioning control assembly - removal and installation

Warning: *The models covered by this manual are equipped with airbags. Always disable the airbag system before working in the vicinity of the impact sensors, steering column or instrument panel to avoid the possibility of accidental deployment of the airbag(s), which could cause personal injury (see Chapter 12). The yellow wires and connectors routed through the instrument panel are for this system. Do not use electrical test equipment on these yellow wires or tamper with them in any way while working under the instrument panel.*

Removal

Refer to illustrations 12.3a, 12.3b, 12.3c, 12.3d, 12.3e and 12.3f

1   Disconnect the battery cable from the negative battery terminal. **Caution:** *On models equipped with a Delco Loc II or Theftlock audio system, be sure the lockout feature is turned off before performing any procedure which requires disconnecting the battery.*

2   Remove the main instrument panel bezel to allow access to the heater/air conditioning control mounting screws (see Chapter 11).

3   Remove the control assembly retaining screws and pull the unit from the dash **(see illustrations)**. It can be pulled out just far enough to allow disconnecting the control cable end, electrical connections and vacuum harness (on air-conditioned models) from the control head. Use a small screwdriver to release the clips retaining the control cable.

Installation

4   To install the control assembly, reverse the removal procedure. **Caution:** *When reconnecting vacuum harness to the control assembly, do not*

use any lubricant to make them slip on easier; it can affect vacuum operation. If necessary, use a drop of plain water to make reconnection easier.

13 Air conditioning and heating system - check and maintenance

Warning: *The air conditioning system is under high pressure. DO NOT loosen any fittings or remove any components until after the system has been discharged. Air conditioning refrigerant should be properly discharged into an EPA-approved recovery container at a dealership service department or an automotive air conditioning repair facility. Always wear eye protection when disconnecting air conditioning system fittings.*

1   The following maintenance steps should be performed on a regular basis to ensure that the air conditioner continues to operate at peak efficiency:

a) *Check the drivebelt (see Chapter 1).*

b) *Check the condition of the hoses. Look for cracks, hardening and deterioration. Look at potential leak areas (hoses and fittings) for signs of refrigerant oil leaking out.* **Warning:** *Do not replace air conditioning hoses until the system has been discharged by a dealership or air conditioning repair facility.*

c) *Check the fins of the condenser for leaves, bugs and other foreign material. A soft brush and compressed air can be used to remove them.*

d) *Check the wire harness for correct routing, broken wires, damaged insulation, etc. Make sure the electrical connectors are clean and tight.*

e) *Maintain the correct refrigerant charge.*

12.3e Disconnect the lighting harness assembly

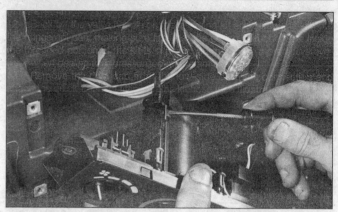

12.3f Use a small screwdriver to pry apart the clips that retain the air conditioner control cable (If equipped)

13.9 Install an automotive thermometer into the cooling duct closest to the evaporator core (right side duct)

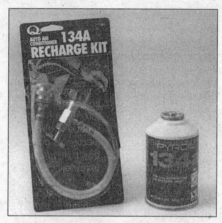

13.10 A basic charging kit for R-134a systems is available at most auto parts stores

13.13 Add refrigerant to the low-side port only - the procedure is easier if you wrap the can with a warm, wet towel to prevent icing

2   The system should be run for about 10 minutes at least once a month. This is particularly important during the winter months because long-term non-use can cause hardening of the internal seals.

3   Because of the complexity of the air conditioning system and the special equipment required to effectively work on it, accurate troubleshooting of the system should be left to a certified air conditioning technician.

4   If the air conditioning system doesn't operate at all, check the fuse panel. Check the HVAC fuse and the air conditioning compressor relay.

5   The most common cause of poor cooling is simply a low system refrigerant charge. If a noticeable drop in cool air output occurs, the following quick check will help you determine if the refrigerant level is low. For more complete information on the air conditioning system, refer to the *Haynes Automotive Heating and Air Conditioning Manual*.

Checking the refrigerant charge

Refer to illustration 13.9

6   Warm the engine up to normal operating temperature.

7   Place the air conditioning temperature selector at the coldest setting and the blower at the highest setting. Open the doors (to make sure the air conditioning system doesn't cycle off as soon as it cools the passenger compartment).

8   With the compressor engaged - the clutch will make an audible click and the center of the clutch will rotate - feel the surface of the accumulator and the evaporator inlet pipe. If there's no perceptible difference between the inlet pipe and the accumulator, the system is properly charged. If there's a difference, there's something wrong with the system. It might be low charge, but it might be something else. To be sure, take the vehicle to a dealer service department or other qualified repair facility for further diagnosis.

9   Place a thermometer in the dashboard vent nearest the evaporator **(see illustration)** and add refrigerant to the system until the indicated temperature is around 40 to 45-degrees F. If the ambient (outside) air temperature is very high, say 110-degrees F, the duct air temperature will probably be higher, but generally the air conditioning is 30 to 40-degrees F cooler than the ambient air. **Note:** *Humidity of the ambient air also affects the cooling capacity of the system. Higher ambient humidity lowers the effectiveness of the air conditioning system.*

Adding refrigerant

Refer to illustrations 13.10 and 13.13

10   Buy an automotive charging kit at an auto parts store. A charging kit includes a 14-ounce can of refrigerant, a tap valve and a short section of hose that can be attached between the tap valve and the system low side service valve **(see illustration)**. Because one can of refrigerant may not be sufficient to bring the system charge up to the proper level, it's a good idea to buy an extra can. Make sure that one of the cans contains red refrigerant dye. If the system is leaking, the red dye will leak out with the refrigerant and help you pinpoint the location of the leak.

11   Hook up the charging kit by following the manufacturer's instructions. **Warning:** *DO NOT hook the charging kit hose to the system high side!* The fittings on the charging kit are designed to fit **only** on the low side of the system.

12   Back off the valve handle on the charging kit and screw the kit onto the refrigerant can, making sure first that the O-ring or rubber seal inside the threaded portion of the kit is in place. **Warning:** *Wear protective eyewear when dealing with pressurized refrigerant cans.*

13   Remove the dust cap from the low-side charging connection and attach the quick-connect fitting on the kit hose **(see illustration)**.

14   Warm up the engine and turn on the air conditioner. Keep the charging kit hose away from the fan and other moving parts. **Note:** *The charging process requires the compressor to be running. Your compressor may cycle off if the pressure is low due to a low charge. If the clutch cycles off, you can disconnect the air conditioning compressor clutch connector near the compressor and apply battery voltage (+) using a jumper wire. This will keep the compressor ON.*

15   Turn the valve handle on the kit until the stem pierces the can, then back the handle out to release the refrigerant. You should be able to hear the rush of gas. Add refrigerant to the low side of the system until both the accumulator surface and the evaporator inlet pipe feel about the same temperature . Allow stabilization time between each addition.

16   If you have an accurate thermometer, you can place it in the center air conditioning duct inside the vehicle and keep track of the "conditioned" air temperature. A charged system that is working properly should put out air that is 40-degrees F. If the ambient (outside) air temperature is very high, say 110-degrees F, the duct air temperature will probably be higher, but generally the air conditioning is 30 to 40-degrees-F cooler than the ambient air.

17   When the can is empty, turn the valve handle to the closed position and release the connection from the low-side port. Replace the dust cap. **Warning:** *Never add more than one can of refrigerant to the system (if more than one can is required, the system should be evacuated and leak tested).*

18   Remove the charging kit from the can and store the kit for future use with the piercing valve in the UP position, to prevent inadvertently piercing the can on the next use.

Heating systems

19   If the carpet under the heater core is damp, or if antifreeze vapor or steam is coming through the vents, the heater core is leaking. Remove it (see Section 11) and install a new unit (most radiator shops will not repair a leaking heater core).

20   If the air coming out of the heater vents isn't hot, the problem could stem from any of the following causes:

  a) *The thermostat is stuck open, preventing the engine coolant from warming up enough to carry heat to the heater core. Replace the thermostat (see Section 3).*

  b) *A heater hose is blocked, preventing the flow of coolant through the heater core. Feel both heater hoses at the firewall. They should be hot. If one of them is cold, there is an obstruction in one of the hoses or in the heater core, or the heater control valve is shut. Detach the hoses and back flush the heater core with a water hose. If the heater core is clear but circulation is impeded, remove the two hoses and flush them out with a water hose.*

  c) *If flushing fails to remove the blockage from the heater core, the core must be replaced (see Section 11).*

14   Air conditioning accumulator/drier - removal and installation

Removal

Refer to illustration 14.2

Warning: *The air conditioning system is under high pressure. DO NOT loosen any fittings or remove any components until after the system has been discharged. Air conditioning refrigerant should be properly discharged into an EPA-approved container at a dealership service department or an automotive air conditioning repair facility. Always wear eye protection when disconnecting air conditioning system fittings.*

1   Have the air conditioning system discharged (see **Warning** above). Disconnect the cable from the negative terminal of the battery. **Caution:** *On models equipped with a Delco Loc II or Theftlock audio system, be sure the lockout feature is turned off before performing any procedure which requires disconnecting the battery.*

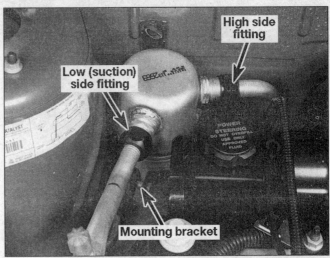

14.2 Disconnect the inlet and outlet lines, then remove the mounting bracket bolt

2   Disconnect the refrigerant inlet and outlet lines **(see illustration)**, using back-up wrenches. Cap or plug the open lines immediately to prevent the entry of dirt or moisture.

3   Loosen the clamp bolt on the mounting bracket and slide the accumulator/drier assembly up and out of the compartment.

Installation

4   If you are replacing the accumulator/drier with a new one, add one ounce of fresh refrigerant oil to the new unit (oil must be R-134a compatible).

5   Place the new accumulator/drier into position in the bracket.

6   Install the inlet and outlet lines, using clean refrigerant oil on the new O-rings. Tighten the mounting bolt securely.

7   Connect the cable to the negative terminal of the battery.

8   Have the system evacuated, recharged and leak tested by a dealership service department or an automotive air conditioning repair facility.

15   Air conditioning compressor - removal and installation

Note: *Whenever the compressor is replaced because of internal damage, the expansion (orifice) tube should also be replaced (see Section 18).*

Removal

Refer to illustrations 15.5 and 15.6

Warning: *The air conditioning system is under high pressure. DO NOT loosen any fittings or remove any components until after the system has been discharged. Air conditioning refrigerant should be properly discharged into an EPA-approved container at a dealership service department or an automotive air conditioning repair facility. Always wear eye protection when disconnecting air conditioning system fittings.*

Note: *The accumulator/drier (see Section 14) should be replaced whenever the compressor is replaced.*

1   Have the air conditioning system discharged (see **Warning** above). Disconnect the cable from the negative terminal of the battery. **Caution:** *On models equipped with a Delco Loc II or Theftlock audio system, be sure the lockout feature is turned off before performing any procedure which requires disconnecting the battery.*

2   Clean the compressor thoroughly around the refrigerant line fittings.

3   Remove the serpentine drivebelt (see Chapter 1).

4   Raise the vehicle and support it securely on jackstands.

5   Disconnect the electrical connector from the air conditioning

3

15.5 Disconnect the electrical connector (arrow) from the air conditioning compressor

15.6 Remove the retaining bolt (arrow) securing the refrigerant lines to the back of the compressor

17.3 Disconnect the refrigerant lines (arrows) leading to the evaporator core

17.5a Remove the heater core shroud mounting bolt and then drop the assembly using the hinge and . . .

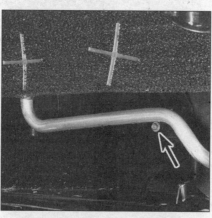

17.5b . . . access the rear heater core shroud bolt by pulling down the assembly to expose the rear section

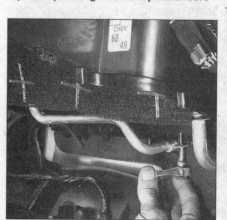

17.6a Remove the evaporator core inlet line clamp and bolt

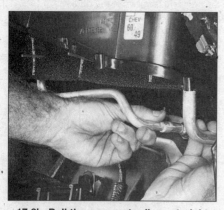

17.6b Pull the evaporator lines straight back and out of the firewall

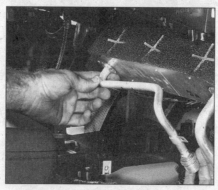

17.6c Angle the evaporator core inlet and outlet pipes down to separate the core from the heater core box

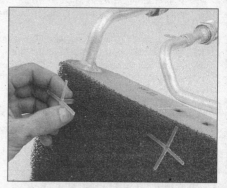

17.7a Pull the filter pad retainers straight out and lift the filter pad from the evaporator core

compressor **(see illustration)**.

6   Disconnect the suction and discharge lines from the compressor. Both lines are mounted to the back of the compressor with a plate secured by one bolt. Plug the open fittings to prevent the entry of dirt and moisture, and discard the seals between the plate and compressor **(see illustration)**.

7   Unbolt and remove the rear compressor mount. **Note:** *On later models it may necessary to remove the right side engine mount bolt and raise the engine to access the rear compressor mount bolts.*

8   Remove the compressor-to-front-bracket bolts and nuts and remove the compressor from the engine compartment.

Installation

9   If a new compressor is being installed, pour the oil from the old compressor into a graduated container and add that exact amount of new refrigerant oil to the new compressor. Also follow any directions included with the new compressor. **Note:** *Some replacement compressors come with refrigerant oil in them. Follow the directions with the compressor regarding the draining of excess oil prior to installation.* **Caution:** *The oil used must be labeled as compatible with R-134a refrigerant systems.*

10   Installation is the reverse of the disassembly. When installing the line fitting bolt to the compressor, use new seals lubricated with clean refrigerant oil, and tighten the bolt securely.

11   Reconnect the battery cable to the negative battery terminal.

12   Have the system evacuated, recharged and leak tested by a dealership service department or an automotive air conditioning repair facility.

17.7b Check the evaporator core for leaks, damage or cracks

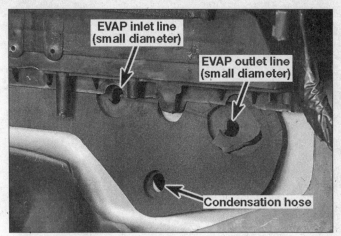

17.9 Before installing the evaporator core, make sure the pad is intact and the holes are not plugged or damaged

16 Air conditioning condenser - removal and installation

Warning 1: *The models covered by this manual are equipped with airbags. Always disable the airbag system before working in the vicinity of the impact sensors, steering column or instrument panel to avoid the possibility of accidental deployment of the airbag(s), which could cause personal injury (see Chapter 12).*
Warning 2: *The air conditioning system is under high pressure. DO NOT loosen any fittings or remove any components until after the system has been discharged. Air conditioning refrigerant should be properly discharged into an EPA-approved container at a dealership service department or an automotive air conditioning facility. Always wear eye protection when disconnecting air conditioning system fittings.*

Removal

1   Have the air conditioning system discharged (see **Warning** above). Disconnect the cable from the negative terminal of the battery. **Caution:** *On models equipped with a Delco Loc II or Theftlock audio system, be sure the lockout feature is turned off before performing any procedure which requires disconnecting the battery.*
2   Disconnect the refrigerant line fittings from the right side of the condenser and cap the open fittings to prevent the entry of dirt and moisture.
3   Disconnect the upper engine compartment panel (see Chapter 11). It will be necessary to remove the hood latch from the panel (see Chapter 11) without disconnecting the cable.
4   Remove the right and left headlamp assemblies (see Chapter 12).
5   Raise the vehicle and support it securely on jackstands.
6   Disable the airbag system and remove the front airbag (crash) sensor from the mounting brackets (see Chapter 12).
7   Working in the engine compartment, remove the right side radiator mount. Make sure the hood latch mechanism and the airbag sensor harness are secured, out of the way.
8   Remove the condenser mounting bolts from the radiator (see **illustration 5.9**).
9   Tilt the upper half of the radiator forward and dislodge the condenser insulator mounts from the radiator support.
10   Pull the condenser up between the radiator and the radiator support to remove it from the vehicle. **Caution:** *The condenser is made of aluminum - be careful not to damage it during removal.*

Installation

11   Installation is the reverse of removal. Be sure to use new, compatible O-rings on the refrigerant line fittings (lubricate the O-rings with clean refrigerant oil. If a new condenser is installed, add 1 ounce of new refrigerant oil to the system (oil must be R-134a compatible).
12   Have the system evacuated, recharged and leak tested by a dealership service department or an automotive air conditioning repair facility.

17 Air conditioning evaporator - removal and installation

Warning 1: *The models covered by this manual are equipped with airbags. Always disable the airbag system before working in the vicinity of the impact sensors, steering column or instrument panel to avoid the possibility of accidental deployment of the airbag(s), which could cause personal injury (see Chapter 12). The yellow wires and connectors routed through the instrument panel are for this system. Do not use electrical test equipment on these yellow wires or tamper with them in any way while working under the instrument panel.*
Warning 2: *The air conditioning system is under high pressure. DO NOT loosen any fittings or remove any components until after the system has been discharged. Air conditioning refrigerant should be properly discharged into an EPA-approved container at a dealership service department or an automotive air conditioning repair facility. Always wear eye protection when disconnecting air conditioning system fittings.*

Removal

Refer to illustrations 17.3, 17.5a, 17.5b, 17.6a, 17.6b, 17.6c, 17.7a and 17.7b
1   Have the air conditioning system discharged (see **Warning** above). Disconnect the cable from the negative terminal of the battery. **Caution:** *On models equipped with a Delco Loc II or Theftlock audio system, be sure the lockout feature is turned off before performing any procedure which requires disconnecting the battery.*
2   Drain the cooling system (see Chapter 1).
3   Disconnect the air conditioning lines at the passenger side of the firewall **(see illustration)**.
4   Follow the procedures in Section 11 for removing the heater core.
Note: *Remember, this will require complete instrument panel removal, heater housing disassembly and heater core removal to gain access to the evaporator core.*
5   Remove the evaporator protective shroud **(see illustrations)**.
6   Remove the evaporator core clamp bolt and clamp **(see illustration)**, then slide the evaporator core out carefully **(see illustrations)**.
7   Check the core over carefully for signs of leaks **(see illustrations)**.
8   If a new evaporator core is to be installed, save all of the sealing gaskets from the original unit and transfer them.

Installation

Refer to illustration 17.9
9   Installation is the reverse of the removal procedure. Lubricate all O-rings with clean refrigerant oil **(see illustration)**.
10   If a new evaporator has been installed, add 1 ounce of refrigerant oil (oil must be R-134a compatible). Have the system evacuated, recharged and leak tested by a dealership service department or an automotive air conditioning repair facility.

3

18.3 Working from under the engine compartment, disconnect the high pressure line near the bottom of the engine compartment (arrow)

18.4 Carefully remove the expansion tube using needle-nose pliers

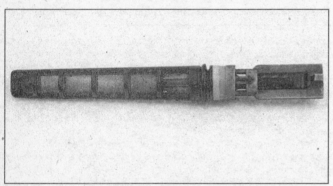

18.5 The expansion tube is equipped with a tapered mesh screen that must be cleaned and not have any holes or damage

18 Air conditioning expansion (orifice) tube - removal and installation

Refer to illustrations 18.3, 18.4 and 18.5

Warning 2: *The air conditioning system is under high pressure. DO NOT loosen any fittings or remove any components until after the system has been discharged. Air conditioning refrigerant should be properly discharged into an EPA-approved container at a dealership service department or an automotive air conditioning repair facility. Always wear eye protection when disconnecting air conditioning system fittings.*

1   Have the air conditioning system discharged and the refrigerant recovered (see **Warning** above). Disconnect the cable from the negative terminal of the battery. **Caution:** *On models equipped with a Theft-lock audio system, be sure the lockout feature is turned off before performing any procedure which requires disconnecting the battery.*

2   Remove the air cleaner and duct (see Chapter 4).

3   Disconnect the refrigerant high-pressure line at the fitting at the bottom of the engine compartment **(see illustration)**.

4   The expansion tube is a tube with a fixed-diameter orifice and a mesh filter at each end **(see illustration)**. When you separate the pipe at the fitting you will see one end of the orifice tube inside the pipe leading to the evaporator. Use needle-nose pliers to remove the orifice tube.

5   The orifice tube acts to meter the refrigerant, changing it from high-pressure liquid to low-pressure liquid. It is possible to reuse the orifice tube if **(see illustration)**:

a) *The screens aren't plugged with grit or foreign material*
b) *Neither screen is torn*
c) *The plastic housing over the screens is intact*
d) *The brass orifice inside the plastic housing is unrestricted*

6   Installation is the reverse of removal. Be sure to insert the expansion tube with the shorter end in first, toward the evaporator. **Caution:** *Always use a new O-ring when installing the expansion (orifice) tube.*

7   Retighten the fitting and refrigerant line, then have the system evacuated, recharged and leak-tested by the shop that discharged it.

Chapter 4
Fuel and exhaust systems

Contents

4

Specifications

General

Fuel pressure
Key On, engine not running .... 41 to 47 psi
Engine idling
   With pressure regulator vacuum hose connected .... 31 to 44 psi
   With pressure regulator vacuum hose disconnected .... 42 to 50 psi
Injector resistance .... 11 to 13 ohms

Torque specifications
Ft-lbs (unless otherwise indicated)

Air intake plenum bolts (2.2L engine) 22
Exhaust pipe-to-manifold nuts .... 15 to 22
Fuel rail bracket
   Nuts .... 89 in-lbs
   Bolts 18
Fuel rail mounting bolts
   1995 through 1997 .... 31 in-lbs
   1998 and later 18
IAC valve screws .... 27 in-lbs
Throttle body bolts/nuts 16

1   General information

These models are equipped with Multiport Fuel Injection (MFI) system. These modern fuel injection systems are also equipped with the updated OBD II self-diagnosis system (see Chapter 6).

The fuel system consists of a fuel tank, an electric fuel pump and fuel pump relay, an air cleaner assembly and a fuel injection system.

Multiport Fuel Injection (MFI) system

This system utilizes injectors mounted above each intake port. The throttle body on the MFI system serves only to control the amount of air passing into the system. Because each cylinder is equipped with an injector mounted immediately adjacent to the intake valve, much better control of the fuel/air mixture ratio is possible.

These models are equipped with a CLEAR FLOOD MODE built into the PCM. If the engine is flooded (excess fuel in combustion chambers), completely depress the accelerator pedal while cranking the engine over. This will cancel the fuel injector ON Time signals from the PCM and allow the engine to clear the excess fuel through the exhaust system. Once the system is cleared, release the accelerator pedal to 70-percent to start the engine.

Fuel pump and lines

Fuel is circulated from the fuel tank to the fuel injection system, and back to the fuel tank, through a pair of lines running along the underside of the vehicle. An electric fuel pump is attached to the fuel sending unit inside the fuel tank. A return system routes all vapors and excess fuel back to the fuel tank through separate return lines.

The fuel pump system is controlled by a fuel pump relay that is mounted on the inner fender panel in the engine compartment. When the ignition is turned ON without the engine running, the PCM energizes the fuel pump relay for two seconds. When the ignition is cranking the engine, the PCM supplies power to the fuel pump relay as long as it receives reference pulses from the crankshaft position sensor. An inoperative fuel pump relay can result in long cranking times or a no-start condition. **Note:** *Early models are equipped with a fuel pump relay/oil pressure indicator switch. The switch is normally OPEN until oil pressure reaches six psi, causing it to close. If oil pressure falls below two psi, the switch will open and simultaneously, shut down the fuel pump. This switch is a fail-safe device used to shut down the engine if oil pressure fails. If the fuel pump relay fails, the switch will provide battery voltage to the fuel pump directly from terminal 87 on the fuel pump relay (ignition switch circuit). In the event of fuel pump relay problems and/or switch problems, be sure to check both components. Because of the intricate fuel pump electrical system, it will be necessary to relieve the fuel pressure by disconnecting the fuel pump harness connector to disable the fuel pump.*

Exhaust system

The exhaust system includes an exhaust manifold fitted with an exhaust oxygen sensor, a catalytic converter, an exhaust pipe, and a muffler. The catalytic converter is an emission control device added to the exhaust system to reduce pollutants. A single-bed converter is used in combination with a three-way (reduction) catalyst. Refer to Chapter 6 for more information regarding the catalytic converter.

---

2   Fuel pressure relief procedure

Warning: *Gasoline is extremely flammable, so take extra precautions when you work on any part of the fuel system. Don't smoke or allow open flames or bare light bulbs near the work area, and don't work in a garage where a natural gas-type appliance (such as a water heater or a clothes dryer) with a pilot light is present. Since gasoline is carcinogenic, wear latex gloves when there's a possibility of being exposed to fuel, and, if you spill any fuel on your skin, rinse it off immediately with soap and water. Mop up any spills immediately and do not store fuel-soaked rags where they could ignite. The fuel system is under constant pressure, so, if any fuel lines are to be disconnected, the fuel pressure in the system must be relieved first. When you perform any kind of work on the fuel system, wear safety glasses and have a Class B type fire extinguisher on hand.*
Note: *After the fuel pressure has been relieved, it's a good idea to lay a shop towel over any fuel connection to be disassembled, to absorb the residual fuel that may leak out when servicing the fuel system.*

1   Before servicing any fuel system component, you must relieve the fuel pressure to minimize the risk of fire or personal injury.
2   Remove the fuel filler cap - this will relieve any pressure built up in the tank.
3   Raise the vehicle and secure it on jackstands.
4   Disconnect the fuel pump electrical connector located near the fuel tank.
5   Start the engine and allow it to run until the remaining fuel is consumed and the engine stalls. The fuel pressure within the fuel lines is

now relieved. It is a good idea to place shop towels around the fuel fitting to be disconnected to absorb any residual fuel that spills out.
6   Unless this procedure is followed before servicing fuel lines or connections, fuel spray (and possible injury) may occur.

---

3   Fuel pump/fuel pressure - check

Warning: *Gasoline is extremely flammable, so take extra precautions when you work on any part of the fuel system. See the **Warning** in Section 2.*
Note: *The following checks assume the fuel filter is in good condition. If you doubt its condition, install a new one (see Chapter 1).*
1   Check that there is adequate fuel in the fuel tank. Relieve the fuel pressure (see Section 2).

Fuel pump output and pressure check

Refer to illustrations 3.2, 3.5 and 3.6
2   Connect a fuel pressure gauge between the fuel rail and inlet line located at the rear of the engine compartment. You will need a fuel pressure gauge capable of measuring high fuel pressure and equipped with the correct adapters to install onto the quick-connect adapters on the fuel lines. On some models the fuel pressure gauge can be connected in line with the fuel rail and fuel inlet line using a T-fitting **(see illustration)**. On others it will be necessary to install special fuel line adapters to the fuel line quick-connect fittings (see Section 4). Such adapters can be fabricated from fuel line and the necessary quick-connect fittings.
3   Turn the ignition switch ON (engine not running). The fuel pump should run for about two seconds - note the reading on the gauge. After the pump stops running the pressure should hold steady. It should be within the range listed in this Chapter's Specifications. **Note:** *If there is no response from the fuel pump, use a jumper wire attached to the positive terminal of the battery and apply battery voltage to the test terminal (red wire) located on the driver's side of the engine compartment, near the power brake booster.*
4   Start the engine and let it idle at normal operating temperature. The pressure should be lower by 3 to 10 psi. Now detach the vacuum hose from the fuel pressure regulator - the pressure should increase by 3 to 10 psi. If all the pressure readings are within the limits listed in this Chapter's Specifications, the system is operating properly.
5   If the pressure did not drop by 3 to 10 psi after starting the engine, or if it didn't increase when the vacuum hose was disconnected from the fuel pressure regulator, apply 12 to 14 inches of vacuum to the pressure regulator **(see illustration)**. If the pressure drops, repair the vacuum source to the regulator. If the pressure does not drop, replace the regulator. **Note:** *This test will work only on engines where the fuel pressure regulator is easily accessible.*
6   If the fuel pressure is not within specifications, check the following:

a) *If the pressure is higher than specified, check for vacuum to the fuel pressure regulator. Vacuum must fluctuate with the increase or decrease in engine rpm. If vacuum is present, check for a pinched or clogged fuel return hose or pipe. If the return line is OK, replace the regulator.*

b) *If the pressure is lower than specified, change the fuel filter to rule out the possibility of a clogged filter. If the pressure is still low, install a fuel line shut-off adapter between the pressure regulator and the return line (this can be fabricated from fuel line, a shut off valve and the necessary fittings to mate with the pressure regulator and the return line, or, instead of a shut-off valve, use fuel hose that can be pinched with a pair of pliers) **(see illustration)**. With the valve open (or the hose not pinched), start the engine (if possible) and slowly close the valve or pinch the hose (only pinch the hose on the adapter you fabricated). If the pressure rises above 47 psi, replace the regulator (see Section 13). **Warning:** Don't allow the fuel pressure to exceed 60 psi. Also, don't attempt to restrict the return line by pinching it, as the nylon fuel line will be damaged.*

3.2 Attach a fuel pressure gauge between the fuel rail and the inlet fuel line using special fuel line adapters that can be coupled with the factory fuel line connections

3.5 Connect a vacuum pump to the fuel pressure regulator, apply vacuum to the fuel pressure regulator and check the fuel pressure - the fuel pressure should decrease as the vacuum increases

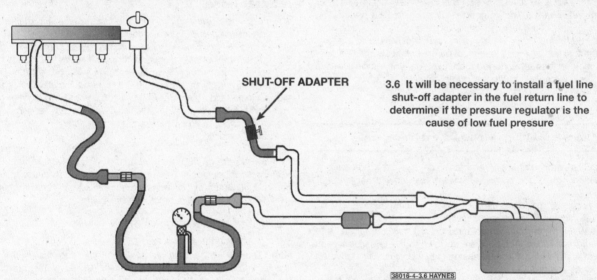

SHUT-OFF ADAPTER

3.6 It will be necessary to install a fuel line shut-off adapter in the fuel return line to determine if the pressure regulator is the cause of low fuel pressure

38016-4-3.6 HAYNES

4

c) *If the pressure is still low with the fuel return line restricted, an injector (or injectors) may be leaking (see Section 13) or the fuel pump may be faulty.*

7   After the testing is done, relieve the fuel pressure (see Section 2) and remove the fuel pressure gauge.

8   If there are no problems with any of the above-listed components, check the fuel pump electrical circuits (see below).

Fuel pump electrical circuit check

Refer to illustration 3.12

9   If you suspect a problem with the fuel pump, verify the pump actually runs. Have an assistant turn the ignition switch to ON - you should hear a brief whirring noise as the pump comes on and pressurizes the system. Have the assistant start the engine. This time you should hear a constant whirring sound from the pump (but it's more difficult to hear with the engine running).

10   If the pump does not come on (makes no sound), proceed to the next Step.

11   Check the ignition fuse (see Chapter 12). The ignition fuse supplies battery voltage to terminal 87 on the fuel pump relay. If the fuse is blown, replace the fuse and see if the pump works. If the pump still does not work, go to the next Step.

12   Check the fuel pump relay circuit. If the pump does not run, check for an open circuit between the relay and the fuel pump. With the ignition key ON (engine not running), check for battery voltage at the relay

connector **(see illustration)**. The fuel pump relay is located in the fuse/relay control box in the engine compartment. **Note:** *If oil pressure drops below the specified pressure level, the oil pressure switch will act as a fuel pressure cut-off device on early models. Be sure to check the oil pressure switch and circuit in the event of a problem diagnosing the fuel pump circuit.*

86		30
87	87A	85

B+

3.12 The fuel pump relay is located in the engine compartment fuse/relay block, near the air cleaner housing

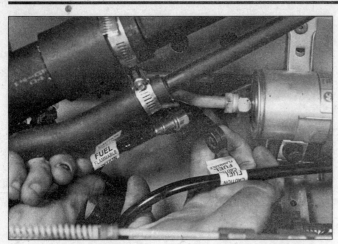

4.7a Disconnecting the fuel line using the special tool

4.7b Location of the inlet and return lines (arrows) on the fuel rail

13   If battery voltage exists, check the relay (see Chapter 12) or replace the relay with a known good relay and retest. If necessary, have the relay checked by a qualified automotive electrical specialist.

14   If the fuel pump does not activate, check for power to the fuel pump at the fuel tank. Access to the fuel pump is difficult, but it is possible to check for battery voltage at the electrical connector near the tank. If voltage is present at the fuel pump connector, replace the fuel pump.

4   Fuel lines and fittings - repair and replacement

Warning: *Gasoline is extremely flammable, so take extra precautions when you work on any part of the fuel system. See the* **Warning** *in Section 2.*

1   Always relieve the fuel pressure before servicing fuel lines or fittings on fuel-injected vehicles (see Section 2).

2   The fuel feed, return and vapor lines extend from the fuel tank to the engine compartment. The lines are secured to the underbody with clip and screw assemblies. These lines must be occasionally inspected for leaks, kinks and dents.

3   If evidence of dirt is found in the system or fuel filter during disassembly, the line should be disconnected and blown out. Check the fuel strainer on the fuel level sending unit (see Section 8) for damage and deterioration.

Steel and nylon tubing

Refer to illustrations 4.7a and 4.7b

4   Because fuel lines used on fuel-injected vehicles are under high pressure, they require special consideration.

5   If replacement of a metal fuel line or emission line is called for, use welded steel tubing meeting GM specification 124-M or its equivalent. Don't use copper or aluminum tubing to replace steel tubing. These materials cannot withstand normal vehicle vibration.

6   If is becomes necessary to replace a section of nylon fuel line, replace it only with the correct part number - don't use any substitutes.

7   Some fuel lines have threaded fittings with O-rings. Any time the fittings are loosened to service or replace components:

a) *Use a backup wrench while loosening and tightening the fittings.*

b) *Check all O-rings for cuts, cracks and deterioration. Replace any that appear worn or damaged.*

c) *If the lines are replaced, always use original equipment parts, or parts that meet the manufacturer's standards.*

d) *Use the proper fuel line disconnect tools to release the spring-lock couplings inside the quick-disconnect fittings present in many locations throughout the system (fuel filter, fuel rail, fuel pressure regulator, etc.)* **(see illustrations)**.

Rubber hose

Warning: *These models are equipped with electronic fuel injection - use only original equipment replacement hoses or their equivalent. Others may fail from the high pressures of this system.*

8   When rubber hose is used to replace a metal line, use reinforced, fuel resistant hose (GM Specification 6163-M). Hose other than this could fail prematurely and could fail to meet Federal emission standards. Hose inside diameter must match line outside diameter. **Warning:** *Don't substitute rubber hose for metal line on high-pressure systems. Use only genuine factory replacement lines or lines meeting factory specifications.*

9   Don't use rubber hose within four inches of any part of the exhaust system or within ten inches of the catalytic converter. Metal lines and rubber hoses must never be allowed to chafe against the frame. A minimum of 1/4-inch clearance must be maintained around a line or hose to prevent contact with the frame.

Removal and installation

Refer to illustrations 4.10, 4.12a, 4.12b and 4.13

Note: *The following procedure and accompanying illustrations are typical for vehicles covered by this manual. On quick-disconnect (non-threaded) fittings, clean off the fittings before disconnection to prevent dirt from getting in the fittings. After disconnection, clean the fittings with compressed air and apply a few drops of oil.*

10   Relieve the fuel pressure (see Section 2) and disconnect the cable from the negative terminal of the battery. **Caution:** *If the vehicle is equipped with a Delco Loc II or Theftlock audio system, make sure you*

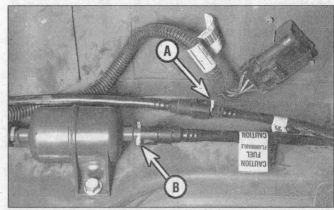

4.10 Some models are equipped with fuel lines that can be disconnected by pinching the tabs and separating each connector

A   Fuel return line   B   Fuel feed line

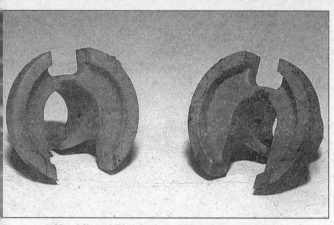

4.12a 3/8 and 5/16-inch fuel line disconnect tools

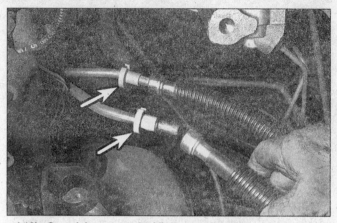

4.12b On quick-connect fuel lines, use the proper tool (arrows) and push them into the connector to separate the fuel lines

have the correct activation code before disconnecting the battery. Dis-connect the fuel feed, return or vapor line at the fuel tank **(see illustra-tion).** **Note:** *Some fuel line connections may be threaded. Be sure to use a back-up wrench when separating the connections.*

11   Remove all fasteners attaching the lines to the vehicle body.

12   Detach the fitting(s) that attach the fuel hoses to the engine com-partment metal lines **(see illustrations).** Twisting them back and forth will allow them to separate more easily.

13   Installation is the reverse of removal. Be sure to use new O-rings at the threaded fittings, if equipped **(see illustration).**

5   Fuel tank - removal and installation

Refer to illustration 5.6

Warning: *Gasoline is extremely flammable, so take extra precautions when you work on any part of the fuel system. See the* **Warning** *in Sec-tion 2.*

Note: *Don't begin this procedure until the fuel gauge indicates the tank is empty or nearly empty. If the tank must be removed when it isn't empty (for example, if the fuel pump malfunctions), siphon any remain-ing fuel from the tank prior to removal.*

1   Unless the vehicle has been driven far enough to completely empty the tank, it's a good idea to siphon the residual fuel out before removing the tank from the vehicle. **Warning:** *DO NOT start the siphoning action by mouth! Use a siphoning kit, available at most auto parts stores.*

2   Relieve the fuel system pressure (see Section 2).

3   Detach the cable from the negative terminal of the battery. **Cau-tion:** *If the vehicle is equipped with a Delco Loc II or Theftlock audio*

system, make sure you have the correct activation code before discon-necting the battery.

4   Raise the vehicle and support it securely on jackstands placed underneath the jacking points.

5   Disconnect the exhaust rubber hangers and allow the rear portion of the exhaust system to rest on the rear axle. Remove the fuel tank heat shield on 1999 models.

6   Disconnect the fuel filler hose and vapor lines from the fuel tank **(see illustration).**

7   Disconnect the fuel feed and return lines and the vapor return line from the fuel pump. **Note:** *Use the proper fuel line disconnect tools* (see Section 4).

8   Disconnect the bolts from the fuel filler neck. Angle the filler neck down and away from the tank and remove it from the vehicle.

9   Support the fuel tank with a floor jack.

10   Disconnect both fuel tank retaining straps.

11   Lower the tank enough to disconnect the electrical connectors from the fuel pump/fuel level sending unit, and the fuel tank pressure sensor (if equipped).

12   Remove the tank from the vehicle.

13   Installation is the reverse of removal.

6   Fuel tank cleaning and repair - general information

1   All repairs to the fuel tank or filler neck should be carried out by a professional who has experience in this critical and potentially danger-ous work. Even after cleaning and flushing of the fuel system, explo-sive fumes can remain and ignite during repair of the tank.

2   If the fuel tank is removed from the vehicle, it should not be

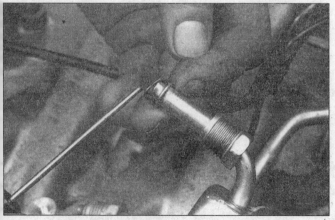

4.13 Always replace the fuel line O-rings (if equipped)

5.6 Remove the fuel filler and the vapor return hose clamps (arrows)

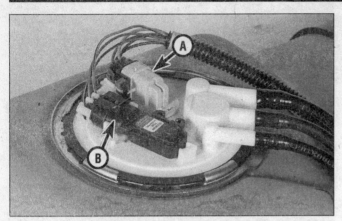

7.4 Location of the fuel pump/fuel level sending unit (A) and the fuel pressure sensor (B) harness connectors

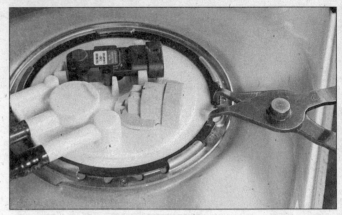

7.5 Use a pair of snap-ring pliers to remove the retaining collar from the fuel pump assembly

placed in an area where sparks or open flames could ignite the fumes coming out of the tank. Be especially careful inside garages where a natural gas-type appliance is located, because the pilot light could cause an explosion.

7 Fuel pump - removal and installation

Refer to illustrations 7.4, 7.5, 7.6, 7.8a and 7.8b

Warning: *Gasoline is extremely flammable, so take extra precautions when you work on any part of the fuel system. See the* **Warning** *in Section 2.*

1   Relieve the fuel system pressure (see Section 2).
2   Disconnect the cable from the negative battery terminal. **Caution:** *If the vehicle is equipped with a Delco Loc II or Theftlock audio system, make sure you have the correct activation code before disconnecting the battery.*
3   Remove the fuel tank (see Section 5).
4   Disconnect the fuel pump/sending unit harness connector and the fuel tank pressure sensor from the fuel tank **(see illustration)**.
5   The fuel pump/sending unit assembly is located inside the fuel tank. Using a pair of snap-ring pliers, contract the retaining collar and remove it from the top of the fuel pump assembly **(see illustration)**.
6   Lift the fuel pump/sending unit assembly from the fuel tank **(see illustration). Caution:** *The fuel level float and sending unit are delicate. Do not bump them against the tank during removal or the accuracy of the sending unit may be affected.*
7   Inspect the condition of the O-ring around the opening of the tank. If it is dried, cracked or deteriorated, replace it.

8   Remove the strainer from the lower end of the fuel pump **(see illustrations)**. If it is dirty, clean it with a suitable solvent and blow it out with compressed air. If it is too dirty to be cleaned, replace it.
9   Remove the fuel pressure sensor mounting bolts and separate the sensor from the top of the fuel pump assembly.
10   If it is necessary to separate the fuel pump and sending unit, disconnect the electrical connectors at the pump, noting their positions. Follow the procedure in Section 8.
11   Disconnect the fuel line from the pump.
12   Installation is the reverse of removal. Reassemble the fuel pump to the sending unit bracket assembly and insert the assembly into the fuel tank.
13   Install the fuel tank (see Section 5).

8 Fuel level sending unit - check and replacement

Warning: *Gasoline is extremely flammable, so take extra precautions when you work on any part of the fuel system. See the* **Warning** *in Section 2.*

Check

1   Raise the vehicle and support it securely on jackstands. **Note:** *Because the harness connector is not accessible with the fuel tank in the vehicle, remove the fuel tank (see Section 5) to gain access to the fuel pump/fuel level sending unit.*
2   Position the probes of an ohmmeter onto the terminals of the sending unit electrical connector and check for resistance. Refer to the wiring diagrams at the end of this manual for the proper terminals, if necessary.

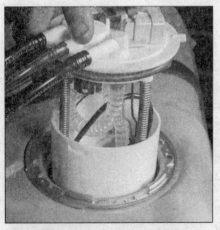

7.6 Lift the fuel pump assembly from the fuel tank

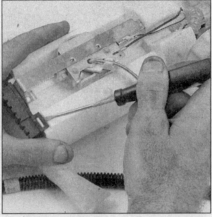

7.8a Pry on the plastic tab to remove the protective shield from the foot of the assembly

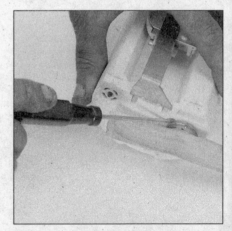

7.8b Carefully pry the fuel strainer from the inlet pipe

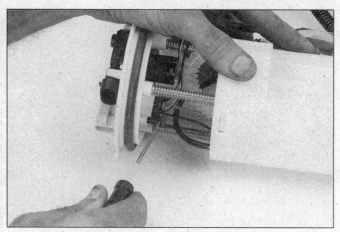

8.6 Remove the fuel level electrical connector from the fuel pump assembly

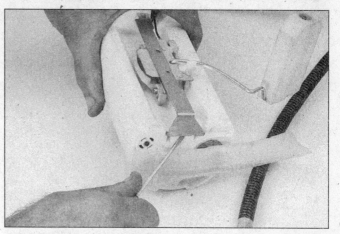

8.7 On 1998 and earlier models, pry the bracket from the base of the fuel pump assembly

3   First, check the resistance of the sending unit with the fuel tank completely full. The resistance of the sending unit should be about 90 ohms. Then check for resistance with the float at the bottom (empty). The resistance should be about 2 ohms.

4   If the readings are incorrect or there is very little change in resistance as the float travels from full to empty, replace the fuel level sending unit assembly.

Replacement

Refer to illustrations 8.6 and 8.7

5   Remove the fuel tank (see Section 5) and the fuel pump (see Section 7) from the vehicle.

6   Disconnect the sending unit electrical connector from the assembly **(see illustration)**.

7   On 1998 and earlier models, carefully separate the sending unit bracket from the base of the fuel pump assembly **(see illustration)**. On 1999 models, detach the sending unit retaining clip and release the sending unit from the fuel pump module.

8   Installation is the reverse of removal.

9   Air cleaner assembly - removal and installation

Refer to illustration 9.3

1   Disconnect the negative battery cable, then the positive cable from the battery and remove the battery (see Chapter 5). **Caution:** *If the vehicle is equipped with a Delco Loc II or Theftlock audio system,*

make sure you have the correct activation code before disconnecting the battery.

2   Disconnect the electrical connector to the intake air temperature (IAT) sensor (see Chapter 6).

3   Detach the wiring harness clamps from the air cleaner **(see illustration)**.

4   Loosen the clamp that retains the air intake duct to the air cleaner and detach the duct from the housing **(see illustration 9.3)**.

5   Remove the cover and lift the air filter element from the air cleaner housing (see Chapter 1).

6   Use a punch or a narrow screwdriver and push the locking tabs toward the fender to unlock the air cleaner housing from the bracket. The locking tabs are located near the relay center and the battery tray.

7   Lift the lower portion of the air cleaner housing from the engine compartment, while moving the relay center aside.

8   Installation is the reverse of removal.

10   Accelerator cable - removal and installation

Refer to illustrations 10.2, 10.4, 10.5a, 10.5b, 10.6a and 10.6b

Removal

1   Detach the screws and the clip retaining the lower instrument panel trim and lower the trim (if necessary) (see Chapter 11).

2   Detach the accelerator cable from the accelerator pedal **(see illustration)**.

3   Squeeze the accelerator cable casing tangs and push the cable

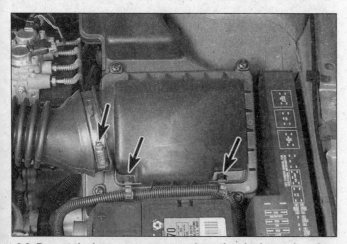

9.3 Pry out the harness connectors from the air cleaner housing and loosen the clamp from the air intake duct

10.2 Slide the accelerator cable through the slot in the pedal arm

4

10.4 Some splash shields must be pried up to release the tangs from the bracket

10.5a Release the locking tab to separate the cruise control cable from the accelerator control bracket

10.5b Unclip the cruise control cable end off the throttle lever

through the firewall into the engine compartment.

4   Remove the accelerator control splash shield **(see illustration)**. Some models have screws, others can be pried off.

5   Using a screwdriver, pry back the locking tab on the cruise control cable **(see illustration)** and detach the cable assembly from the throttle lever **(see illustration)**.

6   First, pry the locking tab on the accelerator cable and detach the casing from the bracket **(see illustration)** then rotate the throttle lever and detach the accelerator cable from the throttle **(see illustration)**.

Installation

7   Installation is the reverse of removal. **Note:** *To prevent possible interference, flexible components (hoses, wires, etc.) must not be routed within two inches of moving parts, unless routing is controlled.*

8   Operate the accelerator pedal and check for any binding condition by completely opening and closing the throttle.

9   Apply sealant around the accelerator cable at the engine compartment side of the firewall.

11 Fuel injection system - general information

Refer to illustration 11.3

1   These models are equipped with a Multiport Fuel Injection (MFI) system.

2   The fuel system consists of a fuel tank, an electric fuel pump and fuel pump relay, an air cleaner assembly and an electronic fuel injection system.

Multiport Fuel Injection (MFI) system

3   Multiport Fuel Injection (MFI) consists of an air intake manifold, the throttle body, the injectors, the fuel rail assembly, an electric fuel pump and associated plumbing **(see illustration)**.

4   Air is drawn through the air cleaner and throttle body. A Manifold Absolute Pressure (MAP) sensors informs the PCM of vacuum and pressure variations.

5   While the engine is running, the fuel constantly circulates through the fuel rail, which removes vapors and keeps the fuel cool while maintaining sufficient pressure to the injectors under all running conditions.

6   Operation of the MFI system is controlled by the PCM so that it works in conjunction with the rest of the vehicle functions to provide optimum driveability and emissions control.

7   Because the MFI system meters fuel and air precisely, it is important that the fuel and air filters be changed at the specified intervals.

12 Fuel injection system - check

Refer to illustrations 12.8 and 12.10

Warning: *Gasoline is extremely flammable, so take extra precautions when you work on any part of the fuel system. See the* **Warning** *in Section 2.*

Note: *The following procedure is based on the assumption that the fuel pump is working and the fuel pressure is adequate* (see Section 3).

1   Check to see that the battery is fully charged, as the control unit and sensors depend on an accurate supply voltage in order to properly meter the fuel.

10.6a Release the locking tab on the accelerator cable casing and slide it out of the bracket

10.6b Pass the accelerator cable end through the slot in the throttle lever

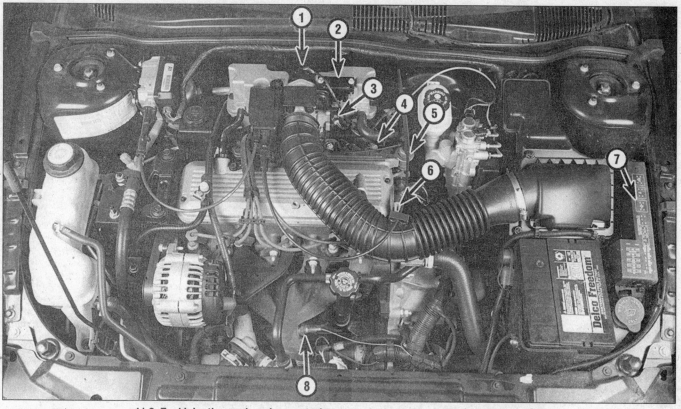

11.3 Fuel injection and engine control component locations (1996 2.2L OHV engine)

1	Idle Air Control (IAC) valve	3	Throttle Position Sensor (TPS)	6	Intake Air Temperature (IAT) sensor
2	Manifold Absolute Pressure (MAP) sensor	4	Fuel injectors	7	Fuel pump relay
		5	Fuel pressure regulator	8	Oxygen sensor

2   Check the air filter element - a dirty or partially blocked filter will severely impede performance and economy (see Chapter 1).

3   Check the fuel filter and replace it if necessary (see Chapter 1).

5   Check the ground wire connections on the intake manifold for tightness. Check all electrical connectors that are related to the system. Loose connectors and poor grounds can cause many problems that resemble more serious malfunctions.

5   If a blown fuse is found, replace it and see if it blows again. If it does, search for a grounded wire in the harness.

6   Check the air intake duct to the intake manifold for leaks, which will result in an excessively lean mixture. Also check the condition of all vac-uum hoses connected to the intake manifold.

7   Remove the air intake duct from the throttle body and check for dirt, carbon or other residue build-up. If it's dirty, clean it with aerosol carburetor cleaner and a rag.

8   With the engine running, place an automotive stethoscope against each injector, one at a time, and listen for a clicking sound, indicating operation **(see illustration)**. If you don't have a stethoscope, place the tip of a screwdriver against the injector and listen through the handle.

9   Unplug the injector electrical connector(s) and test the resistance of each injector. Compare the values to the Specifications listed in this Chapter.

10   Install an injector test light ("noid" light) into each injector electrical connector, one at a time **(see illustration)**. Crank the engine over. Confirm that the light flashes evenly on each connector. This will test

12.8 Use a stethoscope or screwdriver to determine if the injectors are working properly - they should make a steady clicking sound that rises and falls with engine speed changes (MFI system)

12.10 Install a "noid" light into each injector electrical connector and confirm that it blinks when the engine is cranking or running

4

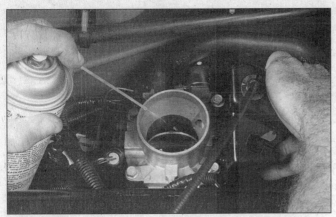

13.2 Spray carburetor cleaner into the throttle body to break away any carbon deposits or sludge that may have collected around the throttle plate

13.9a Disconnect the vacuum lines (arrows) from the throttle body

for voltage to the injectors.

11   The remainder of the system checks can be found in the following Sections.

---

13   Fuel injection system - component check and replacement

Warning 1: *Gasoline is extremely flammable, so take extra precautions when you work on any part of the fuel system. See the* **Warning** *in Section 2.*

Warning 2: *DO NOT use any type of solvents or cleaners containing Methyl Ethyl Ketone (MEK) on the fuel system components as damage may occur to the throttle body and other fuel system components.*

Throttle body
Check

Refer to illustration 13.2

1   Detach the air intake duct from the throttle body and move the duct out of the way.

2   Have an assistant depress the throttle pedal while you watch the throttle valve. Check that the throttle valve moves smoothly when the throttle is moved from closed (idle position) to fully open (wide open throttle). **Note:** *Spray carburetor cleaner into the throttle body, especially around the shaft area* **(see illustration)** *to free-up any binding caused by the accumulation of carbon deposits or sludge buildup.*

3   Wiggle the throttle lever while watching the throttle shaft inside the bore. If it appears worn (loose), replace the throttle body unit.

Removal

Refer to illustrations 13.9a, 13.9b and 13.10

Warning: *Wait until the engine is completely cool before beginning this procedure.*

4   Disconnect the cable from the negative terminal of the battery. **Caution:** *If the vehicle is equipped with a Delco Loc II or Theftlock audio system, make sure you have the correct activation code before disconnecting the battery.*

5   Detach the air intake duct and the air cleaner resonator from the throttle body (if equipped).

6   Unplug the Idle Air Control (IAC) valve and the Throttle Position Sensor (TPS) electrical connectors (see Chapter 6).

7   Mark and disconnect any vacuum hoses connected to the throttle body. Also detach the breather hose, if equipped.

8   Disconnect the accelerator cable from the throttle lever, then detach the cable housing from its bracket (see Section 10). Remove the cruise control cable, if equipped **(see illustration 10.5a and 10.5b)**.

9   Disconnect the vacuum lines from throttle body **(see illustration)**. On some models it will be necessary to remove the vacuum distributor from the throttle body **(see illustration)**.

10   Remove the throttle body bolts and detach the throttle body **(see illustration)**.

Installation

11   Clean off all traces of old gasket material from the throttle body and the plenum.

12   Install the throttle body and a new gasket and tighten the bolts to the torque listed in this Chapter's Specifications.

13   The rest of the procedure is the reverse of removal. Be sure to check the coolant level (see Chapter 1) and add, if necessary.

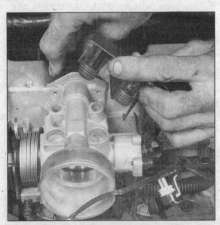

13.9b Some models will require removing the vacuum distributor from the throttle body

13.10 Remove the throttle body mounting bolts (arrows)

13.15 Measure the resistance across terminals A and B, then across terminals C and D

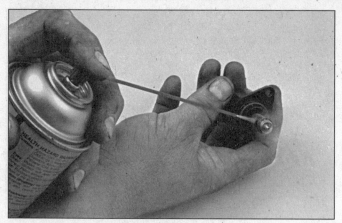

13.17a Clean the IAC valve pintle with aerosol carburetor cleaner to remove carbon deposits

13.17b Spray carburetor cleaner into the IAC valve housing and check for clogged air passages in the air intake plenum

Idle Air Control (IAC) valve

Check

Refer to illustrations 13.15, 13.17a and 13.17b

14   The idle air control valve (IAC) controls the engine idle speed. This output actuator is mounted on the throttle body and is controlled by voltage pulses sent from the PCM (computer). The IAC valve pintle moves in or out allowing more or less intake air into the system according to the engine conditions. To increase idle speed, the PCM retracts the IAC valve pintle away from the seat and allows more air to bypass the throttle bore. To decrease idle speed, the PCM extends the IAC valve pintle towards the seat, reducing the air flow.

15   To check the IAC valve, unplug the electrical connector and, using an ohmmeter, measure the resistance across terminals A and B, then terminals C and D. Each resistance check should indicate 40 to 80 ohms **(see illustration)**. If not, replace the IAC valve.

16   There is an alternate method for testing the IAC valve. Various SCAN tools are available from auto parts stores and specialty tool companies that can be plugged into the DLC (diagnostic connector) for the purpose of monitoring the sensors. Connect the SCAN tool and switch to the Idle Air Motor Position mode and monitor the steps (motor winding position). The SCAN tool should indicate between 10 to 200 steps depending upon the rpm range. Allow the engine to idle for several minutes and while observing the count reading, snap the throttle to achieve high rpm (under 3,500). Repeat the procedure several times and observe the SCAN tool steps (counts) when the engine goes back to idle. The readings should be within 5 to 10 steps each time. If the readings fluctuate greatly, replace the IAC valve. The PCM will set Codes P0506 or P0507 in the event of IAC failure. **Note:** *When the IAC valve electrical connector is disconnected for testing, the PCM will have to "relearn" its idle mode. In other words, it will take a certain amount of time before the idle motor resets for the correct idle speed. Make sure the idle is smooth and not misfiring before plugging in the SCAN tool.* **Note:** *Refer to Chapter 6 for additional information and illustrations concerning SCAN tools.*

17   Next, remove the valve (see Step 18) and inspect it:

   a) *Check the pintle for excessive carbon deposits. If necessary, clean it with aerosol carburetor cleaner* **(see illustration)**. *Also clean the IAC valve housing to remove any deposits* **(see illustration)**.

   b) *Check the IAC valve electrical connections. Make sure the pins are not bent and make good contact with the connector terminals.*

Removal

Refer to illustration 13.19

18   Unplug the electrical connector from the Idle Air Control (IAC) valve.

19   Unscrew the valve or remove the two IAC valve attaching screws and withdraw the valve **(see illustration)**.

20   Check the condition of the rubber O-ring. If it's hardened or deteriorated, replace it. On models equipped with a gasket, remove the gasket.

21   Clean the sealing surface and the bore of the idle air/vacuum signal housing assembly to ensure a good seal. **Caution:** *The IAC valve itself is an electrical component and must not be soaked in any liquid cleaner, as damage may result.*

22   Before installing the IAC valve, the position of the pintle must be checked. If the pintle is extended too far, damage to the assembly may occur.

Installation

23   Measure the distance from the flange or gasket mounting surface of the IAC valve to the tip of the pintle. If the distance is greater than 1-1/8 inch, reduce the distance by applying firm pressure onto the pintle to retract it. Try some side-to-side motion in the event the pintle binds.

24   Position the new O-ring or gasket on the IAC valve. Lubricate the O-ring with a light film of engine oil. Install the IAC valve and tighten the valve or the mounting screws securely.

25   Plug in the electrical connector at the IAC valve assembly. **Note:** *No adjustment is made to the IAC assembly after reinstallation. The IAC resetting is controlled by the PCM when the engine is started.*

Throttle Position Sensor (TPS)

Check

26   Check for stored trouble codes in the PCM using the On Board Diagnosis system (see Chapter 6).

27   To check the operation and replacement of the TPS, refer to the *Information Sensors* in Chapter 6.

13.19 Remove the IAC valve screws (arrows) and separate the IAC from the throttle body

4

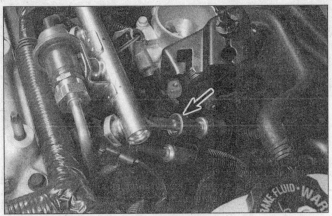

13.29 Use a special fuel line disconnect tool to detach the fuel feed line from the fuel rail

13.32 Remove the fuel rail mounting bolts (arrows) . . .

13.33 . . . and separate the fuel rail from the intake manifold

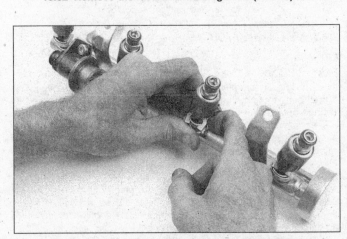

13.34a To remove an injector from the fuel rail, spread the retaining clip with a small screwdriver, then pull the injector from the fuel rail (conventional style injectors)

Fuel rail and injectors

Refer to illustrations 13.29, 13.32, 13.33, 13.34a, 13.34b, 13.34c, 13.34d, 13.35a, 13.35b and 13.35c

Warning: *Before any work is performed on the fuel lines, fuel rail or injectors, the fuel system pressure must be relieved (see Section 2).*

Caution: *There are two different types of fuel rail designs used on these engines. 1995 through 1997 2.2L engines are equipped with bottom-feed fuel injectors that are mounted in the intake manifold rather than in a separate fuel rail. 1998 and later 2.2L engines and all 2.3L and 2.4L engines are equipped with conventional fuel rails and injectors. Be*

sure to follow the correct installation procedure for the bottom-feed style injectors to avoid damaging the engine.

Note: *Refer to Section 12 for the injector checking procedure.*

28   Detach the cable from the negative terminal of the battery. **Caution:** *If the vehicle is equipped with a Delco Loc II or Theftlock audio system, make sure you have the correct activation code before disconnecting the battery.*

29   Using a special tool (as described in Section 4), detach the fuel feed line from the fuel rail **(see illustration)**.

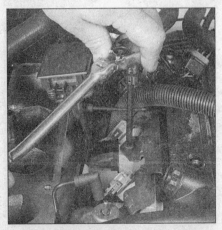

13.34b Remove the fuel injector bracket bolts (bottom-feed style injectors) . . .

13.34c . . . then lift the fuel injector bracket from the intake manifold (bottom-feed style injectors)

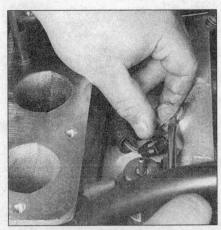

13.34d Carefully pry the fuel injectors out of the manifold (bottom-feed style injectors)

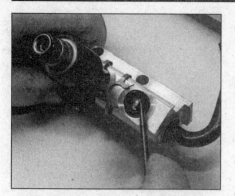

13.35a Remove the injector seal from the fuel rail. The injector seals that are positioned in the fuel rail are black - the seals that go into the manifold are brown (conventional style injectors)

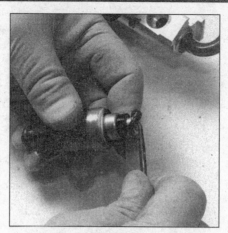

13.35b Carefully pry the seals off the injectors

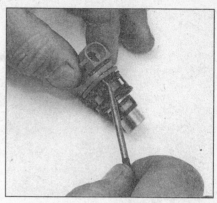

13.35c On bottom-feed style injectors, remove the upper O-ring with a pick or small screwdriver, being careful not to damage the filter screen

30   On all except 1995 through 1997 2.2L engines, detach the fuel return line from the fuel pressure regulator.

31   Remove the air intake plenum on 2.2L engines with the bottom-feed style fuel injectors (see Steps 46 through 54).

32   Label and unplug the injector electrical connectors. Remove the fuel rail retaining bolts **(see illustration)**.

33   Carefully remove the fuel rail with the injectors **(see illustration)**. **Caution 1:** *Use care when handling the fuel rail assembly to avoid damaging the injectors.* **Caution 2:** *On 1995 through 1997 2.2L engines with the bottom-feed injectors, make sure that the O-ring is not stuck inside the intake manifold bore after the injector is removed. These O-rings must not be reused once the injector has been removed.* **Note:** *An identification number is stamped on the side of the fuel rail assembly. Refer to this number if servicing or parts replacement is required.*

34   To remove the fuel injectors on models with a conventional fuel rail, spread the injector retaining clip and pull the injector from the fuel rail **(see illustration)**. To replace bottom-feed type injectors (1995 through 1997 2.2L engines), remove the fuel injector bracket bolts, then remove the bracket and pry the injectors out of the manifold **(see illustrations)**.

35   Remove the injector O-ring seals **(see illustrations)**. **Note:** *On conventional style injectors, be sure to replace the injector seal with the correct color. Brown seals are used on the engine side of the injector, while black seals are used on the fuel rail side of the injector.*

36   Install the new O-ring seal(s), as required, on the injector(s) and lubricate them with a light film of engine oil.

37   Install the injectors on the fuel rail.

38   Secure the injectors with the retainer clips (conventional style only). **Caution:** *1995 through 1997 2.2L engines are equipped with bottom-feed fuel injectors that are mounted in the intake manifold rather than as a separate fuel rail assembly. It is important to make sure that they are* properly seated inside the intake manifold before starting the engine. If they are not seating correctly, the fuel pressure inside the fuel rail portion of the intake manifold will force excessive amounts of fuel into the cylinders and possibly damage the engine. To check the seal, refer to Section 3 and check the fuel pressure with the ignition key ON (engine not running). If the fuel system holds the fuel pressure (gauge remains steady), the fuel system is sealed properly and the engine can be started safely.

39   Installation is the reverse of the removal procedure.

Fuel pressure regulator
Check
40   Refer to Section 3 for the fuel pressure regulator checking procedure.

Replacement
Refer to illustrations 13.43a, 13.43b, 13.45a and 13.45b

41   Relieve the fuel system pressure (see Section 2).

42   Disconnect the cable from the negative terminal of the battery. **Caution:** *If the vehicle is equipped with a Delco Loc II or Theftlock audio system, make sure you have the correct activation code before disconnecting the battery.*

43   Detach the fuel return line from the fuel pressure regulator **(see illustrations)**. Pull off the vacuum hose from the port on the regulator.

44   Remove the pressure regulator mounting screw(s) and detach the fuel pressure regulator. **Note:** *On 1998 and later 2.4L engines it will be necessary to remove the fuel rail assembly, then remove the pressure regulator from the fuel rail.*

45   Reassembly is the reverse of disassembly. Be sure to replace all gaskets and seals **(see illustrations)**, otherwise a dangerous fuel leak may develop. When installing the seals, lubricate them with a light film of engine oil.

13.43a Remove the return line from the fuel pressure regulator, then remove the mounting screw from the regulator (not visible in this photo)

13.43b Remove the fuel pressure regulator mounting screw (arrow) (early type shown)

4

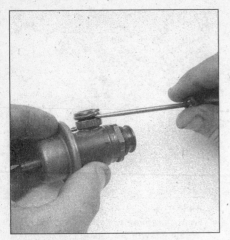

13.45a Replace the fuel pressure regulator-to-fuel rail O-ring

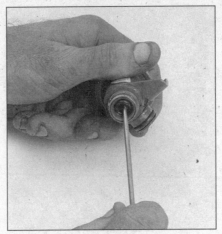

13.45b Also replace the inner return line O-ring

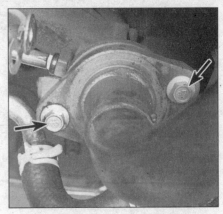

14.1a Be sure to spray penetrating lubricant onto the flange nuts (arrows) before removing them from the exhaust manifold

Air intake plenum (1995 through 1997 2.2L engines)

46   Detach the cable from the negative terminal of the battery. **Caution:** *If the vehicle is equipped with a Delco Loc II or Theftlock audio system, make sure you have the correct activation code before disconnecting the battery.*

47   Remove the air intake duct from the throttle body and the air cleaner assembly (see Section 9).

48   Remove the accelerator cable (see Section 10).

49   Remove the vacuum distributor from the throttle body **(see illustrations 13.9a and 13.9b).**

50   Disconnect the PCV hoses and the power brake vacuum hose.

51   Disconnect the MAP, TPS and IAC valve electrical connectors.

52   Position the fuel injector wiring harness out of the way and remove the air intake plenum mounting bolts.

53   Remove the power steering pump and set it to the side (see Chapter 10).

54   Remove the plenum bolts and the gasket from the intake manifold. Clean the mating surfaces and be sure to remove all traces of the old gasket without chipping the aluminum surface.

55   Installation is the reverse of removal. Tighten the bolts to the torque listed in this Chapter's Specifications in a criss-cross pattern.

14   Exhaust system servicing - general information

Warning: *The vehicle's exhaust system generates very high temperatures and must be allowed to cool down completely before any of the components are touched. Be especially careful around the catalytic converter, where the highest temperatures are generated.*

General Information

Refer to illustrations 14.1a, 14.1b, 14.1c and 14.1d

Caution: *The flex coupling in the front exhaust pipe can be damaged if over-flexed. Always support the exhaust pipe assembly when any components are removed to prevent damage to the flex coupling.*

Replacement of exhaust system components is basically a matter of removing the heat shields, disconnecting the component and installing a new one **(see illustrations)**. The heat shields and exhaust system hangers must be reinstalled in the original locations or damage could result. Due to the high temperatures and exposed locations of the exhaust system components, rust and corrosion can seize parts together. Penetrating oils are available to help loosen frozen fasteners. However, in some cases it may be necessary to cut the pieces apart with a hacksaw or cutting torch. The latter method should be employed only by persons experienced in this work.

14.1b Make sure all exhaust pipe brackets are not cracked or damaged

14.1c Make sure the catalytic converter flange gasket is intact

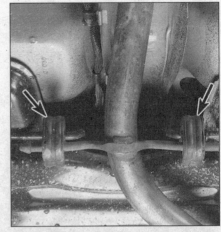

14.1d Check the rubber hangers (arrows) for deterioration or cracks that may cause the exhaust system to drop

Chapter 5
Engine electrical systems

Contents

Specifications

General

Ignition coil resistance	
2.2L engine with DIS	
Primary resistance	0.35 to 1.50 ohms
Secondary resistance	5,000 to 10,000 ohms
2.3L and 2.4L engine with IDI	
Primary resistance	0.40 to 0.80 ohms
Secondary resistance	5,000 to 7,000 ohms
Firing order	1-3-4-2

Charging system

Alternator regulated charging output	13 to 14.5 volts

5

1   General information

Warning: *Because of the very high voltage generated by the ignition system, extreme care should be taken whenever an operation involving ignition components is performed. This not only includes the distributor, coil(s), module and spark plug wires, but related items that are connected to the systems as well, such as the plug connections, tachometer and testing equipment.*

The engines covered by this manual are equipped with either a distributorless Direct Ignition System (DIS) (2.2L engine) or an Integrated Direct Ignition system (IDI) (2.3L and 2.4L OHC engines). The DIS system mounts the coil pack on top of the ignition module with spark plug wires connecting the spark plugs to the coil packs. The IDI system on 2.3L and 2.4L engines incorporates the ignition module, coils and the spark plug boot assembly into one complete unit mounted directly on top of the spark plug assembly. which is mounted directly above the spark plugs to the cylinder head.

DIS and IDI ignition systems are distributorless type ignition systems that use a "waste spark" method of spark distribution. Each cylinder is paired with its opposing cylinder in the firing order, 1-4, 2-3, so that one cylinder on compression fires simultaneously with its opposing cylinder on the exhaust stroke. Since the cylinder on the exhaust stroke requires very little of the available voltage to fire its plug, most of the voltage is used to fire the cylinder on the compression stroke.

The distributorless ignition system includes two coil packs, an ignition control module (ICM), a crankshaft reluctor ring, a magnetic crankshaft sensor and the PCM. The ignition module is located under the coil packs and is connected to the PCM on the DIS type system and is located under the ignition assembly cover on IDI systems.

The crankshaft sensor is located on the side of the engine block. The magnetic crankshaft sensor protrudes through the engine block, within about 0.050-inch of the crankshaft reluctor ring. The reluctor ring is a special disc cast into the crankshaft, which acts as a signal generator for the ignition timing.

The system uses control wires from the PCM, just like conventional distributor systems. The PCM controls timing using crankshaft position, engine rpm, engine temperature and manifold absolute pressure (MAP) sensing.

The charging system consists of a belt-driven alternator with an integral voltage regulator and the battery. These components work together to supply electrical power for the ignition system, the lights and all accessories.

2   Battery - emergency jump starting

Refer to the *Booster battery (jump) starting* procedure at the front of this manual.

3.3 Remove the battery hold-down bolt from the battery carrier

3.6 Check for corrosion on the battery carrier before installing the battery

5.2 The coil packs are arranged 2-3 and 1-4 for companion cylinders (coilpack from a 2.2L engine shown)

3   Battery - removal and installation

Refer to illustrations 3.3 and 3.6
Warning: *Hydrogen gas is produced by the battery, so keep open flames and lighted cigarettes away from it at all times. Always wear eye protection when working around a battery. Rinse off spilled electrolyte immediately with large amounts of water.*

1   The battery is located at the left front corner of the engine compartment.

Removal

2   Detach the cable from the negative terminal of the battery, then from the positive terminal. **Caution 1:** *If the vehicle is equipped with a Delco Loc II or Theftlock audio system, make sure you have the correct activation code before disconnecting the battery.* **Caution 2:** *Always disconnect the negative cable first and hook it up last or the battery may be shorted by the tool being used to loosen the cable clamps.*

3   Detach any brackets or braces that would interfere with the removal of the battery, then remove the hold-down clamp **(see illustration)** from the battery carrier.

4   Carefully lift the battery out of the carrier. **Warning:** *Always keep the battery in an upright position to reduce the possibility of electrolyte spills. If you spill electrolyte on yourself or the vehicle, rinse it off immediately with plenty of water.*

5   If you're installing a new battery, make sure you get one that's identical (same dimensions, amperage rating, cold cranking rating, etc.).

Installation

6   Set the battery in position in the carrier. Don't tilt it. **Note:** *The battery carrier and hold-down clamp should be clean and free from corrosion before installing the battery* **(see illustration)**.

7   Install the hold-down clamp and bolt. The bolt should be snug, but overtightening it may damage the battery case.

8   Install both battery cables - positive first, then negative. **Note:** *The battery terminals and cable ends should be cleaned if necessary* (see Chapter 1).

4   Battery cables - check and replacement

1   Periodically inspect the entire length of each battery cable for damage, cracked or burned insulation and corrosion. Poor battery cable connections can cause starting problems and decreased engine performance.

2   Check the cable-to-terminal connections at the ends of the cables for cracks, loose wire strands and corrosion. The presence of white, fluffy deposits under the insulation at the cable terminal connection is a sign the cable is corroded and should be replaced. Check the terminals for distortion, missing mounting bolts or nuts and corrosion.

3   When removing the cables, always disconnect the negative cable first and hook it up last or the battery may be shorted by the tool used to loosen the cable clamps. Even if only the positive cable is being replaced, be sure to disconnect the negative cable from the battery first (see Chapter 1 for additional information related to battery cable removal).

4   Disconnect the old cables from the battery, then trace each of them to their opposite ends and detach them from the starter solenoid and ground terminals. Note the routing of each cable to ensure correct installation. **Caution:** *If the vehicle is equipped with a Delco Loc II or Theftlock audio system, make sure you have the correct activation code before disconnecting the battery.*

5   If you're replacing either or both cables, take the old ones with you when buying the new ones - the replacements must be identical. Cables have characteristics that make them easy to identify: Positive cables are normally red, larger in diameter and have a larger diameter battery post and clamp; ground cables are normally black, smaller in diameter and have a slightly smaller battery post and clamp.

6   Clean the threads of the solenoid or ground connection with a wire brush to remove rust and corrosion. Apply a light coat of petroleum jelly to the threads to prevent future corrosion.

7   Attach the cable to the solenoid or ground connection and tighten the mounting nut/bolt securely.

8   Before connecting a new cable to the battery, make sure it reaches the battery post without having to be stretched.

9   Connect the positive cable first, followed by the negative cable.

5   Ignition system - general information

Refer to illustration 5.2
Caution: *If the vehicle is equipped with a Delco Loc II or Theftlock audio system, make sure you have the correct activation code before disconnecting the battery.*

1   The engines covered by this manual are equipped with either a distributorless Direct Ignition System (DIS) (2.2L engine) or an Integrated Direct Ignition system (IDI) (2.3L and 2.4L OHC engines). The DIS system mounts the coil packs on top of the ignition module with spark plug wires connecting the spark plugs to the coil packs. The IDI system on 2.3L and 2.4L OHC engines incorporates the ignition module, coils and the spark plug boot assembly into one complete unit which is mounted directly above the spark plugs to the cylinder head. The DIS and IDI ignition systems use a "waste spark" method of spark distribution. Although these two systems differ in the position of the ignition components, they are basically the same type of distributorless ignition system.

6.2 To use a calibrated ignition tester (available at most auto parts stores) on 2.2L engines, simply disconnect a spark plug wire, attach the wire to the tester, clip the tester to a convenient ground and operate the starter - if there's enough power to fire the plug, sparks will be visible between the electrode tip and the tester body

6.5 Check for battery voltage to the ignition module from the ignition key (2.2L engine)

6.6 On 2.2L engines, remove the coil pack from the module assembly and check for a trigger signal between the module terminals with a test light while an assistant cranks the engine over

2   The distributorless ignition system uses a "waste spark" method of spark distribution. Each cylinder is paired with its opposing cylinder in the firing order, 1-4, 2-3 **(see illustration)**, so that one cylinder on compression fires simultaneously with its opposing cylinder on the exhaust stroke. Since the cylinder on the exhaust stroke requires very little of the available voltage to fire its plug, most of the voltage is used to fire the cylinder on the compression stroke.

3   The DIS system includes two coil packs, an ignition control module (ICM), a crankshaft reluctor ring, a magnetic crankshaft sensor and the PCM. The ignition module is located under the coil packs and is connected to the PCM.

4   The crankshaft sensor is located on the side of the engine block near the coil packs. The magnetic crankshaft sensor protrudes through the engine block, within about 0.050-inch of the crankshaft reluctor ring. The reluctor ring is a special disc cast into the crankshaft, which acts as a signal generator for the ignition timing.

5   The system uses control wires from the PCM, just like conventional distributor systems. The PCM controls timing using crankshaft position, engine rpm, engine temperature and manifold absolute pressure (MAP) sensing.

6   Ignition system - check

Warning: *Because of the very high voltage generated by the ignition system, extreme care should be taken whenever an operation is performed involving ignition components. This not only includes the coils, control module and spark plug wires (2.2L engine), but related items connected to the system as well, such as the plug connections, tachometer and any test equipment.*

1   If the engine turns over, but won't start, check for spark at the spark plug by installing a calibrated ignition system tester to one of the spark plug wire/boot terminals. Calibrated ignition testers are available at most auto parts stores in several different types. One type of spark tester has a short center electrode extending from the insulator and is intended for use on breaker-point and older electronic ignitions. The other type does not have a visible center electrode and is for use on high-voltage ignition systems. Be sure to use the high-voltage ignition tester on these models.

2.2L OHV engine

Refer to illustrations 6.2, 6.5, 6.6, 6.7a and 6.7b

2   Insert the ignition tester into the number one spark plug boot terminal. Connect the clip on the tester to a ground such as a metal bracket or valve cover bolt, crank the engine and watch the end of the tester for bright blue, well defined sparks **(see illustration)**.

3   If sparks occur, sufficient voltage is reaching the plugs to fire the engine. However the plugs themselves may be fouled, so remove and check them as described in Chapter 1 or replace them with new ones.

4   If no sparks or intermittent sparks occur, check for a bad spark plug wire by swapping wires or by using an ohmmeter to measure the resistance between the spark plug wire terminal ends. There should be approximately 5000 ohms of resistance per foot.

5   Check the primary wire connections at the ignition control module to make sure they are clean and tight. Refer to the wiring diagrams (see Chapter 12) and check for battery voltage to the ignition control module from the IGN-MOD fuse **(see illustration)**. Battery voltage should be available to the ignition control module with the ignition switch ON. **Note:** *If no voltage is available check the ignition system fuses (see Chapter 12). If the reading is 7 volts or less, repair the primary circuit from the ignition switch to the ignition control module.*

6   If voltage is available at the ignition control module and there's still no spark, check for a trigger signal from the ignition module. Remove the coil pack from the ignition module to expose the module terminals (see Section 7). Connect a test light between each of the module terminals and have an assistant crank the engine over **(see illustration)**. The test light should blink quickly and constantly as each coil pack is triggered to fire by the switching signal from the ignition module. This test checks for the trigger signal (ground) from the computer and Ignition Control Module. **Caution:** *Use only an LED test light to avoid damaging the PCM.*

5

6.7a Check the secondary coil resistance across the towers on each coil pack (2.2L engine)

6.7b Checking the coil primary resistance (2.2L engine)

6.12 To use a calibrated ignition tester on 2.3L and 2.4L engines it will necessary to flip the coil assembly over and bolt it to the engine while grounding the ignition tester – then fabricate three spark plug wires which connect from the remaining spark plug boot terminals on the coil assembly and to spark plugs not being tested

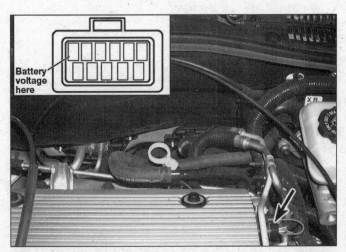

6.15 Checking for battery voltage to the ignition module from the ignition key (2.3L and 2.4L engines)

7   If a trigger signal is present at the coil, the computer and the Ignition Control Module are functioning properly and the problem lies in the ignition coils. Check the primary and secondary resistance of the ignition coils and compare it to the Specifications at the beginning of this Chapter **(see illustrations)**.

8   If the test light does not flash, the ignition module is most likely the problem but not always. Check the crankshaft and camshaft sensors for proper operation. **Note:** *Refer to Chapter 6 for additional information and testing procedures on the crankshaft and the camshaft sensors. It will be necessary to verify that the crankshaft and camshaft sensors are operating correctly before changing the ignition module. A defective ignition module can only be diagnosed by process of elimination.*

9   If the crankshaft and the camshaft sensors check out OK, have the PCM checked by a dealer service department or other qualified automotive repair shop.

2.3L and 2.4L OHC engines

Refer to illustrations 6.12, 6.15, 6.16a, 6.16b, 6.17a and 6.17b

Note: *This procedure is difficult to perform on 2.3L and 2.4L OHC engines, but it is possible with the correct set-up.*

10   Remove the ignition coil/module housing from the cylinder head (see Section 7). Turn the ignition coil/module assembly upside down and bolt it (module cover) to the engine block using the bolt hole in the

cover that the ground strap is attached too, then reconnect the ignition module connector. This will ground the assembly to the engine for testing purposes.

11   Insert the ignition tester into the number one spark plug terminal on the coil assembly, then connect a ground wire from the ignition tester to the engine block.

12   Construct three makeshift spark plug wires from a spare wire set that can be connected to the remaining spark plugs and to each of their corresponding spark plug terminals on the coil assembly **(see illustration)**.

13   Disable the fuel pump (see Chapter 4) and crank the engine over to check for spark at the ignition tester. Test each cylinder one-by-one making sure to ground the remaining spark plugs not being tested before cranking the engine over. Failure to ground the three spark plugs not being tested may result in ignition control module damage!

14   If sparks occur, sufficient voltage is reaching the plugs to fire the engine. However the plugs themselves may be fouled, so remove and check them as described in Chapter 1 or replace them with new ones.

15   If no sparks or intermittent sparks occur, check the primary wire connections at the ignition control module to make sure they are clean and tight. Refer to the wiring diagrams (see Chapter 12) and check for battery voltage to the ignition control module from the IGN-MOD fuse **(see illustration)**. Battery voltage should be available to the ignition control module with the ignition switch ON. **Note:** *If no voltage is available check the ignition system fuses (see Chapter 12). If the reading is 7 volts or less, repair the primary circuit from the ignition switch to the ignition control module.*

6.16a On 2.3L and 2.4L engines, first disconnect the ignition coil-to-ignition module connector and check for battery voltage to the coil with the ignition key On . . .

6.16b . . . then connect an LED test light to the positive terminal of the battery and to each of the coil negative (-) terminals on the ignition module and watch for a blinking light when the engine is cranked

6.17a To check the primary resistance of the ignition coil on 2.3L and 2.4L engines, connect the positive probe to the positive (+) terminal and the negative probe to each of the negative (-) terminals of the coil connector - The resistance should be the same for each check

6.17b Checking the secondary coil resistance (2.3L and 2.4L engines) – be sure the probes touch the contact in the center of each socket

7.4a Remove the bolts from the coil pack assembly

7.4b Some coil pack assemblies are mounted on the side of the engine block

5

16   If voltage is available at the ignition control module and there's still no spark, check for a trigger signal from the ignition module. Remove the coil housing cover and the ignition coil to ignition module electrical connector to expose the module output terminals (see Section 7). First check for battery voltage to the coils with the ignition key On, then connect a test light to the positive terminal of the battery and to each of the coil/module negative terminals and have an assistant crank the engine over **(see illustrations)**. The test light should blink quickly and constantly as each coil pack is triggered to fire by the switching signal from the ignition module. This test checks for the trigger signal (ground) from the computer and Ignition Control Module. **Caution:** *Use only an LED test light to avoid damaging the PCM*

17   If a trigger signal is present at the coil, the computer and the Ignition Control Module are functioning properly and the problem lies in the ignition coils or the wiring to the ignition coils. Check the primary and secondary resistance of the ignition coils and compare it to the Specifications at the beginning of this Chapter **(see illustrations)**. **Note:** *On 2.3L and 2.4L engines, primary resistance can be checked at the coil/ignition module connector or at the bottom of each coil. Always verify that the wires leading to the coils are not defective (have continuity) before replacing an ignition coil.*

18   If the test light does not flash, the ignition module is most likely the problem but not always. Check the crankshaft and camshaft sensors for proper operation. **Note:** *Refer to Chapter 6 for additional infor-*

mation and testing procedures on the crankshaft and the camshaft sensors. It will be necessary to verify that the crankshaft sensors and camshaft sensor is operating correctly before changing the ignition module. A defective ignition module can only be diagnosed by process of elimination.

19   If the crankshaft and the camshaft sensors check out OK, have the PCM checked by a dealer service department or other qualified automotive repair shop.

---

7   Ignition coil and module - replacement

2.2L OHV engine

Refer to illustrations 7.4a, 7.4b, 7.5a and 7.5b

1   Detach the cable from the negative terminal of the battery. **Caution:** *If the vehicle is equipped with a Delco Loc II or Theftlock audio system, make sure you have the correct activation code before disconnecting the battery.*

2   Unplug the electrical connectors from the module.

3   If the plug wires are not numbered, label them and detach the plug wires at the coil assembly.

4   Remove the module/coil assembly mounting bolts and lift the assembly from the vehicle **(see illustrations)**.

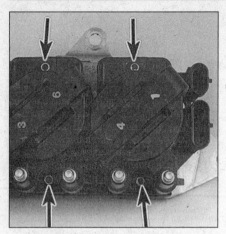

7.5a To detach the coil packs from the ignition module, remove the screws (arrows) . . .

7.5b . . . and pull straight up - when installing a coil pack, line up the blade terminals and press downward - make sure its fully seated before tightening the retaining screws

7.9 Ignition coil and module housing retaining bolts (arrows)

5   To remove the coils from the ignition module, simply remove the mounting screws and unplug the coil from the ignition module terminals **(see illustrations)**.

6   Installation is the reverse of removal. **Note:** *When installing the coils, make sure they are connected properly to the module and all plug wires are fully seated*

2.3L and 2.4L OHC engines

Refer to illustrations 7.9 and 7.10

7   Detach the cable from the negative terminal of the battery. **Caution:** *If the vehicle is equipped with a Delco Loc II or Theftlock audio system, make sure you have the correct activation code before disconnecting the battery.*

8   Unplug the electrical connectors from the module.

9   Remove the ignition coil/module housing retaining bolts **(see illustration)** and lift the housing assembly from the engine.

10   Detach the ignition coil cover screws and remove the coil cover from the housing assembly **(see illustration)**.

11   Detach the coil harness connector.

12   Remove the ignition coils.

13   Remove the ignition module-to-housing screws and detach the module from the housing.

14   Installation is the reverse of removal.

8   Charging system - general information and precautions

The charging system consists of a belt-driven alternator with an integral voltage regulator and the battery. These components work together to supply electrical power for the ignition system, the lights and all accessories.

All models are equipped with the CS type alternator. All CS models have special bolts or rivets instead of screws. CS alternators are rebuildable once the rivets are drilled out. However, we don't recommend this practice. For all intents and purposes, CS types should be considered non-serviceable and, if defective, exchanged as cores for new or rebuilt units.

The purpose of the voltage regulator is to limit the alternator's voltage to a preset value. This prevents power surges, circuit overloads, etc., during peak voltage output. On all models with which this manual is concerned, the voltage regulator is mounted inside the alternator housing.

The charging system doesn't ordinarily require periodic maintenance. However, the drivebelt, battery and wires and connections should be inspected at the intervals outlined in Chapter 1.

The dashboard warning light should come on when the ignition key is turned to START, then go off immediately after the engine has started. If it stays on or comes on when the engine is running, a charg-

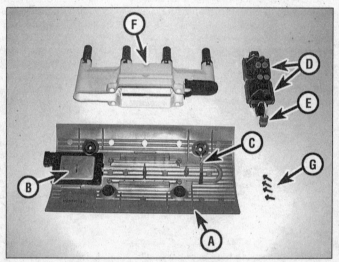

7.10 Exploded view of the ignition coil/module assembly on the 2.3L and 2.4L OHC engines

A	Ignition coil/module housing	E	Ignition coil connector
B	Ignition module	F	Ignition coil cover
C	Ground strap	G	Ignition coil cover
D	Ignition coils		mounting screws

ing system problem has occurred (see Section 9).

Be very careful when making electrical circuit connections to a vehicle equipped with an alternator and note the following:

a) *When reconnecting wires to the alternator from the battery, be sure to note the polarity.*

b) *Before using arc welding equipment to repair any part of the vehicle, disconnect the wires from the alternator and the battery terminals.* **Caution:** *If the vehicle is equipped with a Delco Loc II or Theftlock audio system, make sure you have the correct activation code before disconnecting the battery.*

c) *Never start the engine with a battery charger connected.*

d) *Always disconnect both battery leads before using a battery charger.*

e) *The alternator is turned by an engine drivebelt which could cause serious injury if your hands, hair or clothes become entangled in it with the engine running.*

f) *Because the alternator is connected directly to the battery, it could arc or cause a fire if overloaded or shorted out.*

g) *Wrap a plastic bag over the alternator and secure it with rubber bands before steam cleaning the engine.*

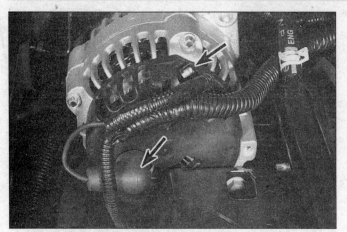

10.3 Disconnect the wires from the rear of the alternator

**10.4 Remove the mounting bolts (arrows) from the alternator
(2.2L engine shown)**

9 Charging system - check

1   If a malfunction occurs in the charging circuit, don't automatically assume the alternator is causing the problem. First check the following items:

   a) *Check the drivebelt tension and condition (Chapter 1). Replace it if it's worn or deteriorated.*

   b) *Make sure the alternator mounting and adjustment bolts are tight.*

   c) *Inspect the alternator wiring harness and the connectors at the alternator. They must be in good condition and tight.*

   d) *Check the fusible link (if equipped) located between the starter solenoid and alternator. If it's burned, determine the cause, repair the circuit and replace the link (the engine won't start and/or the accessories won't work if the fusible link blows). Sometimes a fusible link may look good, but still be bad. If in doubt, remove it and check it for continuity.*

   e) *Start the engine and check the alternator for abnormal noises (a shrieking or squealing sound indicates a bad bearing).*

   f) *Check the specific gravity of the battery electrolyte. If it's low, charge the battery (doesn't apply to maintenance-free batteries).*

   g) *Make sure the battery is fully charged (one bad cell in a battery can cause overcharging by the alternator).*

   h) *Disconnect the battery cables (negative first, then positive). Inspect the battery posts and the cable clamps for corrosion. Clean them thoroughly if necessary (see Chapter 1). Reconnect the cable to the positive terminal.* **Caution:** *If the vehicle is equipped with a Delco Loc II or Theftlock audio system, make sure you have the correct activation code before disconnecting the battery.*

   i) *With the key off, connect a test light between the negative battery post and the disconnected negative cable clamp.*

     1) *If the test light doesn't come on, reattach the clamp and proceed to the next Step.*

     2) *If the test light is brightly illuminated, there's a short (drain) in the electrical system of the vehicle. The short must be repaired before the charging system can be checked.* **Note:** *The test light will glow dimly because of the parasitic drain of the PCM, radio, clock, etc.*

     3) *Disconnect the alternator wiring harness.*

      (a) *If the light goes out, the alternator is bad.*

      (b) *If the light stays on, pull each fuse until the light goes out (this will tell you which component is shorted).*

2   Using a voltmeter, check the battery voltage with the engine off. It should be approximately 12-volts.

3   Start the engine and check the battery voltage again. It should now be as listed in this Chapter's Specifications.

4   Further testing of this type of alternator must be done by a service station, dealer service department or auto electric shop.

5   If the voltmeter indicates low battery voltage, the alternator is faulty and should be replaced with a new one or there is an open circuit between the alternator and the battery.

6   If the voltage reading is 15-volts or higher and a no charge condition exists, the regulator or field circuit is the problem. Remove the alternator (see Section 10) and have it checked by a service station, dealer service department or auto electric shop.

10 Alternator - removal and installation

Refer to illustrations 10.3 and 10.4

1   Detach the cable from the negative terminal of the battery. **Caution:** *If the vehicle is equipped with a Delco Loc II or Theftlock audio system, make sure you have the correct activation code before disconnecting the battery.*

2   Remove the drivebelt (see Chapter 1).

3   Label and detach the wires from the backside of the alternator **(see illustration).**

4   Remove the mounting bolts and separate the alternator from the engine **(see illustration).**

5   If you're replacing the alternator, take the old one with you when purchasing the new one. Make sure the new/rebuilt unit is identical to the old alternator. Look at the terminals - they should be the same in number, size and location as the terminals on the old alternator. Finally, look at the identification numbers - they'll be stamped into the housing or printed on a tag attached to the housing. Make sure the numbers are the same on both alternators.

6   Many new/rebuilt alternators DO NOT have a pulley installed, so you may have to switch the pulley from the old one to the new/rebuilt one.

7   Installation is the reverse of removal.

8   Check the charging voltage to verify proper operation of the alternator (see Section 9).

11 Starting system - general information and precautions

The sole function of the starting system is to turn over the engine quickly enough to allow it to start. The starting system consists of the battery, the starter motor, the starter solenoid and the wires connecting them. The solenoid is mounted directly on the starter motor. The solenoid/starter motor assembly is installed on the lower part of the engine, next to the transmission bellhousing.

When the ignition key is turned to the Start position, the starter solenoid is actuated through the starter control circuit. The starter solenoid then connects the battery to the starter. The battery supplies the electrical energy to the starter motor, which does the actual work of cranking the engine.

The starter motor on a vehicle equipped with a manual transaxle

5

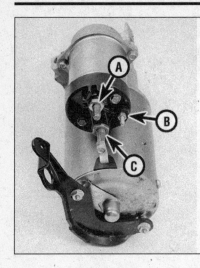

13.5 There are three terminals on the end of a typical starter solenoid

A   *Battery terminal*
B   *Switch terminal (S)*
C   *Motor terminal (M)*

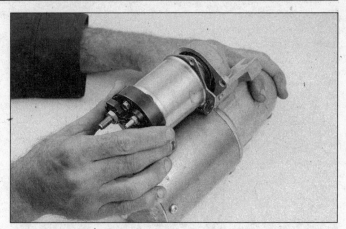

14.5 To remove the solenoid housing from the starter motor, remove the screws and turn it clockwise

can only be operated when the clutch pedal is depressed; the starter on a vehicle equipped with an automatic transaxle can only be operated when the selector lever is in Park or Neutral.

Always observe the following precautions when working on the starting system:

a) *Excessive cranking of the starter motor can overheat it and cause serious damage. Never operate the starter motor for more than 15 seconds at a time without pausing to allow it to cool for at least two minutes.*

b) *The starter is connected directly to the battery and could arc or cause a fire if mishandled, overloaded or shorted out.*

c) *Always detach the cable from the negative terminal of the battery before working on the starting system.*

12   Starter motor - testing in vehicle

Note: *Before diagnosing starter problems, make sure the battery is fully charged.*

1   If the starter motor doesn't turn at all when the switch is operated, make sure the shift lever is in Neutral or Park (automatic transaxle) or the clutch pedal is depressed (manual transaxle).

2   Make sure the battery is charged and all cables, both at the battery and starter solenoid terminals, are clean and secure.

3   If the starter motor spins but the engine isn't cranking, the overrunning clutch in the starter motor is slipping and the starter motor must be replaced.

4   If, when the switch is actuated, the starter motor doesn't operate at all but the solenoid clicks, then the problem lies with either the battery, the main solenoid contacts or the starter motor itself (or the engine is seized).

5   If the solenoid plunger can't be heard when the switch is actuated, the battery is bad, the fusible link is burned (the circuit is open) or the solenoid itself is defective.

6   To check the solenoid, connect a jumper lead between the battery (+) and the ignition switch wire terminal (the small terminal) on the solenoid. If the starter motor now operates, the solenoid is okay and the problem is in the ignition switch, neutral start switch or the wiring.

7   If the starter motor still doesn't operate, remove the starter/solenoid assembly for disassembly, testing and repair.

8   If the starter motor cranks the engine at an abnormally slow speed, first make sure the battery is charged and all terminal connections are clean and tight. If the engine is partially seized, or has the wrong viscosity oil in it, it may crank slowly as well.

9   Run the engine until normal operating temperature is reached, then disable the ignition system by removing the ignition fuse.

10   Connect a voltmeter positive lead to the positive battery post and connect the negative lead to the negative post.

11   Crank the engine and take the voltmeter readings as soon as a

steady figure is indicated. DO NOT allow the starter motor to turn for more than 15 seconds at a time. A reading of 9-volts or more, with the starter motor turning at normal cranking speed, is normal. If the reading is 9-volts or more but the cranking speed is slow, the motor is faulty. If the reading is less than 9-volts and the cranking speed is slow, the solenoid contacts are probably burned, the starter motor is bad, the battery is discharged or there's a bad connection.

13   Starter motor - removal and installation

Refer to illustration 13.5
Note: *On some vehicles, it may be necessary to remove the exhaust pipe(s) or frame crossmember to gain access to the starter motor. In extreme cases it may even be necessary to unbolt the mounts and raise the engine slightly to get the starter out.*

1   Detach the cable from the negative terminal of the battery. **Caution:** *If the vehicle is equipped with a Delco Loc II or Theftlock audio system, make sure you have the correct activation code before disconnecting the battery.*

2   Raise the front of the vehicle and support it securely on jackstands. Apply the parking brake and block the rear wheels to keep the vehicle from rolling off the jackstands.

3   On 2.2L engines with a manual transaxle, remove the brace between the engine and the transaxle. On 2.3L and 2.4L OHC engines, remove the air intake ducts to gain access to the top starter motor bolt located on the transaxle.

4   Remove the mounting bolts and detach the starter. Note the locations of the spacer shims (if used) - they must be reinstalled in the same positions.

5   Working under the vehicle, clearly label, then disconnect the wires from the terminals on the starter motor and solenoid **(see illustration)**.

6   Installation is the reverse of removal.

14   Starter solenoid - removal and installation

Refer to illustration 14.5
1   Detach the cable from the negative terminal of the battery. **Caution:** *If the vehicle is equipped with a Delco Loc II or Theftlock audio system, make sure you have the correct activation code before disconnecting the battery.*

2   Remove the starter motor (Section 13).

3   Disconnect the strap from the solenoid to the starter motor terminal.

4   Remove the screws that secure the solenoid to the starter motor.

5   Twist the solenoid in a clockwise direction to disengage the flange from the starter body **(see illustration)**.

6   Installation is the reverse of removal.

Chapter 6
Emissions and engine control systems

Contents

Specifications

Torque specification

Crankshaft sensor bolt .... 60 in-lbs

---

1 General information

Refer to illustrations 1.5

To prevent pollution of the atmosphere from burned and evaporating gases, a number of emissions control systems are incorporated on the vehicles covered by this manual. The combination of systems used depends on the year in which the vehicle was manufactured, the locality to which it was originally delivered and the engine type. The major systems incorporated on the vehicles with which this manual is concerned include the:

Fuel Control System
Exhaust Gas Recirculation (EGR) system
Evaporative Emissions Control (EVAP) system
Positive Crankcase Ventilation (PCV) system
Catalytic converter

All of these systems are linked, directly or indirectly, to the On Board Diagnostic (OBD) system. The Sections in this Chapter include general descriptions, checking procedures (where possible) and component replacement procedures (where applicable) for each of the systems listed above.

Before assuming that an emissions control system is malfunctioning, check the fuel and ignition systems carefully. In some cases special tools and equipment, as well as specialized training, are required to accurately diagnose the causes of a rough running or difficult to start engine. If checking and servicing become too difficult, or if a procedure is beyond the scope of the home mechanic, consult your dealer service department or other qualified repair shop. This does not necessarily mean, however, that the emissions control systems are particularly difficult to maintain and repair. You can quickly and easily perform many checks and do most (if not all) of the regular maintenance at home with common tune-up and hand tools. **Note:** *The most frequent cause of emissions system problems is simply a loose or broken vacuum hose or wiring connection. Therefore, always check the hose and wiring connections first.*

Pay close attention to any special precautions outlined in this Chapter. It should be noted that the illustrations of the various systems may not exactly match the system installed on your particular vehicle due to changes made by the manufacturer during production or from year to year.

A Vehicle Emissions Control Information (VECI) label is located in the engine compartment of all vehicles with which this manual is concerned **(see illustration)**. This label contains important emissions specifications and setting procedures, as well as a vacuum hose schematic with emissions components identified. When servicing the engine or emissions systems, the VECI label in your particular vehicle should always be checked for up-to-date information. **Note:** *Because of a federally mandated extended warranty which covers the emission control system components (and any components which have a primary purpose other than emission control but have significant effects on emissions), check with your dealer about warranty coverage before*

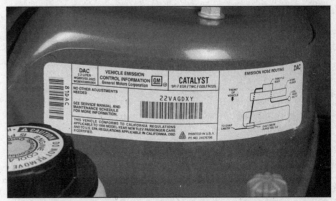

1.5 A Vehicle Emissions Control Information (VECI) label will be found in the engine compartment of all vehicles - if it's missing, obtain a new one from a dealer parts department

6

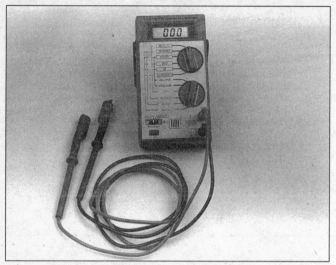

2.1 Digital multimeters can be used for testing all types of circuits; because of their high impedance, they are much more accurate than analog meters for measuring millivolts in low-voltage computer circuits

2.2 Scanners like the Actron Scantool and the AutoXray XP240 are powerful diagnostic aids - programmed with comprehensive diagnostic information, they can tell you just about anything you want to know about your engine management system, but they are expensive

2.4 Trouble code tools simplify the task of extracting the trouble codes

working on any emission related systems.

The number of emissions control system components on later model fuel-injected vehicles has actually decreased due to the high efficiency of the new fuel injection and ignition systems. These models are equipped with a three way catalytic converter containing beads which are coated with a catalyst material containing platinum, palladium and rhodium to reduce the level of nitrogen oxides.

2   On Board Diagnostic (OBD) system and trouble codes

Note: *The diagnostic system and trouble codes are only accessible using expensive, specialized equipment. The codes indicated in the text are designed and mandated by the EPA for all 1994 and later OBD II vehicles produced by automobile manufacturers. These generic trouble codes do not include the manufacturer's specific trouble codes. Consult a dealer service department or other qualified repair shop for additional information. General information on the system sensors and actuators for all models is described in the following text. See the Troubleshooting section at the beginning of this manual for some basic diagnostic aids. Because the OBD II system requires a special SCAN tool to access the trouble codes, have the vehicle diagnosed by a dealer service department or other qualified automotive repair facility if the proper SCAN tool is not available.*

Diagnostic tool information

Refer to illustrations 2.1, 2.2 and 2.4

1   A digital multimeter is necessary for checking fuel injection and emission related components **(see illustration)**. A digital volt-ohmmeter is preferred over the older style analog multimeter for several reasons. The analog multimeter cannot display the volts-ohms or amps measurement in hundredths and thousandths increments. When working with electronic circuits which are often very low voltage, this accurate reading is most important. Another good reason for the digital multimeter is the high impedance circuit. The digital multimeter is equipped with a high resistance internal circuitry (10 million ohms). Because a voltmeter is hooked up in parallel with the circuit when testing, it is vital that none of the voltage being measured should be allowed to travel the parallel path set up by the meter itself. This dilemma does not show itself when measuring larger amounts of voltage (9 to 12 volt circuits) but if you are measuring a low voltage circuit such as the oxygen sensor signal voltage, a fraction of a volt may be a significant amount when diagnosing a problem.

2   Hand-held scanners are the most powerful and versatile tools for analyzing engine management systems used on later model vehicles **(see illustration)**. Early model scanners handle codes and some diagnostics for many OBD I systems. Each brand scan tool must be examined carefully to match the year, make and model of the vehicle you are working on. Often interchangeable cartridges are available to access the particular manufacturer (Ford, GM, Chrysler, etc.). Some manufacturers will specify by continent (Asia, Europe, USA, etc.).

3   With the arrival of the Federally mandated emission control system (OBD II), a specially designed scanner must be used. At this time, several manufacturers plan to release OBD II scan tools for the home mechanic. Ask the parts salesman at a local auto parts store for additional information concerning dates and costs.

4   Another type of code reader and less expensive is available at parts stores **(see illustration)**. These tools simplify the procedure for extracting codes from the engine management computer by simply "plugging in" to the diagnostic connector on the vehicle wiring harness. **Note:** *Some diagnostic connectors are located under the dash, kick panel or glovebox while others are located in the engine compartment.*

OBD system general description

5   Beginning in 1994, General Motors Company began to manufacture a second generation self diagnosis system specified by the CARB and EPA regulations called On Board Diagnosis (OBD) II. This system incorporates a series of diagnostic monitors that detect and identify emissions systems faults and store the information in the computer

memory. This updated system also tests sensors and output actuators, diagnoses drive cycles, freezes data and clears codes. This powerful diagnostic computer must be accessed using the new OBD II SCAN tool and 16 pin Data Link Connector (DLC) located under the driver's dash area. All engines and powertrain combinations described in this manual are equipped with the On Board Diagnosis II (OBD II) system. This system consists of an onboard computer, known as the Powertrain Control Module (PCM), and information sensors, which monitor various functions of the engine and send data to the PCM. Based on the data and the information programmed into the computer's memory, the PCM generates output signals to control various engine functions via control relays, solenoids and other output actuators.

6   The PCM, located in the right, front corner of the engine compartment, is the "brain" of the EFI system. It receives data from a number of sensors and other electronic components (switches, relays, etc.). Based on the information it receives, the PCM generates output signals to control various relays, solenoids and other actuators. The PCM is specifically calibrated to optimize the emissions, fuel economy and driveability of the vehicle.

7   Because of a federally mandated extended warranty which covers the emissions control system components and because any owner-induced damage to the PCM, the sensors and/or the control devices may void the warranty, it isn't a good idea to attempt diagnosis or replacement of the PCM at home while the vehicle is under warranty. Take the vehicle to a dealer service department or other qualified repair shop if the PCM or a system component malfunctions.

Information sensors

8   When battery voltage is applied to the **air conditioning compressor solenoid**, a signal is sent to the PCM, which interprets the signal as an added load created by the compressor and increases engine idle speed accordingly to compensate.

9   The **Intake Air Temperature sensor (IAT)**, positioned in the air intake duct, provides the PCM with intake air temperature information. The PCM uses this information to control fuel flow, ignition timing and EGR system operation.

10   The **Engine Coolant Temperature (ECT) sensor**, which is threaded into a coolant passage in the thermostat housing (2.4L) or the engine block (2.2L), monitors engine coolant temperature. The ECT sends the PCM a voltage signal which influences PCM control of the fuel mixture, ignition timing and EGR operation.

11   The **Heated Exhaust Gas Oxygen (HEGO) sensors**, which are threaded into the exhaust system before and after the catalytic converter, constantly monitor the oxygen content of the exhaust gases. A voltage signal which varies in accordance with the difference between the oxygen content of the exhaust gases and the surrounding atmosphere is sent to the PCM. The PCM converts this exhaust gas oxygen content signal to the fuel/air ratio, compares it to the ideal ratio for current engine operating conditions and alters the signal to the injectors accordingly. **Note:** *Early models are equipped with single wire O2 sensors.*

12   The **Throttle Position Sensor (TPS)**, which is mounted on the side of the throttle body and connected directly to the throttle shaft, senses throttle movement and position, then transmits an electrical signal to the PCM. This signal enables the PCM to determine when the throttle is closed, in its normal cruise condition or wide open.

13   The **Manifold Absolute Pressure (MAP) sensor**, which is mounted on the throttle body, measures the amount and pressure of air entering the engine. The MAP sensor also measure changes in altitude as related to the volume and pressure changes for purposes of accuracy at low and high altitudes.

Output actuators

14   The **fuel pump relay**, which is activated by the ignition switch, supplies battery voltage to the fuel pump in the Start position. Early models are equipped with a fail-safe oil pressure circuit wired into the fuel pump to shut the engine down in the event of oil pressure loss. **Note:** *The fuel pump relay is located in the Relay Distribution Box in the*

engine compartment. Refer to Chapter 12 or your owner's manual for additional information for relay location.

15   The **purge control valve** switches manifold vacuum to operate the purge system. This system allows fuel vapor to flow from the canister to the intake manifold to be burned in the combustion process.

16   The **solenoid-operated fuel injectors** are located above the intake ports (see Chapter 4). The PCM controls the length of time the injector is open. The "open" time of the injector determines the amount of fuel delivered. For information regarding injector replacement, refer to Chapter 4.

17   The **A/C relay** is activated by the PCM with the ignition switch in the On position and the A/C selected. This allows the PCM to monitor engine rpm and air conditioning system pressure changes and shut the system down in the event of overcharge or excessive engine speed. Refer to Chapter 3.

18   The **ignition module** (see Chapter 5) is mounted external to the PCM and the coil packs for the Electronic Integrated (distributorless) ignition system. For further information regarding the ignition system, refer to the appropriate Section in Chapter 5.

Obtaining OBD II system codes

Refer to illustration 2.20
Note: *1995 2.2L and 2.3L engines are equipped with the OBD II 16-pin connector under the dash but the codes are not the updated type for the OBD II systems. These trouble codes are two digit codes similar to many GM OBD I codes from previous years. This will be important when accessing the On Board Diagnostic system using an aftermarket SCAN tool. If the correct tool is not available, have the system diagnosed by a dealer service department or other qualified repair shop.*

19   On OBD II systems, the PCM will illuminate the Malfunction Indicator Light on the dash if it recognizes a component fault for two consecutive drive cycles. It will continue to set the light until the PCM does not detect any malfunction for three or more consecutive drive cycles. Because the OBD II system requires a SCAN tool to reset the light, if the tool is not available for diagnostics, have the system checked by a dealer service department or other qualified repair facility.

20   The diagnostic codes for the OBD II systems can be extracted from the PCM using a special SCAN tool that is programmed to interface with this new system by plugging into the DLC **(see illustration)**. If the tool is not available, have the vehicle checked at a dealer service department or other qualified repair shop.

Clearing codes

21   To clear the codes from the PCM memory, install the OBD II SCAN tool, scroll the menu for the function that describes "CLEARING CODES" and follow the prescribed method for that particular SCAN tool. If necessary, have the codes cleared by a dealer service department or other qualified repair facility. **Caution:** *Do not disconnect the battery from the vehicle to clear the codes. This will erase stored operating parameters from the KAM (Keep Alive Memory) and cause the engine to run rough for a period of time while the computer relearns the information.*

6

2.20 The 16 pin DLC terminal on a 1996 2.2L engine with OBD II

1995 Trouble Codes

Trouble Code	Circuit or system	Probable cause
13	Oxygen sensor circuit	Check the wiring and connectors from the oxygen sensor. Replace oxygen sensor (see Section 5).*
14	Coolant sensor circuit (high temperature indicated)	If the engine is experiencing overheating problems, the problem must be rectified before continuing (see Chapters 1 and 3). Check all wiring and connectors associated with the sensor. Replace the coolant sensor (see Section 5).*
15	Coolant sensor circuit (low temperature indicated)	See above. Also, check the thermostat for proper operation.
19	7X reference signal circuit	The crankshaft position sync pulse is malfunctioning. Have the ignition system checked by a dealer service department or other qualified repair shop.
21	TPS circuit (signal voltage high)	Check for sticking or misadjusted TPS. Check all wiring and connections at the TPS and at the PCM. Replace TPS* (see Section 5).
22	TPS circuit (signal voltage low)	See above.
23	Intake Air Temperature (IAT) sensor	Low temperature indicated (see Sections 4 and 5).
24	Vehicle Speed sensor (VSS)	A fault in this circuit should be indicated only while the vehicle is in motion. Disregard code 24 if set when drive wheels are not turning. Check connections at the PCM. Check the TPS setting (see Section 4).
25	Intake Air Temperature (IAT) sensor*	Check the resistance of the IAT sensor. Check the wiring and connections to the sensor. Replace the (IAT) sensor.
27	Quad driver module	Quad drivers may be faulty. Have the PCM checked by a dealer service department or other qualified repair shop.
28	Quad driver module circuit	Quad driver circuits may be faulty. Have the PCM checked by a dealer service department.
31	PRNDL error (2.2L engine)	PCM recognizes an invalid or incorrect PRNDL parameter. Check the Park/Neutral switch (see Chapter 7) adjustment.
32	Exhaust Gas Recirculation (EGR) failure	Check the vacuum source and all vacuum lines. Check the system electrical connectors at the PCM and EGR valve. Replace the EGR valve or PCM as necessary (see Section 6).*
33	Manifold Absolute Pressure (MAP) signal voltage high	Check vacuum hose(s) from MAP sensor. Check electrical sensor or circuit connections at the PCM. Replace MAP sensor (see Section 5).*
34	Manifold Absolute Pressure (MAP) signal voltage low	Check vacuum hose(s) from MAP sensor. Check electrical sensor or circuit connections at the PCM. Replace MAP sensor (see Section 5).*
43	Knock Sensor (KS) circuit	Check the PCM for an open or short to ground; if necessary, reroute the harness away from other wires such as spark plugs, etc. Replace the knock sensor (see Section 5).*
44	Lean exhaust	Check the wiring and connectors from the oxygen sensor to the PCM. Check the PCM ground terminal. Check the fuel pressure (Chapter 4). Replace the oxygen sensor (see Section 5).*
45	Rich exhaust	Check the evaporative charcoal canister and its components for the presence of fuel. Check for fuel or contaminated oil. Check the fuel pressure regulator. Check for a leaking fuel injector. Check for a sticking EGR valve. Replace the oxygen sensor (see Section 5).*
51	PROM (MEMCAL)	Faulty or incorrect PROM (MEMCAL). Diagnosis should be performed by a dealer service department or other qualified repair shop.
53	System voltage high	Code 53 will set if the voltage at the PCM is greater than 17.1-volts. Check the charging system (see Chapter 5).
55	Fuel system lean	Fuel system lean under heavy load or acceleration. Check the fuel filter and the fuel pressure regulator (see Chapter 4).

Trouble Code	Circuit or system	Probable cause
66	A/C refrigerant pressure sensor circuit	Short in circuit indicates the A/C pressure is below -8 psi or above 448 psi See Chapter 3.
72	Loss of serial data	Have the PCM checked by a dealer service department or other qualified repair shop.

Component replacement may not cure the problem in all cases. For this reason, you may want to seek professional advice before purchasing replacement parts.

OBD II Trouble Codes

Code	Code Definition	Location
P0105	Manifold Absolute Pressure (MAP) and or Throttle Position Sensor (TPS) performance/range fault	See Section 4
P0106	Manifold Absolute Pressure (MAP) system performance fault	See Section 4
P0107	Manifold Absolute Pressure (MAP) sensor circuit low input	See Section 4
P0108	Manifold Absolute Pressure (MAP) sensor circuit high input	See Section 4
P0112	Intake Air Temperature (IAT) sensor circuit low input	See Section 4
P0113	Intake Air Temperature (IAT) sensor circuit high input	See Section 4
P0117	Electronic Coolant Temperature (ECT) sensor circuit low input	See Section 4
P0118	Electronic Coolant Temperature (ECT) sensor circuit high input	See Section 4
P0121	Throttle Position Sensor (TPS) range/performance fault	See Section 4
P0122	Throttle Position Sensor (TPS) circuit low input	See Section 4
P0123	Throttle Position Sensor (TPS) circuit high input	See Section 4
P0125	Engine Coolant Temperature (ECT) excessive time to closed loop	See Section 4
P0131	Upstream heated O_2 sensor circuit low voltage (Bank 1, Sensor 1)	See Section 4
P0132	Oxygen sensor (O2S) circuit high voltage (Sensor 1)	See Section 4
P0133	Upstream heated O_2 sensor circuit high voltage (Bank 1, Sensor 1)	See Section 4
P0134	Oxygen sensor (O2S) circuit insufficient activity (Sensor 1)	See Section 4
P0137	Downstream heated O_2 sensor circuit low voltage (Bank 1, Sensor 2)	See Section 4
P0138	Downstream heated O_2 sensor circuit high voltage (Bank 1, Sensor 2)	See Section 4
P0140	Oxygen sensor (O2S) circuit insufficient activity (Sensor 1)	See Section 4
P0141	O_2 sensor heater circuit fault (Bank 1, Sensor 2)	See Section 4
P0171	System Adaptive fuel too lean	See Chapter 4
P0172	System Adaptive fuel too rich	See Chapter 4
P0191	Injector Pressure sensor system performance	See Chapter 4
P0192	Injector Pressure sensor circuit low input	See Chapter 4
P0193	Injector Pressure sensor circuit high input	See Chapter 4
P0200	Fuel injector control circuit fault	See Chapter 4
P0300	Engine misfire detected	See Chapter 5
P0301	Cylinder number 1 misfire detected	See Chapter 5
P0302	Cylinder number 2 misfire detected	See Chapter 5
P0303	Cylinder number 3 misfire detected	See Chapter 5
P0304	Cylinder number 4 misfire detected	See Chapter 5
P0325	Knock sensor circuit 1 fault	See Section 4
P0326	Knock sensor circuit performance	See Section 4
P0335	Crankshaft Position (CKP) sensor circuit fault	See Section 4
P0341	Camshaft Position (CMP) sensor performance fault	See Section 4
P0342	Camshaft Position (CMP) sensor circuit low input	See Section 4
P0351	COP ignition coil 1 primary circuit fault	See Chapter 5
P0352	COP ignition coil 2 primary circuit fault	See Chapter 5
P0353	COP ignition coil 3 primary circuit fault	See Chapter 5
P0354	COP ignition coil 4 primary circuit fault	See Chapter 5
P0400	EGR flow fault	See Section 6
P0401	EGR insufficient flow detected	See Section 6
P0402	EGR excessive flow detected	See Section 6
P0420	Catalyst system efficiency below threshold (Bank 1)	See Section 9
P0421	Catalyst system efficiency below threshold (Bank 1)	See Section 9
P0430	Catalyst system efficiency below threshold (Bank 2)	See Section 9
P0431	Catalyst system efficiency below threshold (Bank 2)	See Section 9

6

OBD II Trouble Codes (continued)

Code	Code Definition	Location
P0440	Evaporative emissions (EVAP) system fault	See Section 7
P0441	EVAP incorrect purge flow	See Section 7
P0442	Evaporative emissions (EVAP) control small leak detected	See Section 7
P0443	EVAP VMV circuit fault	See Section 7
P0446	Evaporative emissions (EVAP) system performance fault	See Section 7
P0452	EVAP fuel tank pressure sensor low input	See Section 4
P0453	EVAP fuel tank pressure sensor high input	See Section 4
P0461	Fuel level sensor performance fault	See Chapter 4
P0480	Cooling fan relay control circuit fault	See Chapter 3
P0502	VSS circuit low input	See Section 4
P0503	VSS circuit range performance	See Section 4
P0506	IAC system rpm lower than expected	See Chapter 4
P0507	IAC system rpm higher than expected	See Chapter 4
P0530	Air Conditioning (A/C) pressure sensor circuit fault	See Chapter 3
P0562	Ignition system voltage low	See Chapter 5
P0563	Ignition system voltage high	See Chapter 5
P0601	Powertrain Control Module (PCM) read only memory fault	See Section 3
P0602	PCM control module programming error	See Section 4
P0705	Transaxle Range sensor circuit malfunction	See Section 4
P1171	Inadequate fuel flow	See Chapter 4

3 Powertrain Control Module (PCM) - check and replacement

Note: *These models are equipped with an (non-replaceable) EEPROM that must be recalibrated with a special factory SCAN tool (TECH 1) after replacement of the PCM. The PCM is located in the engine compartment near the right front fender.*

Check

Refer to illustration 3.1

1   The PCM is located in the right corner of the engine compartment, under the air cleaner assembly and bolted to the chassis **(see illustration)**.

2   Using the tips of your fingers, tap vigorously on the side of the computer while the engine is running. If the computer is not functioning properly, the engine may stumble or stall and display glitches on the engine data stream obtained using a SCAN tool or other diagnostic equipment.

3   If the PCM fails this test, check the electrical connectors. Each connector is color coded to fit the respective slot in the computer body. If their are no obvious signs of damage, have the unit checked at a dealer service department or other qualified repair shop.

Replacement

Refer to illustration 3.9a, 3.9b and 3.10

Caution: *To prevent damage to the PCM, the ignition switch must be turned Off when disconnecting or connecting the PCM connectors.*

4   The PCM is located in the right corner of the engine compartment under the air cleaner assembly.

5   Disconnect the cable from the negative battery terminal. **Caution:** *On models equipped with a Delco-Loc II or Theftlock audio system, be sure the lockout feature is turned off before disconnecting the battery cable.*

6   Remove the splash shield from the right side of the engine compartment below the fender area (see Chapter 11).

7   Remove the horn electrical connector and the horn assembly from the chassis (see Chapter 11).

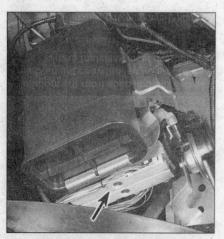

3.1 The PCM (arrow) is located in the right front corner of the engine compartment (fender removed for clarity)

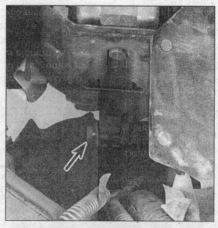

3.9a Remove the bolt (arrow) from the bracket housing to release the PCM (view from top of engine compartment - headlight removed for clarity)

3.9b Remove the bottom retaining bolt (arrow) to release the lower section of the PCM

3.10 Lower the PCM out from the bottom of the fenderwell

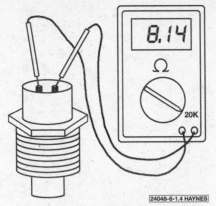

4.1a Be sure the meter probes make clean contact with the terminals of the sensor when checking the resistance (thermister type sensor shown)

Temperature (degrees-F)	Resistance (ohms)
212	176
194	240
176	332
158	458
140	668
122	972
112	1182
104	1458
95	1800
86	2238
76	2795
68	3520
58	4450
50	5670
40	7280
32	9420

4.1b Coolant temperature and intake air temperature sensors approximate temperature vs. resistance relationships

8   Unplug the electrical connectors from the PCM. Each connector is color coded to fit its respective receptacle in the PCM.
9   Remove the bolts from the PCM retainer brackets **(see illustrations)**.
10   Carefully lift the PCM out the bottom of the engine compartment without damaging the electrical connectors and wiring harness to the computer **(see illustration)**.
11   Installation is the reverse of removal.

EEPROM reprogramming

12   These models are equipped with Electrical Erasable Programmable Read Only Memory (EEPROM) chip that is permanently soldered to the PCM circuit board. The EEPROM can be reprogrammed using the GM TECH 1 SCAN tool. Do not attempt to remove this component from the PCM. Have the EEPROM reprogrammed at a dealer service department or other qualified repair shop.

4   Information sensors - general information and testing

Caution 1: *When performing the following tests, use only a high-impedance (10 mega-ohms) digital multi-meter to prevent damage to the PCM.*
Caution 2: *On models equipped with a Delco-Loc II or Theftlock audio system, be sure the lockout feature is turned off before disconnecting the battery cable.*

Thermister (two-wire) sensors (coolant temperature, intake air temperature, air charge sensors, etc.)

Refer to illustrations 4.1a and 4.1b
1   Thermisters are variable resistors that sense temperature level changes and convert them to a voltage signal. The coolant temperature sensor (ECT) and the intake air temperature sensor (IAT) are thermister type sensors. Select the ohms range with the rotary switch, disconnect the harness connector at the sensor and connect the test probes to the sensor terminals **(see illustration)**. The resistance reading (ohms) should should be high when the sensor is cold and low when the sensor is hot **(see illustration)**. Be sure the tips of the probe make clean contact with the terminals inside the sensor to ensure an accurate reading.
2   After the resistance of the sensor has been checked, test the system for the proper reference voltage from the computer. Simply disconnect the harness connector at the sensor, select the voltage range on the rotary switch on the meter and probe the correct terminals on the harness for the voltage signal. Reference voltage should be

approximately 5 volts. The ignition switch must be in the ON position (engine not running). If there is no reference voltage available to the sensor, then the circuit and the computer must be checked.

Potentiometers (three-wire) sensors (TPS, EGR position sensors, etc.)

Refer to illustration 4.3
3   The potentiometer is a variable resistor that converts the voltage signal by varying the resistance according to the driving situations. The signal the potentiometer generates is used by the computer to determine position and direction of the device within the component. Although it is possible to measure resistance changes of the sensor, it is best to measure voltage changes as the end result of the potentiometer's performance. Select the DC volts function on the multimeter, carefully backprobe the harness connector using pins inserted into the correct terminals and connect the meter probes to the pins. Connect the negative probe (-) to the ground terminal and the positive probe to the SIGNAL terminal. It will be necessary to refer to the wiring diagrams at the end of this manual for the correct terminals. Observe

6

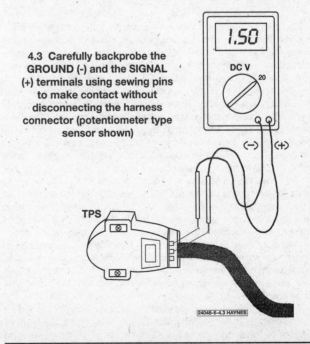

4.3 Carefully backprobe the GROUND (-) and the SIGNAL (+) terminals using sewing pins to make contact without disconnecting the harness connector (potentiometer type sensor shown)

VOLTAGE RANGE	ALTITUDE
3.8 - 5.5V	Below 1,000
3.6 - 5.3V	1,000 - 2,000
3.5 - 5.1V	2,000 - 3,000
3.3 - 5.0V	3,000 - 4,000
3.2 - 4.8V	4,000 - 5,000
3.0 - 4.6V	5,000 - 6,000
2.9 - 4.5V	6,000 - 7,000
2.8 - 4.3V	7,000 - 8,000
2.6 - 4.2V	8,000 - 9,000
2.5 - 4.0V	9,000 - 10,000 FEET

LOW ALTITUDE = HIGH PRESSURE = HIGH VOLTAGE

24071-6-4.09 HAYNES

4.5a Typical Manifold Absolute Pressure (MAP) sensor altitude (pressure) vs. voltage values

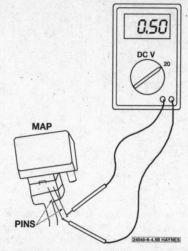

4.5b It will be necessary to run the engine to check for voltage fluctuations in order to create pressure and vacuum changes within the system (pressure/vacuum sensor shown)

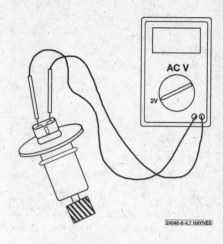

4.7 Connect the probes of the voltmeter directly to the VSS and observe A/C voltage fluctuations as the drive gear is slowly rotated

the meter as the signal arm is moved through its complete range (sweep). The voltage should vary as the arm is moved through its range. These sensors are often used as the throttle position sensor, the EGR valve position sensor or VAF sensor in the intake duct. Most of these type sensors will vary from as little as 0.2 volts, up to 5 volts (closed to open) **(see illustration)**.

4   After the SIGNAL voltage has been checked, test the system for the proper reference voltage from the computer. Simply disconnect the harness connector at the sensor, select the voltage range on the multimeter and probe the correct terminals on the harness for the voltage signal. Reference voltage should be approximately 5 volts. The ignition switch must be in the ON position (engine not running). If there is no reference voltage available to the sensor, then the circuit and the computer must be checked.

Pressure/vacuum (three-wire) sensors (MAP, BARO, vacuum sensors, etc.)

Refer to illustrations 4.5a and 4.5b

5   The pressure/vacuum sensors monitor the intake manifold pressure changes resulting from changes in engine load and speed and converts the information into a voltage output. The PCM uses the sensor to control fuel delivery and ignition timing. The sensor SIGNAL voltage to the PCM varies from below 2 volts at idle (high vacuum) to above 4 volts with the ignition key ON (engine not running) or wide open throttle (WOT) (low vacuum). These values correspond with the altitude and pressure changes the vehicle experiences while driving **(see illustration)**. Select the volts function on the meter, carefully backprobe the harness connector terminals using straight pins and connect the voltmeter probes to the pins **(see illustration)**. Connect the negative probe (-) to the ground terminal and the positive probe to the SIGNAL terminal. It will be necessary to refer to the manufacturer's schematic or wiring diagram for the correct terminals. Observe the meter as the engine idles and then slowly raise the engine rpm to wide open throttle. The voltage should increase from approximately 0.5 to 1.5 volts (high vacuum) to 5.0 volts (low vacuum). If the engine stalls or runs roughly, it is possible to simulate conditions by attaching a hand-held vacuum pump to the pressure/vacuum sensor to simulate running conditions. Some manufacturers equip the fuel injection system with a voltage varying sensor while others use a frequency varying sensor. The latter type must be checked using the frequency (Hz) scale on the meter.

6   If SIGNAL voltage does not exist, test the system for the proper reference voltage from the computer. Simply disconnect the harness connector at the sensor, select the voltage range on the meter and probe the correct terminals on the harness for the voltage signal. Reference voltage should be approximately 5 volts. The ignition switch must be in the ON position (engine not running). If there is no reference voltage available to the sensor, then the circuit and the computer must be checked.

Magnetic reluctance sensors (crankshaft, vehicle speed, camshaft sensors, etc.)

Refer to illustration 4.7

7   Magnetic reluctance type sensors consist of a permanent magnet with a coil wire wound around the assembly. Crankshaft sensors and vehicle speed sensors are common applications of this type sensor. A steel disk mounted on a gear (crankshaft, input shaft, etc.) has tabs that pass between the pole pieces of the magnet causing a break in the magnetic field when passed near the sensor. This break in the field causes a magnetic flux producing reluctance (resistance) thereby, changing the voltage signal. This voltage signal is used to determine crankshaft position, vehicle speed, etc. Because magnetic energy is a free-standing source of energy and does not require battery power to produce voltage, this type of sensor must be checked by observing voltage fluctuations with an AC volt meter. Simply switch the voltmeter to the AC scale and connect the probes to the sensor. On vehicle speed sensors it will be necessary to place the transmission in neutral, then with the help of an assistant, hold one tire still while spinning the other tire (approximately 2 MPH or faster) and, observe the voltage fluctuations. This test can also be performed with the sensor removed from the vehicle, by turning the sensor's drive gear **(see illustration)**. On crankshaft position sensors it will be necessary to have an assistant crank the engine over in short bursts at the ignition key while you observe the voltage fluctuations. The meter should register slight voltage fluctuations that are constant and relatively the same range. These small voltage fluctuations indicate that the magnetic portion of the sensor is producing a magnetic field and "sensing" engine parameters for the computer.

Hall effect (three-wire) sensors (crankshaft, camshaft, etc.)

Refer to illustration 4.9

8   These sensors consist of a single hall effect switch, and a magnet which are separated by an air gap. The hall effect switch completes the ground circuit when a magnetic field is present. As the engine rotates, these sensors will produce ON OFF signals to the PCM to determine crankshaft and camshaft position and speed. Hall effect crankshaft

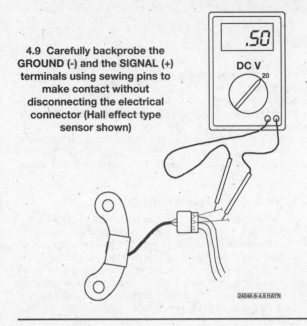

4.9 Carefully backprobe the GROUND (-) and the SIGNAL (+) terminals using sewing pins to make contact without disconnecting the electrical connector (Hall effect type sensor shown)

sensors generally use an interrupter ring located on the back of the harmonic balancer to disrupt the ground signal from the sensor. Always be sure to check the interrupter ring for damage while diagnosing a faulty crankshaft position sensor. Hall effect camshaft sensors typically use a rotating magnet on the camshaft sprocket or in the sensor housing to complete the hall effect switch ground signal. The PCM uses the sensors to control fuel delivery and ignition timing. The camshaft sensor typically produces only one ON OFF signal per camshaft revolution while the crankshaft sensor produces as many as thirty ON OFF signals per crankshaft revolution depending on the make and manufacturer of the vehicle.

9   To check Hall effect type sensors, select the DC volts function on the multi-meter, carefully backprobe the harness connector terminals using pins and connect the voltmeter probes to the pins **(see illustration)**. Connect the negative probe (-) to the ground connection and the positive probe to the SIGNAL terminal. It will be necessary to refer to the wiring diagrams at the back of this manual for the correct terminals. With the ignition key in the ON position and the engine OFF, rotate the engine crankshaft by hand (one full turn) and observe the voltmeter. If the sensor is operating properly it should trigger ON OFF voltage readings on the voltmeter while the engine is being rotated. Depending on the make and model of the vehicle, there should be 5 to 12 volts of signal voltage. **Note:** *When testing camshaft sensors it may be necessary to have an assistant crank the engine over in short bursts at the ignition key (remove the primary [low voltage] wires from the ignition module to prevent the engine from starting) while observing the voltmeter. It could take as many as two complete crankshaft revolutions before a camshaft sensor signal is detected.*

10   If SIGNAL voltage does not exist, test the system for the proper reference voltage from the computer. Simply disconnect the harness connector at the sensor, select the voltage range on the rotary switch on the volt/ohmmeter and probe the correct terminals on the harness for the reference signal. Depending on the make and model of the vehicle there should be 5 to 12 volts of reference voltage (5 volts is the most common). The ignition switch must be in the ON position (engine not running). If there is no reference voltage available to the sensor, the circuit and computer must be checked.

Dual Hall effect (four-wire) sensors (crankshaft)

11   Dual Hall effect sensors consist of two hall-effect switches and a shared magnet mounted between them. The magnet and each Hall-effect switch are separated by an air gap. Each hall effect switch completes the ground circuit when a magnetic field is present. Dual hall effect sensors are often used as a crankshaft sensor in place of two

single hall effect switches described above. Typically the harmonic balancer is equipped with two "interrupter rings" (each having a different number of interrupter blades) which create two separate ON OFF signal patterns that can be used for the timing and the ignition sequence.

12   To check dual hall effect type sensors, select the DC volts function with the rotary switch, carefully backprobe the harness connector terminals using pins and position the voltmeter probes onto the pins **(see illustration 4.9)**. Connect the negative probe (-) to the ground connection and the positive probe to one SIGNAL terminal. It will be necessary to refer to the manufacturer's schematic or wiring diagram for the correct terminals. With the ignition key in the ON position and the engine OFF, rotate the engine crankshaft by hand (one full turn) and observe the volt meter. Next, connect the positive probe to the second SIGNAL terminal. If the sensor is operating properly it should trigger ON OFF voltage readings on the voltmeter from both SIGNAL terminals while the engine is being rotated. Depending on the make and model of the vehicle, there should be 5 to 12 volts of signal voltage at each terminal.

13   If SIGNAL voltage does not exist, test the system for the proper reference voltage from the computer. Simply disconnect the harness connector at the sensor, select the voltage range on the rotary switch on the volt/ohmmeter and probe the correct terminals on the harness for the reference signal. Depending on the make and model of the vehicle there should be 5 to 12 volts of reference voltage (5 volts is the most common). The ignition switch must be in the ON position (engine not running). If there is no reference voltage available to the sensor, the circuit and computer must be checked.

Oxygen (O₂) sensors

14   The oxygen sensor(s) monitors the oxygen content of the exhaust gas stream. The oxygen content in the exhaust reacts with the oxygen sensor to produce a voltage output which varies from 0.1-volt (high oxygen, lean mixture) to 0.9-volts (low oxygen, rich mixture). The PCM constantly monitors this variable voltage output to determine the ratio of oxygen to fuel in the mixture. The PCM alters the air/fuel mixture ratio by controlling the pulse width (open time) of the fuel injectors. The PCM and the oxygen sensor(s) attempt to maintain a mixture ratio of 14.7 parts air to 1 part of fuel at all times. The oxygen sensor produces no voltage when it is below its normal operating temperature of about 600-degrees F. During this initial period before warm-up, the PCM operates in OPEN LOOP mode. When checking the oxygen sensor system, it will be necessary to test all oxygen sensors. **Note:** *Because the oxygen sensor(s) are difficult to access, probing the harness electrical connectors for testing purposes will require patience. The exhaust manifolds and pipes are extremely hot and will melt stray electrical leads that touch the surface during testing. If possible, use a SCAN tool that plugs into the DLC (diagnostic link). This tool will access the PCM data stream and indicates the millivolt changes for each individual oxygen sensor.*

15   Check the oxygen sensor millivolt signal. Locate the oxygen sensor electrical connector and carefully backprobe it using a long pin(s) into the appropriate wire terminals. In most models, connect the positive probe (+) of a voltmeter onto the SIGNAL wire and the negative probe (-) to the ground wire. Consult the wiring diagrams at the end of Chapter 12 for additional information on the oxygen sensor electrical connector wire color designations. **Note:** *Downstream oxygen sensors will produce much slower fluctuating voltage values to reflect the results of the catalyzed exhaust mixture from rich or lean to less presence of CO, HC and Nox molecules. Here the CO_2 and H_2O gaseous forms do not register or react with the oxygen sensors to such a large degree.* Monitor the SIGNAL voltage (millivolts) as the engine goes from cold to warm.

16   The oxygen sensor will produce a steady voltage signal of approximately 0.1 to 0.2 volts (100 to 200 millivolts) with the engine cold (open loop). After a period of approximately two minutes, the engine will reach operating temperature and the oxygen sensor will start to fluctuate between 0.1 to 0.9 volts (100 to 900 millivolts) (closed loop). If the oxygen sensor fails to reach the closed loop mode or there is a very long period of time until it does switch into closed loop mode,

6

replace the oxygen sensor with a new part. **Note:** *Downstream oxygen sensors will not change voltage values as quickly as upstream oxygen sensors. Because the downstream oxygen sensors detect oxygen content after the exhaust has been catalyzed, voltage values should fluctuate much slower and deliberate.*

17   Also inspect the oxygen sensor heater. Disconnect the oxygen sensor electrical connector and working on the oxygen sensor side, connect an ohmmeter between the black wire (-) and brown wire (+). It should measure approximately 5 to 7 ohms. **Note:** *The wire colors often change from the harness connector to the oxygen sensor connector wires according to manufacturer's specifications. Follow the wire colors to the oxygen sensor electrical connector and determine the matching wires and their colors before testing the heater resistance.*

18   Check for proper supply voltage to the heater. Disconnect the oxygen sensor electrical connector and working on the engine side of the harness, measure the voltage between the black wire (-) and brown wire (+) on the oxygen sensor electrical connector. There should be battery voltage with the ignition key ON (engine not running). If there is no voltage, check the circuit between the main relay, the fuse and the sensor. **Note:** *It is important to remember that supply voltage will only reach the O_2 sensor with the ignition key ON (engine not running).*

19   If the oxygen sensor fails any of these tests, replace it with a new part.

Knock sensors

20   Knock sensors detect abnormal vibration in the engine. The knock control system is designed to reduce spark knock during periods of heavy detonation. This allows the engine to use maximum spark advance to improve driveability. Knock sensors produce AC output voltage which increases with the severity of the knock. The signal is fed into the PCM and the timing is retarded to compensate for the severe detonation.

21   To check a knock sensor, disconnect the electrical connector and drain the engine coolant as described in Chapter 1. Remove the sensor from the engine block. **Note:** *Most knock sensors have a 1/4 inch pipe thread, use a pipe plug of the same size to thread into the engine block as the sensor is removed. This will relieve the task of draining the coolant with minimal coolant loss.* These type of sensors must be checked by observing voltage fluctuations with a voltmeter. Simply switch the voltmeter to the lowest voltage scale and connect the negative probe (-) to the sensor body (ground) and the positive probe to the sensor terminal. With the voltmeter connected to the sensor, gently tap on a surface near the knock sensor with a hammer or similar device (this simulates the knock from the engine) and observe voltage fluctuations on the meter. If no voltage fluctuations can be detected, the sensor is bad and should be replaced with a new part.

Mass Air Flow (MAF) sensors

20   The MAF sensor measures the amount of air passing through the sensor body and ultimately entering the engine through the throttle body. The PCM uses this information to control fuel delivery - the more air entering the engine (acceleration), the more fuel needed.

21   A SCAN tool is necessary to check the output of the MAF sensor. The SCAN tool displays the sensor output in grams per second. With the engine idling at normal operating temperature, the display should read 4 to 7 grams per second. When the engine is accelerated the values should raise and remain steady at any given RPM. A failure in the MAF sensor or circuit will also set a diagnostic trouble code.

Neutral Start switch

22   The Neutral Start switch or Transaxle range sensor, located on the rear upper part of the automatic transaxle, indicates to the PCM when the transaxle is in Park or Neutral. This information is used for Transaxle Converter Clutch (TCC), Exhaust Gas Recirculation (EGR) and Idle Air Control (IAC) valve operation. **Caution:** *The vehicle should not be driven with the Neutral Start switch disconnected because idle quality will be adversely affected.*

23   For more information regarding the Neutral Start switch, which is part of the Neutral start and back-up light switch assembly, see Chapter 7.

Air conditioning control

24   During air conditioning operation, the PCM controls the application of the air conditioning compressor clutch. The PCM controls the air conditioning clutch control relay to delay clutch engagement after the air conditioning is turned ON to allow the IAC valve to adjust the idle speed of the engine to compensate for the additional load. The PCM also controls the relay to disengage the clutch on WOT (wide open throttle) to prevent excessively high rpm on the compressor. Be sure to check the air conditioning system as detailed in Chapter 3 before attempting to diagnose the air conditioning clutch or electrical system.

Power steering pressure sensor

25   Turning the steering wheel increases power steering fluid pressure and engine load. The pressure switch will close before the load can cause an idle problem. A pressure switch that will not open or an open circuit from the PCM will cause timing to retard at idle and this will affect idle quality. A pressure switch that will not close or an open circuit may cause the engine to die when the power steering system is used heavily. Any problems with the power steering pressure switch or circuit should be repaired by a dealer service department or other qualified repair shop.

Fuel Tank Pressure (FTP) sensor

26   The fuel tank pressure (FTP) sensor is used to monitor the fuel tank pressure or vacuum during the OBD II test portion for emissions integrity. This test scans various sensors and output actuators to detect abnormal amounts of fuel vapors that may not be purging into the canister and/or the intake system for recycling. The FTP sensor helps the PCM monitor this pressure differential (pressure vs. vacuum) inside the fuel tank. Any problems with the fuel tank pressure (FTP) sensor or circuit should be repaired by a dealer service department or other qualified repair shop.

Transaxle Converter Clutch (TCC) system

27   The purpose of the Torque Converter Clutch (TCC) system, equipped in automatic transaxles, is to eliminate the power loss of the torque converter stage when the vehicle is in the cruising mode (usually above 35 mph). This economizes the automatic transaxle to the fuel economy of the manual transaxle. The lock-up mode is controlled by the PCM through the activation of the TCC apply solenoid which is built into the automatic transaxle. When the vehicle reaches a specified speed, the PCM energizes the solenoid and allows the torque converter to lock-up and mechanically couple the engine to the transaxle, under which conditions emissions are at their minimum. However, because of other operating condition demands (deceleration, passing, idle, etc.), the transaxle must also function in its normal, fluid-coupled mode. When such latter conditions exist, the solenoid de-energizes, returning the torque converter to normal operation. The converter also returns to normal operation whenever the brake pedal is depressed.

28   Due to the requirement of special diagnostic equipment for the testing of this system, and the possible requirement for dismantling of the automatic transaxle to replace components of this system, Checking and replacing of the components should be handled by a dealer service department or other qualified repair facility.

---

5   Information sensors - replacement

Engine Coolant Temperature (ECT) sensor

Refer to illustrations 5.1, 5.2a and 5.2b

Warning: *Wait until the engine is completely cool before beginning this procedure.*

1   Before installing the new sensor, wrap the threads with Teflon sealing tape to prevent leakage and thread corrosion **(see illustration).**

5.1 To prevent coolant leakage, be sure to wrap the temperature sensor threads with Teflon tape before installation

5.2a Location of the engine coolant temperature (ECT) sensor (arrow) - 1997 and earlier 2.2L engine

5.2b Location of the engine coolant temperature (ECT) sensor (arrow) - 1998 and later 2.2L engine

5.5 On 1997 and earlier 2.2L engines, the MAP sensor is mounted on the plenum

5.7 Location of the Intake Air Temperature (IAT) sensor (arrow) - 1998 2.2L engine shown, others similar

5.9a Location of the upstream oxygen sensor on the 2.2L engine (arrow)

6

2   To remove the sensor, release the locking tab, unplug the electrical connector, then carefully unscrew the sensor. **Caution:** *Handle the coolant sensor with care. Damage to this sensor will affect the operation of the entire fuel injection system.* **Note:** *On 1997 and earlier 2.2L engines the coolant temperature sensor is mounted on the left (driver's) end of the cylinder head, just behind the water outlet* **(see illustration)**. *On 2.3L engines it's installed in the left (driver's side) rear corner of the cylinder head. On 1998 and later 2.2L engines it's installed in the coolant bypass pipe at the right front corner of the cylinder head, near the alternator* **(see illustration)**. *On 2.4L engines it's located on the thermostat housing cover.*

3   Installation is the reverse of removal. Check the coolant level and add some, if necessary (see Chapter 1).

Manifold Absolute Pressure (MAP) sensor

Refer to illustration 5.5

4   If you're working on a 1998 and later 2.2L engine, remove the throttle body (see Chapter 4). If you're working on a 2.4L engine, remove the air intake duct from the throttle body.

5   Detach the vacuum hose, unplug the electrical connector and remove the mounting screw(s) (all except 1998 and later 2.2L engines) **(see illustration)**. **Note:** *On 1998 and later 2.2L engines, the MAP sensor plugs directly into the intake manifold.*

6   Installation is the reverse of removal.

Intake Air Temperature (IAT) sensor

Refer to illustration 5.7

7   The Intake Air Temperature (IAT) sensor is located inside the air duct **(see illustration)** directly downstream of the air filter housing. To remove an IAT sensor, unplug the electrical connector and remove the sensor from the air intake duct. Carefully twist the sensor to release it from the rubber boot.

8   Installation is the reverse of removal.

Oxygen sensors

Refer to illustrations 5.9a, 5.9b and 5.13

Note: *Because it is installed in the exhaust manifold or pipe, which contracts when cool, the oxygen sensor may be very difficult to loosen when the engine is cold. Rather than risk damage to the sensor (assuming you are planning to reuse it in another manifold or pipe) or the threads which it screws into, start and run the engine for a minute or two, then shut it off. Be careful not to burn yourself during the following procedure.*

9   Most of the models covered by this manual are equipped with the OBD II engine management system, which employs two oxygen sensors - an upstream sensor (mounted in the exhaust manifold) and a downstream sensor (mounted in the exhaust pipe, after the catalytic converter) **(see illustrations)**.

5.9b Location of the downstream oxygen sensor. Some later downstream O2 sensors are equipped with heaters (4-wire harness connector)

5.13 Use a special slotted socket to unscrew the oxygen sensor

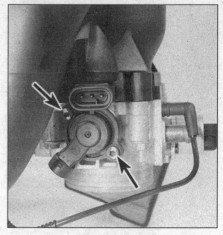

5.19 Remove the TPS mounting screws (arrows) (1998 2.2L engine shown, others similar)

10   Disconnect the cable from the negative terminal of the battery. **Caution:** *On models equipped with a Delco-Loc II or Theftlock audio system, be sure the lockout feature is turned off before disconnecting the battery cable.*

11   For access to the downstream oxygen sensor, raise the vehicle and place it securely on jackstands.

12   Follow the harness from the sensor to the electrical connector, then unplug the electrical connector from the vehicle harness.

13   Carefully unscrew the sensor from the exhaust manifold **(see illustration)**.

14   Anti-seize compound must be used on the threads of the sensor to facilitate future removal. The threads of new sensors will already be coated with this compound, but if an old sensor is removed and reinstalled, recoat the threads.

15   Install the sensor and tighten it securely.

16   Reconnect the electrical connector of the pigtail lead to the main engine wiring harness.

17   Take the vehicle on a test drive and check to see that no trouble codes set.

Throttle Position Sensor (TPS)

Refer to illustration 5.19

18   Disconnect the electrical connector from the TPS.

19   Remove the mounting screws from the TPS **(see illustration)** and remove the TPS from the throttle body.

20   When installing the TPS, be sure to align the socket locating tangs on the TPS with the throttle shaft in the throttle body.

21   Installation is the reverse of removal.

Neutral Start switch (automatic transaxle)

22   Refer to Chapter 7B for the neutral start switch replacement procedure.

Vehicle Speed Sensor (VSS)

Refer to illustration 5.23

23   The Vehicle Speed Sensor (VSS) is located in the transaxle housing at the rear section near the right driveaxle **(see illustration)**.

24   To replace the VSS, detach the sensor retaining screw and bracket, unplug the sensor and remove it from the transaxle.

25   Installation is the reverse of removal.

Crankshaft sensor

Refer to illustration 5.27

26   Disconnect the negative terminal from the battery.

27   Disconnect the electrical connector from the crankshaft sensor. **Note:** *The crankshaft sensor is mounted to either the front side or the back side of the engine block, depending on engine type* **(see illustration)**.

28   Remove the bolt from the crankshaft sensor and remove the sensor.

29   Installation is the reverse of removal. Be sure to use a new O-ring and tighten the bolt to the torque listed in this Chapter's Specifications.

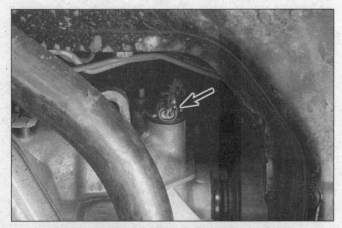

5.23 Location of the Vehicle Speed Sensor (VSS)

5.27 Locations of the crankshaft sensor (A) and the knock sensor (B) - viewed from above, cylinder head removed for clarity

5.33a Remove the bolts and lift the cover to expose the knock sensor module

5.33b Pinch the tabs and lift the knock sensor module from the PCM

Knock sensor

Warning: *Wait for the engine to cool completely before performing this procedure.*

30   The knock sensor is located on the side of the engine block near the bellhousing **(see illustration 5.27)**. It is threaded into the engine block coolant passage; when it is removed, the coolant will drain from the engine block.

31   Drain the cooling system (see Chapter 1). Place a drain pan under the sensor, disconnect the electrical connector and unscrew the knock sensor. A new sensor is pre-coated with thread sealant - do not apply any additional sealant or the operation of the sensor may be effected. Install the knock sensor and tighten it securely (approximately 14 ft-lbs). Don't overtighten the sensor or damage may occur. Plug in the electrical connector, refill the cooling system (see Chapter 1) and check for leaks.

Knock sensor module (1997 and earlier)

Refer to illustrations 5.33a and 5.33b

32   Remove the PCM (see Section 3).

33   Remove the bolts from the knock sensor module **(see illustration)** and lift the module from the PCM **(see illustration)**.

34   Installation is the reverse of removal. Be sure the module is placed in position perfectly straight to avoid bending the mating pins.

6   Exhaust Gas Recirculation (EGR) system

General description

1   The EGR system meters exhaust gases into the engine induction system through passages cast into the intake manifold. From there the exhaust gases pass into the fuel/air mixture for the purpose of lowering combustion temperatures, thereby reducing the amount of oxides of nitrogen (NOx) formed.

2   The linear EGR valve feeds small amounts of exhaust gas back into the intake manifold and then into the combustion chamber independent of intake manifold vacuum. The linear EGR valve operates without intake manifold vacuum thereby allowing accurate (finer) amounts of exhaust gas to be recirculated back into the intake system. A PCM controlled pintle within the linear EGR valve passes exhaust gasses through a small orifice upon command from the PCM. This EGR valve operates similar to the stepper motor type particular to IAC valves that control idle quality. By sending a 5 volt reference signal to the EGR valve, the PCM detects and sets the pintle position with the feedback signal from the EGR valve.

3   Common engine problems associated with the EGR system are rough idling or stalling at idle, rough engine performance during light throttle application and stalling during deceleration.

4   If the EGR system will not allow the PCM to control the position of the EGR valve pintle, it will set a diagnostic trouble code. Have the EGR system checked by a dealer service department or qualified repair facility in the event of EGR system failure.

Check

5   Special electronic diagnostic equipment is needed to check this valve and should be left to a dealer service department or other qualified repair facility.

Replacement

Refer to illustration 6.7a and 6.7b

6   Disconnect the electrical connector from the EGR valve.

7   Remove the two mounting bolts and remove the EGR valve from the intake manifold or EGR valve adapter **(see illustrations)**.

8   Remove the EGR valve and gasket.

6.7a Location of the Linear EGR valve on a 2.2L engine (arrow)

6.7b Remove the Linear EGR valve mounting bolts (arrows) (cylinder head removed for clarity)

9   Clean the mounting surface of the EGR valve. Remove all traces of gasket material from the intake manifold and from the valve if it is to be reinstalled. Clean both mating surfaces with a cloth dipped in lacquer thinner or acetone.
10   Install a new gasket and the EGR valve and tighten the bolts securely.
11   Connect the electrical connector onto the EGR valve.

7   Evaporative Emissions Control (EVAP) System

Note: *These models are equipped with an enhanced Evaporative Emission (EVAP) system. This EVAP system will conduct up to eight different tests using the On Board Diagnostic system to detect leaks, pressure variations or electrical problems within this closed network. Because it is governed by the PCM, have the system checked by a dealer service department or other qualified repair shop if the simple checks which follow don't rectify a problem.*

General description

1   This system is designed to trap and store fuel that evaporates from the throttle body and fuel tank which would normally enter the atmosphere and contribute to hydrocarbon (HC) emissions.
2   The system consists of a charcoal-filled canister and lines running to and from the canister. These lines include a vent line from the gas tank, a vent line from the throttle body, an EVAP vent solenoid (electronic), a charcoal canister and the intake manifold. In addition, there is an EVAP canister purge valve in the canister. The PCM controls the vacuum to the purge valve with an electrically operated solenoid. The fuel tank cap is also an integral part of the system. An indication that the system is not operating properly is a strong fuel odor.
3   The PCM monitors the vacuum level through the fuel tank pressure sensor signal (mounted on the tank). At the appropriate time, the EVAP canister purge valve and the EVAP vent solenoid are turned on which allows the engine to draw a small amount of vacuum on the entire EVAP system. Once the proper vacuum level is set, the PCM turns off the purge valve sealing the system closed. The PCM then monitors the system for leaks and sets a diagnostic code (P0442) if trouble is detected.

Check

Refer to illustration 7.4
4   Check all lines in and out of the canister for kinks, leaks and breaks along their entire lengths **(see illustration)**. Repair or replace as necessary.
5   Check the gasket in the gas cap for signs of drying, cracking or breaks. Replace the gas cap with a new one if defects are found.
6   The computer operates the purge valve and the vent solenoid by changing its frequency signal. The EVAP pressure sensor detects

7.4 The charcoal canister is located at the right front of the vehicle, behind the fender (fender removed for clarity)

abnormal high pressure in the purge lines. Check for battery voltage to the purge control solenoid and the pressure sensor with the ignition key ON (engine not running). If no battery voltage is present, have the PCM and the EVAP system circuit checked at a dealer service department or other qualified repair shop. If battery voltage exists, connect the solenoid and the pressure sensor and if there is no obvious sounds from the EVAP vent solenoid (buzzing), have it checked by a dealer service department or other qualified repair shop. Remember, the purge valve will not be activated by the computer until fuel tank pressure exceeds 0.7 psi.

Canister replacement

7   Remove the splash shield under the right-hand side of the engine.
8   Disconnect the hoses from the canister.
9   Remove the mounting bolt and remove the canister and bracket.
10   Installation is the reverse of removal.

8   Positive Crankcase Ventilation (PCV) system

Note: *These models are equipped with an oil/air separator assembly attached to the side of the engine block near the oil filter. This separator helps to filter oil that has become air-born within the crankcase ventilation gasses. This separator will require changing in the event the system becomes clogged and the crankcase vapors are not allowed to pass into the venting tubes and PCV valve. Refer to Chapter 1 for the changing procedures.*

General description

1   The positive crankcase ventilation system reduces hydrocarbon emissions by circulating fresh air through the crankcase to pick-up blow-by gases, which are then rerouted through the throttle body and burned in the engine.
2   The main components of this system are vacuum hoses and a PCV valve, which regulates the flow of gases according to engine speed and manifold vacuum.

Check and component replacement

3   Checking the system and PCV valve replacement are covered in Chapter 1.

9   Catalytic converter

General description

1   The catalytic converter is an emission control device added to the exhaust system to reduce pollutants from the exhaust gas stream. These systems are equipped with a single bed monolith catalytic converter. This monolithic converter contains a honeycomb mesh which is also coated with two types of catalysts. One type is the oxidation catalyst while the other type is a three-way catalyst that contains platinum and palladium. The three-way catalyst lowers the levels of oxides of nitrogen (NOx) as well as hydrocarbons (HC) and carbon monoxide (CO) emissions. The oxidation catalyst lowers the levels of hydrocarbons and carbon monoxide.

Check

2   The test equipment for a catalytic converter is expensive and highly sophisticated. If you suspect the converter is malfunctioning, take it to a dealer service department or authorized emissions inspection facility for diagnosis and repair.
3   Whenever the vehicle is raised for service of underbody components, check the converter for leaks, corrosion and other damage. If damage is discovered, the converter should be replaced.
4   Because the converter is welded to the exhaust system, converter replacement requires removal of the exhaust pipe assembly (see Chapter 4). Take the vehicle, or the exhaust system, to a dealer service department or a muffler shop.

Chapter 7 Part A
Manual transaxle

Contents

Specifications

General

Lubricant type... See Chapter 1

Torque specifications

	Ft-lbs
Back-up light switch	24
Transaxle-to-engine bolts	55

1   General information

The vehicles covered by this manual are equipped with either a 5-speed manual or an automatic transaxle. Information on the manual transaxle is included in this Part of Chapter 7. Information on the automatic transaxle is in Part B.

The 1995 through 1999 models are equipped with a 5-speed Isuzu transaxle, while the 2000 models are equipped with a 5-speed Getrag transaxle. On all models, the transaxle consists of a transmission and differential housed in a single all-aluminum assembly. Because of its complexity, the special tools needed to overhaul it, and the difficulty of obtaining replacement parts, overhauling this unit is beyond the scope of the average home mechanic. The information in this chapter is limited to general diagnosis, external adjustments and removal and installation.

Depending on the cost, it may be a good idea to consider replacing the old unit with either a rebuilt or used transaxle instead of a new one. Your local dealer or transmission shop should be able to supply information concerning cost, availability and exchange policy. Regard-less of how you decide to remedy a transaxle problem, however, you will save money by removing and installing it yourself.

2   Shift cables - removal and installation

1   Disconnect the shift and select cables from the levers on the transaxle. The cables can be pried from the pins on the levers with a screwdriver.
2   Remove the center console (see Chapter 11).
3   Disconnect the shift and select cables from the shift control assembly. The cables can be pried from the pins on the shift control assembly with a screwdriver. To detach the cables from their bracket on the shift control assembly base, pry off the large retaining plate.
4   Remove the screws from the right sill plate, remove the plate, then pull the carpet back for access to the cables.
5   Remove the cable grommet retainer screws (1995 and 1996 models) or pry loose the grommet (1997 and later models).
6   Pull the cables through into the passenger compartment and remove them from the vehicle.
7   Installation is the reverse of removal.

3   Shift control assembly - removal and installation

1   Remove the shift lever knob, console and shift boot (Chapter 11).
2   Disconnect the shift and select cables from the shift control assembly (see Section 2).
3   Remove the large shift cable retainer plate and detach the cables from their bracket on the shift control assembly base.
4   Remove the shift control assembly retaining nuts and remove the shift control assembly.
5   Installation is the reverse of removal.

4   Back-up light switch - check and replacement

Check

1   Turn the ignition key to the On position and move the shift lever to the Reverse position. The switch should close the back-up light circuit and turn on the back-up lights.
2   If it doesn't, check the back-up light fuse (see Chapter 12).
3   If the fuse is okay, verify that there's voltage available on the battery side of the switch (with the ignition turned to On).
4   If there's no voltage on the battery side of the switch, check the wire between the fuse and the switch; if there is voltage, put the shift lever in reverse and see if there's voltage on the ground side of the switch.
5   If there's no voltage on the ground side of the switch, replace the switch (see below); if there is voltage, note whether one or both back-up lights are out.
6   If only one bulb is out, replace it; if they're both out, the bulbs could be the problem, but it's more likely that the wire between the switch and the bulbs has an open somewhere.

Replacement

7   The back-up light switch is located on top of the left end (driver's side) of the transaxle.
8   Unplug the electrical connector from the switch.
9   Unscrew the switch with a *socket*. Do NOT use an open-end wrench.
10   Installation is the reverse of removal. Be sure to tighten the new switch to the torque listed in this Chapter's Specifications.

5   Manual transaxle - removal and installation

Removal

Refer to illustration 5.2

1   Disconnect the negative cable from the battery. **Caution:** *If the vehicle is equipped with a Delco Loc II or Theftlock audio system, make sure you have the correct activation code before disconnecting the battery.*
2   The engine must be supported during transaxle removal with an engine support tool **(see illustration)**. If this tool isn't available, support the engine with a hoist that can hold it high enough to allow the transaxle to be lowered from the engine compartment with the vehicle raised.
3   Working in the passenger compartment, remove the left sound insulator (see Chapter 11).
4   Disconnect the clutch master cylinder pushrod from the clutch pedal (see Chapter 8).
5   Remove the air cleaner and duct assembly from the throttle body (see Chapter 4).
6   Detach the wiring harness from its mounting bracket.
7   Remove the upper transaxle mount-to-transaxle bolts.
8   Disconnect the clutch actuator cylinder line and remove the clutch master cylinder from the clutch actuator cylinder (see Chapter 8).
9   Detach the wire harness from the mount bracket.

5.2 During transmission removal, support the engine with a support fixture designed for this purpose. They are often available from rental yards

10   Disconnect the ground cables from their transaxle mounting studs.
11   Unplug the electrical connector from the back-up light switch.
12   Detach the vent tube from the transaxle.
13   Remove the rear transaxle-to-engine bolts.
14   Lower the engine to facilitate removal and installation of the transaxle.
15   Loosen the front wheel lug nuts, raise the vehicle and place it securely on jackstands. Remove the front wheels.
16   Drain the transaxle fluid (Chapter 1).
17   Remove the left front inner splash shield.
18   Unplug both front ABS wheel speed sensor connectors (see Chapter 9) and set the harnesses aside. It's not necessary to remove the sensor. Make sure the harnesses and connectors are safely out of harm's way where they won't be damaged.
19   Remove the flywheel housing cover bolts.
20   Remove the vehicle speed sensor from the transaxle (see Chapter 6).
21   Remove the left and right balljoint nuts and separate the balljoints from the steering knuckles.
22   Remove the left stabilizer link pin and remove the left U-bolt from the stabilizer bar (see Chapter 10).
23   Remove the left suspension support attaching bolts.
24   Detach the driveaxles from the transaxle (see Chapter 8).
25   Remove the front lower transaxle mount.
26   Support the transaxle with a jack, preferably a transmission jack made for this purpose. Chain the transaxle to the jack to make sure it doesn't fall off during removal.
27   Remove the rest of the transaxle-to-engine bolts.
28   Make a final check that all wiring, cables, etc. are disconnected from the transaxle.
29   Separate the transaxle from the engine by carefully prying the bellhousing away from the engine.
30   Lower the transaxle and remove it from the left side of the engine compartment.
31   The clutch components can now be inspected (Chapter 8). In most cases, new clutch components should be installed as a matter of course when the transaxle is removed.

Installation

32   With the clutch components installed and properly aligned (see Chapter 8), carefully raise the transaxle into place, guide the right side driveaxle into the transaxle and slide the input shaft into place in the clutch hub splines.
33   Install the transaxle-to-engine bolts. Tighten the bolts to the torque listed in this Chapter's Specifications.
34   Install the front transaxle mount and tighten the nuts and bolts securely.

35   Install the flywheel housing cover and tighten the bolts securely.
36   Install the driveaxles (see Chapter 8).
37   Install the left suspension support and tighten the bolts securely.
38   Install the left stabilizer bar U-bolt (see Chapter 10).
39   Attach the left and right balljoints to the steering knuckles (see Chapter 10).
40   Install the left stabilizer bar link (see Chapter 10).
41   Reroute the ABS harnesses and connectors and plug the connectors into the front wheel speed sensors.
42   Install the front inner splash shield.
43   Install the wheels.
44   Install the vehicle speed sensor (see Chapter 6).
45   Lower the vehicle.
46   Connect the ground cables to their transaxle mounting studs.
47   Attach the vent tube to the transaxle.
48   Plug in the connector for the back-up light switch.
49   Install the upper transaxle-to-engine bolts and tighten them to the torque listed in this Chapter's Specifications.
50   Connect the clutch master cylinder to the clutch release cylinder (see Chapter 8).
51   Install the rear transaxle mount and tighten the nuts and bolts securely.
52   Attach the wiring harness to the mount bracket.
53   Remove the engine support.
54   Connect the shift and select cables to the shift levers and to the shift cable bracket.
55   Install the air cleaner and ducting (see Chapter 4).
56   Connect the clutch master cylinder pushrod to the clutch pedal (see Chapter 8).
57   Install the left sound insulator (see Chapter 11).
58   Fill the transaxle with the specified lubricant (see Chapter 1).
59   Connect the negative battery cable.

6   Manual transaxle overhaul - general information

Overhauling a manual transaxle is a difficult job for the do-it-yourselfer. It involves the disassembly and reassembly of many small parts. Numerous clearances must be precisely measured and, if necessary, changed with select fit spacers and snap-rings. As a result, if transaxle problems arise, it can be removed and installed by a competent do-it-yourselfer, but overhaul should be left to a transmission repair shop. Rebuilt transaxles may be available - check with your dealer parts department and auto parts stores. At any rate, the time and money involved in an overhaul is almost sure to exceed the cost of a rebuilt unit.

Nevertheless, it's not impossible for an inexperienced mechanic to rebuild a transaxle if the special tools are available and the job is done in a deliberate, step-by-step manner so nothing is overlooked.

The tools necessary for an overhaul include internal and external snap-ring pliers, a bearing puller, a slide hammer, a set of pin punches, a dial indicator and possibly a hydraulic press. In addition, a large, sturdy workbench and a vise or transaxle stand will be required.

During disassembly of the transaxle, make careful notes of how each piece comes off, where it fits in relation to other pieces and what holds it in place. Note how the parts are installed when you remove them; this will make it much easier to get the transaxle back together.

Before taking the transaxle apart for repair, it will help if you have some idea what area of the transaxle is malfunctioning. Certain problems can be closely tied to specific areas in the transaxle, which can make component examination and replacement easier. Refer to the Troubleshooting section at the front of this manual for information regarding possible sources of trouble.

Notes

Chapter 7 Part B
Automatic transaxle

Contents

Specifications

Torque specifications

	Ft-lbs
Park/Neutral Position switch retaining bolts	18
Transaxle-to-engine bolts	
All bolts except lower bolt	71
Lower bolt	41

1 General information

Due to the complexity of the clutches and the hydraulic control system, and because of the special tools and expertise required to perform an automatic transmission overhaul, it should not be undertaken by the home mechanic. Therefore, the procedures in this Chapter are limited to general diagnosis, routine maintenance and adjustment and transmission removal and installation.

If the transmission requires major repair work it should be left to a dealer service department or an automotive or transmission repair shop. You can, however, remove and install the transmission yourself and save the expense, even if the repair work is done by a transmission specialist.

Adjustments that the home mechanic may perform include those involving the throttle valve cable, the shift linkage and the neutral safety switch. **Caution:** *Never tow a disabled vehicle equipped with an automatic transaxle at speeds greater than 30 mph or distances over 50 miles unless the front wheels are off the ground. Failure to observe this precaution may result in severe transmission damage caused by lack of lubrication.*

2 Diagnosis - general

Note: *Automatic transaxle malfunctions may be caused by five general conditions: poor engine performance, improper adjustments, hydraulic malfunctions, mechanical malfunctions or malfunctions in the computer or its signal network. Diagnosis of these problems should always begin with a check of the easily repaired items: fluid level and condition (see Chapter 1), shift linkage adjustment (see Section 3) and throttle linkage adjustment (see Section 5). Next, perform a road test to determine if the problem has been corrected or if more diagnosis is necessary. If the problem persists after the preliminary tests and corrections are completed, additional diagnosis should be done by a dealer service department or transmission repair shop. Refer to the Troubleshooting section at the front of this manual for transaxle problem diagnosis.*

Preliminary checks

1   Drive the vehicle to warm the transaxle to normal operating temperature.

2   Check the fluid level as described in Chapter 1:

3.2 To disconnect the shift cable from the transaxle manual lever, pry it off the pin on the lever with a prying tool or screwdriver

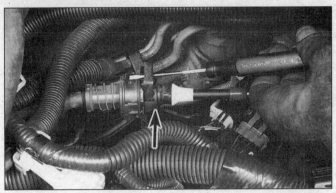

3.3 To disengage the shift cable from the cable bracket on the transaxle, remove this locking clip, then squeeze the two locking tangs (arrow indicates one tang, other tang on other side of cable housing) and pull the cable housing from the bracket

a) *If the fluid level is unusually low, add enough fluid to bring the level within the designated area of the dipstick, then check for external leaks.*

b) *If the fluid level is abnormally high, drain off the excess, then check the drained fluid for contamination by coolant. The presence of engine coolant in the automatic transmission fluid indicates that a failure has occurred in the internal radiator walls that separate the coolant from the transmission fluid (see Chapter 3).*

c) *If the fluid is foaming, drain it and refill the transaxle, then check for coolant in the fluid or a high fluid level.*

3   Check the engine idle speed. **Note:** *If the engine is malfunctioning, do not proceed with the preliminary checks until it has been repaired and runs normally.*

4   Check the Throttle Valve (TV) cable for freedom of movement. Adjust it if necessary (see Section 5). **Note:** *The TV cable may function properly when the engine is shut off and cold, but it may malfunction once the engine is hot. Check it cold and at normal engine operating temperature.*

5   Inspect the shift control cable (see Section 3). Make sure that it's properly adjusted and that the linkage operates smoothly.

Fluid leak diagnosis

6   Most fluid leaks are easy to locate visually. Repair usually consists of replacing a seal or gasket. If a leak is difficult to find, the following procedure may help.

7   Identify the fluid. Make sure it's transmission fluid and not engine oil or brake fluid (automatic transmission fluid is a deep red color).

8   Try to pinpoint the source of the leak. Drive the vehicle several miles, then park it over a large sheet of cardboard. After a minute or two, you should be able to locate the leak by determining the source of the fluid dripping onto the cardboard.

9   Make a careful visual inspection of the suspected component and the area immediately around it. Pay particular attention to gasket mating surfaces. A mirror is often helpful for finding leaks in areas that are hard to see.

10   If the leak still cannot be found, clean the suspected area thoroughly with a degreaser or solvent, then dry it.

11   Drive the vehicle for several miles at normal operating temperature and varying speeds. After driving the vehicle, visually inspect the suspected component again.

12   Once the leak has been located, the cause must be determined before it can be properly repaired. If a gasket is replaced but the sealing flange is bent, the new gasket will not stop the leak. The bent flange must be straightened.

13   Before attempting to repair a leak, check to make sure that the following conditions are corrected or they may cause another leak. **Note:** *Some of the following conditions cannot be fixed without highly specialized tools and expertise. Such problems must be referred to a transmission shop or a dealer service department.*

Gasket leaks

14   Check the pan periodically. Make sure the bolts are tight, no bolts are missing, the gasket is in good condition and the pan is flat (dents in the pan may indicate damage to the valve body inside).

15   If the pan gasket is leaking, the fluid level or the fluid pressure may be too high, the vent may be plugged, the pan bolts may be too tight, the pan sealing flange may be warped, the sealing surface of the transaxle housing may be damaged, the gasket may be damaged or the transaxle casting may be cracked or porous. If sealant instead of gasket material has been used to form a seal between the pan and the transaxle housing, it may be the wrong sealant.

Seal leaks

16   If a transaxle seal is leaking, the fluid level or pressure may be too high, the vent may be plugged, the seal bore may be damaged, the seal itself may be damaged or improperly installed, the surface of the shaft protruding through the seal may be damaged or a loose bearing may be causing excessive shaft movement.

17   Make sure the dipstick tube seal is in good condition and the tube is properly seated. Periodically check the area around the speedometer gear or sensor for leakage. If transmission fluid is evident, check the O-ring for damage. Also inspect the side gear shaft oil seals for leakage.

Case leaks

18   If the case itself appears to be leaking, the casting is porous and will have to be repaired or replaced.

19   Make sure the oil cooler hose fittings are tight and in good condition.

Fluid comes out vent pipe or fill tube

20   If this condition occurs, the transaxle is overfilled, there is coolant in the fluid, the case is porous, the dipstick is incorrect, the vent is plugged or the drain back holes are plugged.

---

3   Shift cable - replacement and adjustment

Replacement

Refer to illustration 3.2, 3.3, 3.5a, 3.5b and 3.6

1   Disconnect the negative battery cable. **Note:** *If the vehicle is equipped with a Delco Loc II or Theftlock system, make sure you have the correct activation code before disconnecting the battery* (see Chapter 12).

2   Disconnect the shift cable from the transaxle manual lever **(see illustration)**.

3   Disengage the shift cable from the cable bracket on the transaxle **(see illustration)**.

4   Remove the shift lever knob (see Section 4) and the console (see Chapter 11).

3.5a To disengage the shift cable from the bracket on the shift lever base, remove this locking clip . . .

3.5b . . . squeeze the locking tangs together, then disengage the cable from the bracket

3.6 To disconnect the shift cable from the pin on the shift lever, simply pry it loose

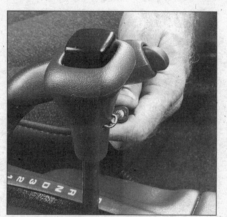

4.2a To remove the shift lever knob, pry out this locking clip . . .

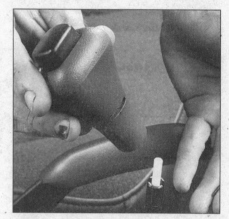

4.2b . . . then pull the knob straight up

5   Pry out the locking clip and disengage the cable from the bracket at the front of the shift lever base **(see illustrations)**.
6   Disconnect the shift cable from the pin on the shift lever **(see illustration)**.
7   Remove the sill plate, pull up the carpet and trace the cable to the cable grommet (the point at which it goes through the firewall). Pry out the grommet and pull the cable through the hole and remove it.
8   Installation is the reverse of removal. When you're done installing the new cable, be sure to adjust it.

Adjustment

9   Place the manual lever on the transaxle in the Neutral position. This is accomplished by rotating the lever clockwise from the Park position, through Reverse and into Neutral.
10   Place the shift lever inside the car in Neutral.
11   Push the tab on the cable adjuster at the cable bracket on the transaxle. The cable will automatically adjust itself.
12   Make sure the engine will start in the Park and Neutral positions only.
13   If the engine can be started in any position other than Park or Neutral, check the adjustment of the Park-Neutral Position Switch (see Section 7), then adjust and check the shift cable again.
14   If the cable still can't be adjusted properly, have the vehicle examined by a dealer service department.

---

4   Shift lever assembly - removal and installation

Removal

Refer to illustrations 4.2a, 4.2b and 4.6
1   Disconnect the negative battery cable. **Note:** *If the vehicle is*

equipped with a Delco Loc II or Theftlock system, make sure you have the correct activation code before disconnecting the battery (see Chapter 12).
2   Remove the shift lever knob **(see illustrations)**.
3   Remove the console (see Chapter 11).
4   Disconnect the shift cable from the shift lever (see Section 3).
5   Unplug the electrical connector.
6   Remove the shift lever assembly retaining nuts **(see illustration)** and remove the shift lever assembly.

Installation

7   Place the shift lever in position on the mounting studs and install the nuts. Tighten the nuts securely.
8   Installation is the reverse of removal. Adjust the shift cable (see Section 3) when you're done.

7B

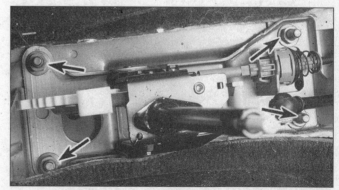

4.6 To detach the shift lever assembly from the floorpan, remove these four nuts (arrows)

5.2 Disconnect the TV cable from the throttle lever at the throttle body

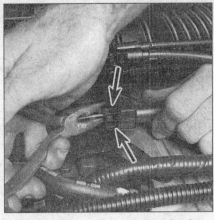

5.3 Disengage the TV cable housing from the cable bracket by compressing the tangs (arrows) and pushing the housing through the bracket

5.5 To detach the TV cable from the transaxle, remove this bolt (arrow)

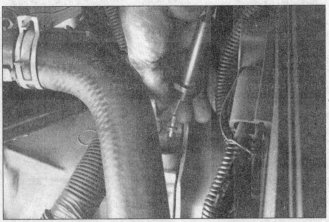

5.6 Pull up on the TV cable until the end of the cable is visible, then disengage it from the transaxle link

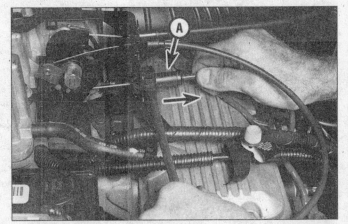

5.9 To adjust the TV cable, depress the adjustment button with a screwdriver or other suitable tool, move the slider (A) back (in direction of arrow) against the fitting until it stops, then release the adjustment button

5   Throttle Valve (TV) cable (3T40 transaxle) - replacement and adjustment

1   The Throttle Valve cable controls transaxle line pressure, which determines the shift "feel" and the timing of part-throttle and detent downshifts. The TV cable is attached to the throttle lever on the throttle body and to a link inside the transaxle.

Replacement
Refer to illustrations 5.2, 5.3, 5.5 and 5.6

2   Disconnect the TV cable from the throttle lever at the throttle body **(see illustration)**.

3   To disengage the TV cable housing from the cable bracket, compress the tangs and push the housing out of the bracket **(see illustration)**.

4   Disconnect any clips or straps retaining the cable to the transaxle.

5   Remove the bolt that secures the TV cable to the transaxle **(see illustration)**.

6   Pull up on the TV cable until the end of the cable is visible, then disengage it from the transaxle link **(see illustration)**.

7   Installation is the reverse of removal. Adjust the cable when you're done.

Adjustment
Refer to illustration 5.9

8   To check TV cable freeplay, pull on the upper end of the cable. It should travel a short distance with light resistance due to the small coiled return spring. To verify that the TV cable, lever and plunger are moving freely, pull the cable farther out to move the lever into contact with the plunger (this compresses the heavier TV spring). When released, the cable should return to its closed position. If the cable doesn't operate as described, turn off the engine (the engine must be turned off during adjustment), and adjust the TV cable as follows.

9   Depress the adjustment button **(see illustration)** and hold it down.

10   Pull the cable housing out until the slider mechanism hits the stop.

11   Release the adjustment button.

12   Remove the floormat.

13   Firmly depress the accelerator pedal all the way to the floor (wide-open throttle).

14   Verify that the cable moves freely.

15   Road test the vehicle. Sometimes, the cable will operate okay when the engine is cold, but will fail to operate properly when the engine is warmed up. If delayed or only full-throttle shifts still occur, try adjusting the cable one more time. If the problem persists, have the vehicle checked by a dealer service department or transmission shop.

6   Park/Lock system - description, adjustment and cable replacement

Warning: *The models covered by this manual are equipped with*

6.4 There must be no gap between the metal terminal stop (A) and the protruding end of the plastic collar (B). If there is a gap between the terminal stop and the collar, as there is in this photo, adjust the cable

airbags. *Always disable the airbag system before working in the vicinity of the impact sensors, steering column or instrument panel to avoid the possibility of accidental deployment of the airbag(s), which could cause personal injury (see Chapter 12). The yellow wires and connectors routed through the console are for this system. Do not use electrical test equipment on these yellow wires or tamper with them in any way while working around the console.*

Description

1   The Park/Lock cable prevents the shift lever from being moved out of Park unless the brake pedal is depressed simultaneously. When the car is started, a solenoid is energized, locking the shift lever in Park; when the brake pedal is depressed, the solenoid is deenergized, unlocking the shift lever so that it can be moved into some other gear.

Check and adjustment

Refer to illustrations 6.4 and 6.9

2   Remove the left sound insulator trim panel from underneath the left end of the dash (see Chapter 11).
3   Locate the forward end of the Park/Lock cable.
4   The terminal stop on the steering column end of the Park-Lock cable must be touching the blue or white plastic collar that protrudes from the ignition switch. To check it:

a) *Turn the ignition key to the Lock position.*
b) *Put the shift lever in the Park position.*
c) *Note the position of the terminal stop (the metal plug on the end of the cable). There must be no gap between the metal terminal stop and the protruding end of the plastic collar* **(see illustration)**. *If there's a gap between the terminal stop and the collar, adjust the cable. The plastic collar must be either flush or recessed about 0.04 inch (1 mm) into the ignition park-lock housing. If the plastic collar isn't in the correct position, adjust the cable.*

5   The plastic collar must have no more than 0.06 inch (1.5 mm) of travel. To check it:

a) *Put the key lock cylinder in the Lock position.*
b) *Put the shift lever in the Park position.*
c) *Gently, squeeze the park lock button on the shift lever until you feel resistance.*
d) *Verify that the plastic collar travels no more than 0.06 inch (1.5 mm) and that the shift lever doesn't come out of the Park position.*

6   With the ignition key turned to the On position, verify that the shift lever moves through all gear positions.
7   While moving the shift lever through all gear positions, verify that the ignition key can't be turned to the Lock position.
8   Verify that the ignition key can be removed when it's in the Lock position and the shift lever is in the Park position.
9   If the Park/Lock cable fails any of the above five tests, adjust it as follows.

6.9 To release the Park/Lock cable, raise the locking tab on the cable housing at the cable bracket on the shift lever base; to lock the cable into place, press the locking tab on the cable housing back into the housing

a) *Put the shift lever in Park.*
b) *Turn the ignition key to Lock.*
c) *Remove the shift lever knob (see Section 3).*
d) *Remove the console (see Chapter 11).*
e) *Raise the locking tab on the cable housing at the cable bracket on the shift lever base* **(see illustration)**. *This releases the Park/Lock cable.*
f) *Adjust the outer cable "conduit" (the cable sheathing) to correctly position the white plastic housing in the ignition switch* **(see illustration 6.4)**. *There must be no gap between the metal terminal stop and the protruding end of the white plastic collar. The white plastic collar must either be flush or recessed about 0.04 inch (1 mm) into the ignition park lock housing.*
g) *Press the locking tab on the cable housing back into the housing.*

10   Following Steps 4 through 8 above, check the cable as described. Make sure it passes all five checks.

Cable replacement

Refer to illustrations 6.18 and 6.19

11   Remove the shift lever knob (see Section 3).
12   Remove the console, left sound insulator and steering column knee bolster panel (see Chapter 11).
13   Unbolt and lower the steering column.
14   Put the shift lever in the Park position.
15   Turn the ignition key to the Run position.
16   Insert a screwdriver blade into the slot in the ignition switch inhibitor, depress the cable latch and detach the cable from the inhibitor.
17   Raise the locking tab on the cable housing at the shift lever base to the up position **(see illustration 6.9)**.
18   Snap the rear end of the cable loose from the pin on the Park/Lock lever **(see illustration)**.

6.18 Pry the cable loose from the pin on the Park/Lock lever

6.19 To disengage the housing from the cable bracket, depress the two cable housing latches on the sides of the cable housing

19   Depress the two cable housing latches on the sides of the cable housing **(see illustration)** and disengage the housing from the cable bracket.
20   Remove any cable clips and remove the Park/Lock cable.
21   Make sure the cable locking tab is in the up position and the shift lever is in Park.
22   With the ignition key in the Run position (this is very important), snap the cable into the inhibitor housing.
23   Turn the ignition key to the Lock position.
24   Snap the rear end of the cable onto the pin on the park/lock lever.
25   Push the nose of the cable connector forward to remove the slack.
26   With no load on the connector nose, snap down the cable housing locking tab.
27   Check the operation of the park/lock cable as follows.

a) *With the shift lever in Park and the key in Lock, make sure the shift lever cannot be moved to another position and the key can be removed.*

b) *With the key in Run and the shift lever in Neutral, make sure the key cannot be turned to Lock.*

28   If it operates as described above, the park/lock cable system is properly adjusted. Proceed to Step 30.
29   If the park/lock system does not operate as described, return the cable connector lock to the up position and repeat the adjustment procedure. Push the cable connector down and recheck the operation.
30   If the key cannot be removed in the Park position, snap the locking tab to the up position and move the nose of the cable connector rearward until the key can be removed from the ignition switch.
31   Install the cable into the retaining clips.
32   Raise the steering column into position, install the bolts and tighten them securely.
33   Install the steering column knee bolster, the left sound insulator and the console (see Chapter 11).
34   Install the shift lever knob (see Section 3).

7   Park/Neutral Position (PNP) switch - check and replacement

Check

1   Refer to the *Information sensor checks* in Chapter 6.

Replacement

Refer to illustrations 7.4 and 7.13

2   Disconnect the negative cable from the battery. **Note:** *If the vehicle is equipped with a Delco Loc II or Theftlock system, be sure you have the correct code before disconnecting the battery. See the information at the front of this manual for the radio re-activation procedure.*

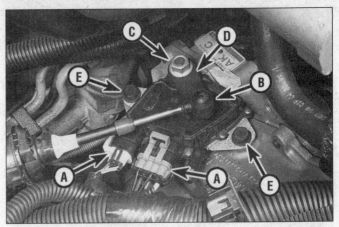

7.4 The Park/Neutral Position switch is located on top of the transaxle; to remove the switch, unplug the electrical connectors (A), disconnect the shift cable (B) from the manual lever, remove the manual lever retaining nut (C) and manual lever (D), and remove the switch retaining bolts (E)

3   Apply the parking brake and put the shift lever in Neutral.
4   Locate the Park/Neutral Position (PNP) switch **(see illustration)**, which is mounted on the transaxle at the manual lever.
5   Disconnect the shift cable from the manual lever (see Section 3).
6   Unplug the electrical connector(s) from the PNP switch **(see illustration 7.4)**.
7   Remove the switch retaining screws or bolts **(see illustration 7.4)** and detach the switch.
8   Remove the manual lever retaining nut and remove the manual lever **(see illustration 7.4)**.
9   Remove the Park/Neutral Position switch.
10   Put the shift shaft in the Neutral position.
11   Align the flats of the shift shaft with the flats of the Park/Neutral Position switch and install the switch.
12   Install the switch mounting bolts. If you're installing the old switch, don't tighten them yet. If you're installing a new switch, tighten the bolts to the torque listed in this Chapter's Specifications.
13   If you're installing the old switch, insert a 3/32-inch gauge pin or drill bit into the service adjustment hole **(see illustration)** and rotate the switch until the pin drops into the second hole; the pin or drill bit should drop into the holes to a depth of 9/64-inch. Tighten the switch mounting bolts to the torque listed in this Chapter's Specifications. Remove the gauge pin or drill bit. Verify that the engine will start only in Park or Neutral. If it starts in any other gear, readjust the switch.
14   If you're installing a new switch, align the service adjustment holes with the mounting boss on the transaxle. Put the manual lever in the Neutral position, but do NOT rotate the switch; the switch is pinned

7.13 To adjust the old Park/Neutral Position switch, insert a 3/32-inch gauge pin or drill bit into the service adjustment hole and rotate the switch until the pin drops into the second hole; the pin or drill bit should drop into the holes to a depth of 9/64-inch

8.4 Rubber-type driveaxle oil seals (arrow) can be pried out of the transaxle housing with a screwdriver; be careful not to damage the splines of the axleshaft (transaxle removed for clarity)

8.5 Dislodge metal-type driveaxle oil seals by working around the outer circumference with a chisel and hammer (transaxle removed for clarity)

in the Neutral position. If it has been rotated and the pin is broken, adjust the switch as described in the previous step. Verify that the engine will start only in Park or Neutral. If it starts in any other gear, readjust the switch.
15   The remainder of installation is the reverse of removal.

8   Oil seal replacement

1   Oil leaks frequently occur due to wear of the driveaxle oil seals. Replacement of these seals is relatively easy, since the repairs can usually be performed without removing the transaxle from the vehicle.

Driveaxle oil seals

Refer to illustrations 8.4 and 8.5

2   The driveaxle oil seals are located in the sides of the transaxle, where the driveaxles are attached. If leakage at the seal is suspected, raise the vehicle and support it securely on jackstands. If the seal is leaking, fluid will be found on the sides of the transaxle.
3   Remove the driveaxles (see Chapter 8).
4   On rubber-type seals, use a screwdriver or prybar to carefully pry the oil seal out of the transaxle bore. Be careful not to damage the splines on the output shaft **(see illustration)**. If the oil seal cannot be removed with a screwdriver or prybar, a special oil seal removal tool (available at auto parts stores) will be required.
5   On metal-type seals, use a hammer and chisel to pry up the outer lip of the seal to dislodge it so it can be pried out of the housing **(see illustration)**.
6   Compare the old seal to the new one to be sure it's the correct one.
7   Coat the outside and inside diameters of the new seal with a small amount of transmission fluid.
8   Using a large section of pipe or a large deep socket as a drift, install the new oil seal. Drive it into the bore squarely and make sure it's completely seated.
9   Install the driveaxle(s). Be careful not to damage the lip of the new seal.

Speedometer gear seal

10   Disconnect the speedometer cable, if equipped, from the transaxle (see Chapter 10). **Note:** *Some later models may use a vehicle speed sensor and an electric speedometer instead of a conventional mechanical speedometer, so there is no cable. However, the drive gear assembly - and the seals themselves - are similar to a conventional mechanical speedometer drive gear assembly and its seals.*
11   On models with a mechanical speedometer, remove the retainer clip that secures the speedometer cable, remove the speedometer driven gear and sleeve assembly and discard the O-ring.
12   On models with an electric speedometer, unplug the VSS connector, remove the governor cover or (on some models) the speed

sensor housing, and discard the O-ring.
13   Installation is the reverse of removal. Be sure to dip the new O-ring into clean automatic transmission fluid before installing it.

9   Transaxle mount - check and replacement

1   Insert a large screwdriver or prybar between the mount bracket and the rubber portion of the mount and pry up.
2   The transaxle should not move excessively away from the mount. If it does, or if the rubber is torn or badly cracked, replace the mount.
3   To replace a mount, support the transaxle with a jack, remove the nuts and bolts and remove the mount. It may be necessary to raise the transaxle slightly to provide enough clearance to remove the mount.
4   Installation is the reverse of removal.

10   Transaxle - removal and Installation

Removal

Refer to illustrations 10.5, 10.9, 10.18a and 10.18b

1   Disconnect the negative battery cable. **Caution:** *If the vehicle is equipped with a Delco Loc II or Theftlock system, make sure you have the correct activation code before disconnecting the battery (see Chapter 12).*
2   Remove the intake air duct (see Chapter 4).
3   Disconnect the shift cable from transaxle (see Section 3).
4   Disconnect the TV cable from the transaxle on models equipped with a 3T40 transaxle (see Section 5).
5   Disconnect the vent hose **(see illustration)**.

10.5 Detach the vent hose (arrow) from the top of the transaxle

7B

10.9 To detach the upper part of the transaxle from the engine, remove these nuts (arrows); be sure to label all ground wires, such as the two shown here, to ensure that they are reattached correctly

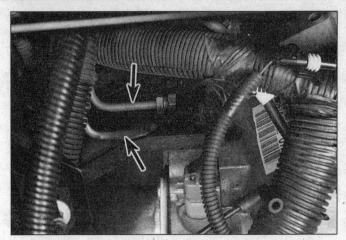

10.18a To detach the oil cooler lines (arrows) from a 3T40 transaxle, unscrew the tube nuts at the transaxle

6   Clearly label, then unplug, all electrical connectors.
7   Detach the power steering pump assembly (see Chapter 10) and set it aside.
8   Attach a suitable lifting device to the engine and raise the engine sufficiently to support the weight of the engine. **Note:** *The engine must remain supported while the transaxle is out of the vehicle.*
9   Remove the upper transmission-to-engine bolts **(see illustration)**.
10   Loosen the front wheel lug nuts, raise the vehicle and support it securely on jackstands. Remove the front wheels.
11   Remove the left splash shield.
12   Remove the front ABS wheel speed sensors and harnesses (see Chapter 9) and set them aside.
13   Remove the driveaxles (see Chapter 8).
14   Remove the engine-to-transaxle brace.
15   Remove the three torque converter cover bolts and remove the cover.
16   Remove the starter motor (see Chapter 5).
17   Mark the relationship of the torque converter to the driveplate and remove the flywheel-to-torque converter bolts.
18   Disconnect and plug the transaxle cooler lines **(see illustrations)**.
19   Disconnect all ground wires.
20   Remove the cooler line bracket.
21   Remove the exhaust brace.
22   Support the transaxle with a jack.
23   Remove the transaxle mount-to-body bolts.
24   Remove the transaxle mount bracket-to-transaxle bolts and remove the mount and bracket assembly.
25   Remove the nut and bolt from the heater core hose pipe-to-transaxle bracket.
26   Remove the remaining transaxle-to-engine bolts.
27   Remove the transaxle from the engine by sliding it toward the left side of the vehicle.

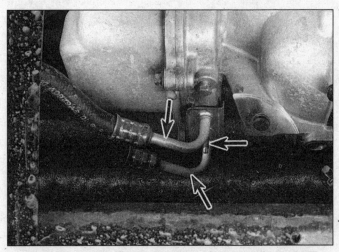

10.18b To detach the oil cooler lines (arrows) from a 4T40E transaxle, remove this nut (arrow)

Installation

28   Installation is the reverse of removal, with attention paid to the following points:

a) *Tighten all transaxle-to-engine bolts to the torque listed in this Chapter's Specifications.*
b) *The front end alignment should be checked by a dealer or alignment shop.*
c) *Adjust the TV cable (see Section 5).*
d) *Check the transaxle fluid level (see Chapter 1).*

Chapter 8
Clutch and driveaxles

Contents

Specifications

Clutch

Fluid type	See Chapter 1
CV joint boot length (see illustrations 11.3u and 11.3v)	
1995 models	
Ball-and-cage type joint	5-1/4 inches
Tripot type joint	
29/32 inch diameter axleshaft	7-1/2 inches
1-1/32 inch diameter axleshaft	6-11/16 inches
1996 and 1997 models (all)	4-29/32 inches
1998 and later models	
2.2L engine	4.0 inches
2.4L engine	4-29/32 inches

Torque specifications
Ft-lbs (unless otherwise indicated)

Clutch master cylinder mounting nuts	180 in-lbs
Clutch pressure plate-to-flywheel bolts	
2.2L	180 in-lbs, then rotate 30-degrees
2.3L, 2.4L	180 in-lbs, then rotate 45-degrees
Driveaxle/hub nut	185
Intermediate shaft support bolts	48
Wheel lug nuts	See Chapter 1

1   General information

The information in this Chapter deals with the components from the rear of the engine to the front wheels, except for the transaxle, which is dealt with in Chapters 7A and 7B. For the purposes of this Chapter, these components are grouped into two categories: clutch and driveaxles. Separate Sections within this Chapter offer general descriptions and checking procedures for both groups.

Since nearly all the procedures covered in this Chapter involve working under the vehicle, make sure it's securely supported on sturdy jackstands or a hoist where the vehicle can be easily raised and lowered.

2   Clutch - description and check

1   All vehicles with a manual transaxle use a single dry plate, diaphragm spring type clutch. The clutch disc has a splined hub which allows it to slide along the splines of the transaxle input shaft. The clutch and pressure plate are held in contact by spring pressure exerted by the diaphragm in the pressure plate.

2   The clutch release system is operated by hydraulic pressure. The hydraulic release system consists of the clutch pedal, a master cylinder and fluid reservoir, the clutch fluid hydraulic line, and an integral clutch release cylinder and release bearing assembly.

3   When pressure is applied to the clutch pedal to release the clutch,

8

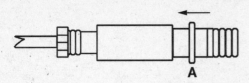

3.2 To disconnect the clutch hydraulic line, push in the release slide (A), hold it there and pull the line and fitting apart

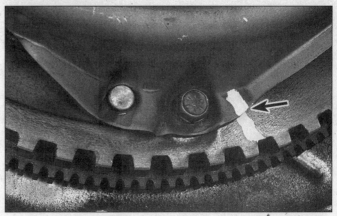

3.6 If you're going to re-use the same pressure plate, mark the relationship of the pressure plate to the flywheel

hydraulic pressure is exerted against the release bearing, which pushes against the fingers of the diaphragm spring of the pressure plate assembly, which in turn releases the clutch disc.

4   Terminology can be a problem regarding the clutch components because common names have in some cases changed from that used by the manufacturer. For example, the driven plate is also called the clutch plate or disc, the pressure plate assembly is also referred to as the clutch cover, the clutch release bearing is also called a throw-out bearing, and the actuator cylinder is also known as a release or slave cylinder.

5   Unless you're replacing components that are obviously damaged, make the following preliminary checks to determine the nature of the clutch system failure.

a) *Check the fluid level in the clutch master cylinder (see Chapter 1). If the fluid level is low, add fluid as necessary and inspect the hydraulic clutch system for leaks. If the master cylinder reservoir has run dry, bleed the system (see Section 6) and retest the clutch operation.*

b) *Check "clutch spin down time": run the engine at normal idle speed with the transaxle in Neutral (not with the clutch pedal depressed), depress the clutch pedal, wait several seconds and shift the transaxle into Reverse. You should not hear a grinding noise, the most likely cause of which is a defective pressure plate or clutch disc.*

c) *Check for complete clutch release: run the engine (with the parking brake applied) and hold the clutch pedal about 1/2-inch from the floor. Shift the transaxle between 1st gear and Reverse several times. If the shift is not smooth, component failure is indicated. Check the release cylinder pushrod travel. With the clutch pedal depressed completely the release cylinder pushrod should extend substantially. If it doesn't, check the fluid level in the clutch master cylinder.*

d) *Visually inspect the clutch pedal bushing at the top of the clutch pedal to make sure there is no sticking or excessive wear.*

3   Clutch components - removal, inspection and installation

Warning: *Dust produced by clutch wear and deposited on clutch components may contain asbestos, which is hazardous to your health. DO NOT blow it out with compressed air and DO NOT inhale it. DO NOT use gasoline or petroleum based solvents to remove the dust. Brake system cleaner should be used to flush the dust into a drain pan. After the clutch components are wiped clean with a rag, dispose of the contaminated rags and cleaner in a labeled, covered container.*

Removal
Refer to illustrations 3.2 and 3.6

1   Access to the clutch components is normally accomplished by removing the transaxle, leaving the engine in the vehicle. If, of course, the engine is being removed for major overhaul, then the opportunity should always be taken to check the clutch for wear and replace worn components as necessary. However, the relatively low cost of the clutch components compared to the time and labor involved in gaining access to them warrants their replacement any time the engine or transaxle is removed, unless they are new or in near-perfect condition.

The following procedures assume that the engine will stay in place.

2   Disconnect the clutch hydraulic line from the clutch release cylinder **(see illustration)**. Have rags handy as some fluid will be lost as the line is disconnected. **Caution:** *Don't allow brake fluid to come into contact with paint, as it will damage the finish.*

3   Remove the clutch master cylinder assembly (see Section 4).

4   Remove the transaxle from the vehicle (see Chapter 7A).

5   To support the clutch disc during removal, install a clutch alignment tool through the clutch disc hub **(see illustration 3.14)**.

6   Carefully inspect the flywheel and pressure plate for indexing marks. The marks are usually an X, an O or a white letter. If they cannot be found, scribe marks yourself so the pressure plate and the flywheel will be in the same alignment during installation **(see illustration)**.

7   Slowly loosen the pressure plate-to-flywheel bolts. Work in a diagonal pattern and loosen each bolt a little at a time until all spring pressure is relieved. Then hold the pressure plate securely and completely remove the bolts, followed by the pressure plate and clutch disc.

Inspection
Refer to illustrations 3.10, 3.12a and 3.12b

8   Ordinarily, when a problem occurs in the clutch, it can be attributed to wear of the clutch driven plate assembly (clutch disc). However, all components should be inspected at this time.

9   Inspect the flywheel for cracks, heat checking, score marks and other damage. If the imperfections are slight, a machine shop can resurface it to make it flat and smooth. Refer to Chapter 2 for the flywheel removal procedure.

10   Inspect the lining on the clutch disc. There should be at least 1/16-inch of lining above the rivet heads. Check for loose rivets, distortion, cracks, broken springs and other obvious damage **(see illustration)**. As mentioned above, ordinarily the clutch disc is replaced as a matter of course, so if in doubt about the condition, replace it with a new one.

11   Inspect the clutch release cylinder and bearing (see Section 5).

12   Check the machined surface and the diaphragm spring fingers of the pressure plate **(see illustrations)**. If the surface is grooved or otherwise damaged, replace the pressure plate assembly. Also check for obvious damage, distortion, cracking, etc. Light glazing can be removed with emery cloth or sandpaper. If a new pressure plate is indicated, new or factory rebuilt units are available.

Installation
Refer to illustrations 3.14 and 3.21

13   Before installation, carefully wipe the flywheel and pressure plate machined surfaces clean. It's important that no oil or grease is on these surfaces or the lining of the clutch disc. Handle these parts only with clean hands.

14   Position the clutch disc and pressure plate with the clutch held in place with an alignment tool **(see illustration)**. Make sure it's installed properly (most replacement clutch plates will be marked "flywheel

3.10 Examine the clutch disc for evidence of excessive wear, such as smeared friction material, chewed-up rivets, worn hub splines and distorted damper cushions or springs

1   **Lining** - wears down in use
2   **Rivets** - secure the lining and can damage the flywheel or pressure plate if allowed to contact the surfaces
3   **Markings** - usually says something like "Flywheel side"

side" or something similar - if not marked, install the clutch disc with the damper springs or cushion toward the transaxle).
15   Tighten the pressure plate-to-flywheel bolts only finger tight, working around the pressure plate.
16   Center the clutch disc by ensuring the alignment tool is through the splined hub and into the recess in the crankshaft. Wiggle the tool up, down or side-to-side as needed to bottom the tool. Tighten the pressure plate-to-flywheel bolts a little at a time, working in a criss-cross pattern to prevent distortion of the cover. After all of the bolts are snug, tighten them to the torque listed in this Chapter's Specifications. Remove the alignment tool.
17   Using high-temperature grease, lubricate the inner groove of the release bearing (see Section 4). Also place grease on the transaxle input shaft bearing retainer.
18   Install the clutch release cylinder and bearing (see Section 5).
19   Install the transaxle (see Chapter 7A).
20   Install the clutch master cylinder (see Section 4).
21   Connect the clutch hydraulic line to the clutch release cylinder **(see illustration)**.
22   Bleed the clutch hydraulic system (see Section 6).

NORMAL FINGER WEAR

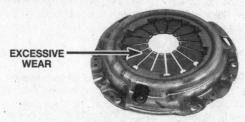

EXCESSIVE WEAR

EXCESSIVE FINGER WEAR

BROKEN OR BENT FINGERS

3.12a Replace the pressure plate if excessive wear is noted

3.12b Examine the pressure plate friction surface for score marks, cracks and evidence of overheating (blue spots)

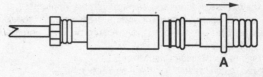

A

`38016-8-3.21 HAYNES`

3.21 To connect the clutch hydraulic line, pull back on the release slide (A), hold it there, and push the line into the fitting until you hear a click

8

3.14 Center the clutch disc in the pressure plate with a clutch alignment tool

4   Clutch master cylinder - removal and installation

1   Remove the sound insulator from under the left side of the dashboard (see Chapter 11).
2   Using a flashlight, locate the clutch master cylinder pushrod. Disconnect the pushrod from the clutch pedal.
3   Remove the two nuts which attach the clutch master cylinder to the firewall. Detach the clutch master cylinder from the firewall.
4   Detach the clutch master cylinder remote reservoir from the firewall.
5   Disconnect the clutch hydraulic line from the clutch release cylinder (see illustration 3.2). Have rags handy as some fluid will be lost as the line is removed. **Caution:** *Don't allow brake fluid to come into contact with paint, as it will damage the finish.*
6   Remove the clutch master cylinder, remote reservoir and clutch hydraulic line as a single assembly.
7   Installation is the reverse of removal. Be sure to tighten the clutch master cylinder nuts to the torque listed in this Chapter's Specifications. To reconnect the clutch hydraulic line at the clutch release cylinder, refer to illustration 3.21.
8   When you're done, bleed the clutch hydraulic system (see Section 6).

5   Clutch release cylinder and bearing - removal, inspection and installation

Warning: *Dust produced by clutch wear and deposited on clutch components may contain asbestos, which is hazardous to your health. DO NOT blow it out with compressed air and DO NOT inhale it. DO NOT use gasoline or petroleum-based solvents to remove the dust. Brake system cleaner should be used to flush it into a drain pan. After the clutch components are wiped clean with a rag, dispose of the contaminated rags and cleaner in a labeled, covered container.*

Removal

1   Remove the clutch master cylinder (see Section 4).
2   Remove the transaxle (see Chapter 7A).
3   Slide the clutch release cylinder and bearing assembly off the transaxle input shaft.

Inspection

4   Hold the bearing by the outer race and rotate the inner race while applying pressure. If the bearing doesn't turn smoothly or if it's noisy, replace the bearing/hub assembly with a new one. Wipe the bearing with a clean rag and inspect it for damage, wear and cracks. Don't immerse the bearing in solvent - it's sealed for life and to do so would ruin it. Also check the release cylinder for leaks. A thin coating of hydraulic fluid near the seal is acceptable, but a liberal amount of fluid indicates a damaged seal. If the release cylinder and bearing assembly is noisy or leaking, replace it.

Installation

5   Fill the inner groove of the release bearing with high-temperature grease. Also apply a light coat of the same grease to the transaxle input shaft splines.
6   Slide the release cylinder and bearing onto the input shaft.
7   Apply a light coat of high-temperature grease to the face of the release bearing where it contacts the pressure plate diaphragm fingers.
8   Install the transaxle (see Chapter 7A).
9   Install the clutch master cylinder (see Section 4).
10   Bleed the clutch hydraulic system (see Section 6).

6   Clutch hydraulic system - bleeding

1   The hydraulic system should be bled of all air whenever any part of the system has been removed or if the fluid level has been allowed to fall so low that air has been drawn into the master cylinder. The procedure is similar to bleeding a brake system.
2   Fill the master cylinder with new brake fluid conforming to DOT 3 specifications. **Caution:** *Do not reuse any of the fluid coming from the system during the bleeding operation or use fluid which has been inside an open container for an extended period of time.*
3   Raise the vehicle and place it securely on jackstands to gain access to the release cylinder, which is located on the left side of the clutch housing.
4   Locate the bleeder valve on the clutch release cylinder (right above the fitting for the hydraulic fluid line). Remove the dust cap which fits over the bleeder valve and push a length of plastic hose over the valve. Place the other end of the hose into a clear container with about two inches of brake fluid in it. The hose end must be submerged in the fluid.
5   Have an assistant depress the clutch pedal and hold it. Open the bleeder valve on the release cylinder, allowing fluid to flow through the hose. Close the bleeder valve when fluid stops flowing from the hose. Once closed, have your assistant release the pedal.
6   Continue this process until all air is evacuated from the system, indicated by a full, solid stream of fluid being ejected from the bleeder valve each time and no air bubbles in the hose or container. Keep a close watch on the fluid level inside the clutch master cylinder reservoir; if the level drops too low, air will be sucked back into the system and the process will have to be started all over again.
7   Install the dust cap and lower the vehicle. Check carefully for proper operation before placing the vehicle in normal service.

7   Clutch start switch - check and replacement

1   Remove the left side under-dash panel.

Check

2   Verify that the engine will not start when the clutch pedal is released. Now, depress the clutch pedal - the engine should start.
3   Locate the switch at the upper end of the clutch pedal and unplug the electrical connector.
4   Using an ohmmeter, verify that there is continuity between the terminals of the clutch start switch when the pedal is depressed. There should be no continuity when the pedal is released.
5   If the switch does not work as described, replace it.

Replacement

6   Unplug the electrical connector from the switch.
7   Detach the clutch pedal position switch from the clutch pedal bracket.
8   Installation is the reverse of removal. The switch is self-adjusting, so there's no need for adjustment.
9   Verify that the engine doesn't start when the clutch pedal is released, and does start when the pedal is depressed.

8   Driveaxles - general information and inspection

1   Power is transmitted from the transaxle to the wheels through a pair of driveaxles. The inner end of each driveaxle is splined into the differential side gears, or onto a splined intermediate axleshaft. The outer ends of the driveaxles are splined to the axle hubs and locked in place by a large nut.
2   The inner ends of the driveaxles are equipped with sliding constant velocity joints, which are capable of both angular and axial motion. Most inner joint assemblies consist of a tripot bearing and a joint housing (outer race) in which the joint is free to slide in and out as the driveaxle moves up and down with the wheel; some inner joints on 1995 models are the "cross-groove," or "ball-and-cage" type, which consists of six ball bearings riding between an inner race and outer race and held in position by a cage. The inner joints can be disassembled, cleaned,

9.5 To hold the hub/disc while breaking loose the driveaxle hub nut, jam a punch into the cooling vanes of the disc

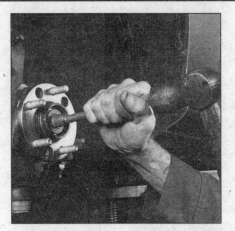

9.6 Using a brass punch, strike the end of the driveaxle sharply with a hammer; when it breaks free, it will move noticeably

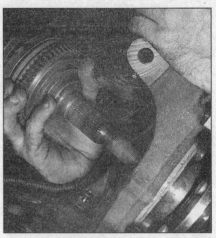

9.9 Pull the steering knuckle out and slide the end of the driveaxle out of the hub

inspected and repacked, but they cannot be overhauled. If any parts are damaged, an inner joint must be replaced as a unit.

3   The outer CV joints are the cross-groove," or "ball-and-cage," type. The outer joints are capable of angular but not axial movement. The outer joints can be disassembled, cleaned, inspected and repacked, but they cannot be overhauled. If any parts are damaged, an outer joint must be replaced as a unit.

4   The boots should be inspected periodically for damage and leaking lubricant. Torn CV joint boots must be replaced immediately or the joints can be damaged. Boot replacement involves removal of the driveaxle (see Section 9). **Note:** *Some auto parts stores carry "split" type replacement boots, which can be installed without removing the driveaxle from the vehicle. This is a convenient alternative; however, the driveaxle should be removed and the CV joint disassembled and cleaned to ensure the joint is free from contaminants such as moisture and dirt which will accelerate CV joint wear.* The most common symptom of worn or damaged CV joints, besides lubricant leaks, is a clicking noise in turns, a clunk when accelerating after coasting and vibration at highway speeds. To check for wear in the CV joints and driveaxle shafts, grasp each axle (one at a time) and rotate it in both directions while holding the CV joint housings, feeling for play indicating worn splines or sloppy CV joints. Also check the driveaxle shafts for cracks, dents and distortion.

9.10 To separate the inner end of the driveaxle from the transaxle, pry on the CV joint housing like this with a large screwdriver or prybar - you may need to give the prybar a sharp rap with a brass hammer

9   Driveaxle - removal and installation

Removal

Refer to illustrations 9.5, 9.6, 9.9 and 9.10

1   Disconnect the cable from the negative terminal of the battery. **Caution:** *If the stereo in your vehicle is equipped with an anti-theft system, make sure you have the correct activation code before disconnecting the battery.*

2   Set the parking brake.

3   Loosen the front wheel lug nuts, raise the vehicle and support it securely on jackstands.

4   Remove the wheel.

5   Remove the driveaxle/hub nut and washer. To prevent the disc/hub from turning, wedge a long punch into the brake disc cooling vanes and allow it to rest against the caliper anchor **(see illustration)**.

6   To loosen the driveaxle from the hub splines, tap the end of the driveaxle with a soft-faced hammer or a hammer and a brass punch **(see illustration)**. **Note:** *Don't attempt to push the end of the driveaxle through the hub yet. Applying force to the end of the driveaxle, beyond just breaking it loose from the hub, can damage the driveaxle or transaxle.* If the driveaxle is stuck in the hub splines and won't move, it

may be necessary to remove the brake disc (see Chapter 9) and push it from the hub with a two-jaw puller after Step 8 is performed.

7   Place a drain pan underneath the transaxle to catch the lubricant that will spill out when the driveaxles are removed.

8   Separate the strut from the steering knuckle (see Chapter 10).

9   Pull out on the steering knuckle and detach the driveaxle from the hub **(see illustration)**. Don't let the driveaxle hang by the inner CV joint after the outer end has been detached from the steering knuckle, as the inner joint could become damaged. Support the outer end of the driveaxle with a piece of wire, if necessary.

10   Carefully pry the inner CV joint out of the transaxle **(see illustration)** or off the splined end of the intermediate shaft (right driveaxle only, on vehicles so equipped).

11   Refer to Chapter 7 for the driveaxle oil seal replacement procedure.

Installation

12   Installation is the reverse of the removal procedure, but with the following additional points:

a)   *Seat the inner CV joint in the differential side gear by positioning the end of a large screwdriver in the groove in the CV joint housing and tapping it into position with a hammer. Once this has been done pull out on the joint housing to make sure the retaining ring has seated.*

11.3a Cut off the boot seal retaining clamps, using wire cutters or a chisel and hammer

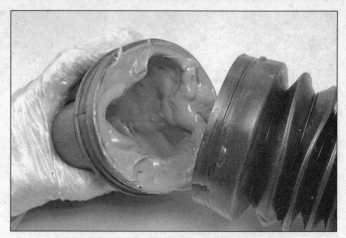

11.3b Slide the housing off the spider assembly

b) Tighten the strut-to-knuckle bolts/nuts to the torque listed in the Chapter 10 Specifications.

c) Install a new driveaxle/hub nut and tighten it to the torque listed in this Chapter's Specifications.

d) Install the wheel and lug nuts, lower the vehicle and tighten the lug nuts to the torque listed in the Chapter 1 Specifications.

e) Check the transaxle lubricant and add, if necessary, to bring it to the proper level (see Chapter 1).

10   Intermediate shaft - removal and installation

1   Loosen the right front wheel lug nuts. Raise the vehicle and support it securely on jackstands. Remove the right wheel.

2   Remove the right driveaxle assembly (see Section 9).

3   Unbolt the intermediate shaft support bracket from the engine block.

4   Carefully pull the intermediate shaft assembly from the transaxle. Make sure the splines on the inner end of the shaft don't damage the side-gear seal lip. Remove the intermediate shaft assembly.

5   Turn the shaft and verify that the bearing is in good condition. If the bearing makes grinding noises when the shaft is turned, have the bearing pressed off and a new one installed by an automotive machine shop.

6   Installation is the reverse of removal. Be sure to tighten the intermediate shaft support bolts to the torque listed in this Chapter's Specifications.

11   Driveaxle boot - replacement

Note: If the CV joint boots must be replaced, explore all options before beginning the job. Complete rebuilt driveaxles are available on an exchange basis, which eliminates much time and work. Whichever route you choose to take, check on the cost and availability of parts before disassembling the vehicle.

1   Remove the driveaxle (see Section 10).

2   Place the driveaxle in a vise lined with rags to avoid damage to the axleshaft. Check the CV joint for excessive play in the radial direction, which indicates worn parts. Check for smooth operation throughout the full range of motion for each CV joint. If a boot is torn, disassemble the joint, clean the components and inspect for damage due to loss of lubrication and possible contamination by foreign matter.

Inner CV joint

Refer to illustrations 11.3a through 11.3w

Note: Some 1995 models use a "ball-and-cage" type inner CV joint instead of the usual tripot design. Aside from a wire retainer ring which must be removed before the ball-and-cage assembly can be removed from the CV joint housing, this unit is similar in construction to the outer CV joint, which is covered in the sequence beginning with illustration 11.4a.

3   To replace the inner boot, refer to the accompanying illustrations **(see illustrations 11.3a through 11.3w)**.

11.3c On models with a ball-and-cage inner joint, pry out the wire ring bearing retainer with a screwdriver

11.3d Slide the boot towards the center of the driveaxle

11.3e If the spider is positioned by a stop ring, spread the ends of the stop ring apart and slide it towards the center of the shaft

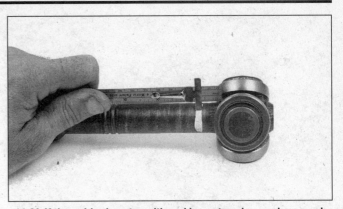

11.3f If the spider is not positioned by a stop ring, make a mark on the axleshaft (such as the white paint mark shown here), measure the distance from your mark to the tripot, and jot down this measurement for reassembly; when it's reinstalled, the spider must be placed in exactly the same position on the axleshaft splines

11.3g Slide the spider assembly back to expose the retaining ring and pry off the ring

11.3h Carefully tap the spider off the axleshaft with a brass punch (but don't hit it so hard that it flies off, or you'll be picking up needle bearings!)

11.3i When you slide the spider off the driveaxle, hold the bearings in place with your hand; even better, use tape or a cloth wrapped around the spider bearing assembly to retain them

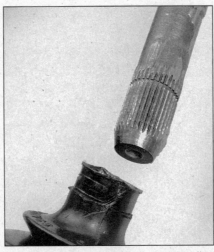

11.3j Slide the boot off the axleshaft (and the stop ring, if equipped)

11.3k Clean all of the old grease out of the housing and spider assembly, then remove each bearing, one at time

8

11.3l Carefully disassemble each section of the spider assembly, clean the needle bearings with solvent and inspect the rollers, spider cross, bearings and housing for scoring, pitting and other signs of abnormal wear

11.3m Apply a coat of CV joint grease to the inner bearing surfaces to hold the needle bearings in place and slide the bearing over them

11.3n Wrap the axleshaft splines with tape to avoid damaging the boot, then slide the small clamp and boot onto the axleshaft

11.3o Slide the spider stop ring, if equipped, onto the axleshaft, past the groove in which it seats

11.3p Install the spider bearing with the recess in the counterbore facing the end of the driveaxle

11.3q Use a screwdriver to install the spider retaining ring, then slide the spider assembly against it and install the stop ring, if equipped, in its groove; on units with no stop ring, position the spider the same distance from your mark that it was at prior to disassembly

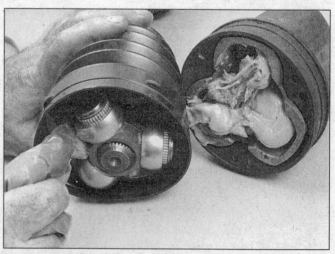

11.3r Pack the housing with half of the grease furnished with the new boot and place the remainder in the boot

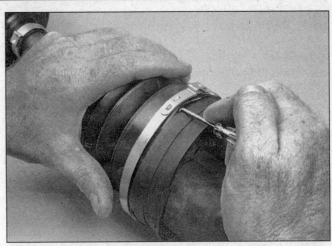

11.3s With the retaining clamps in place (but not tightened), install the tripot housing

11.3t Seat the boot in the housing and axle seal grooves - a small screwdriver can make the job easier (make sure the boot isn't dimpled, stretched or out of shape)

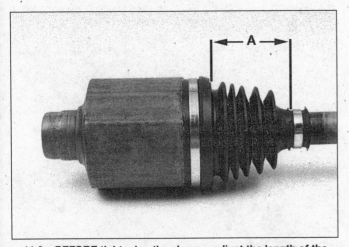

11.3u BEFORE tightening the clamps, adjust the length of the joint so that the indicated distance (measured from the inner ridge for the small clamp to the edge of the end of the boot folds on 1998 and later 2.2L models . . .

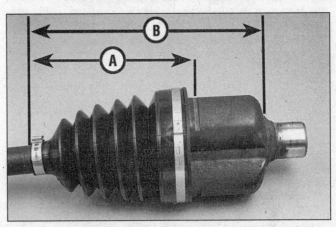

11.3v . . . or from the small end of the boot to the groove in the CV joint housing [A] on all others except 1995 models with tripot inner joints - on those models, measure dimension B) is the same as the dimension listed in this Chapter's Specifications

Outer CV joint

Refer to illustrations 11.4a through 11.4u

4   Refer to the accompanying illustrations and perform the outer CV joint boot replacement procedure (see illustrations 11.4a through 11.4v).

11.3w With the joint at the proper length, equalize the pressure in the boot by inserting a small screwdriver between the boot and the housing, then secure the boot clamps with special pliers (available at auto parts stores)

11.4a Cut off the band retaining the boot to the shaft, then slide the boot toward the center of the shaft

8

11.4b Some outer CV joints can be removed by spreading apart the ends of an internal snap-ring, then sliding off (or tapping off, if necessary) the CV joint assembly (you can identify this type by the large recess into which the snap-ring pliers are inserted to spread the snap-ring; the other type of outer joint, discussed in the next caption, doesn't have this recess, so there's no way to spread the snap-ring)

11.4c On joints that aren't retained by a snap-ring, clean all grease off the axleshaft and paint a mark on the shaft, then measure the distance from your mark to the face of the inner race and record this measurement; the inner race must be installed on the axleshaft in exactly the same position in which it was installed prior to removal

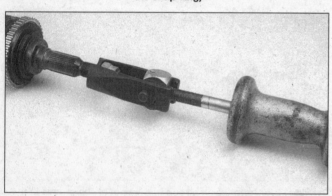

11.4d Outer CV joints without a recess for spreading apart the internal snap-ring must be removed with a slide hammer; you'll need an adapter) and a slide hammer setup such as the one shown here

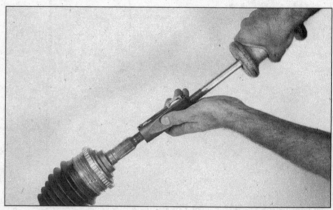

11.4e With the axleshaft firmly clamped down in a bench vise and the adapter gripping the driveaxle/hub nut, carefully extract the outer CV joint from the axleshaft

11.4f Press down on the inner race far enough to allow a ball bearing to be removed - if it's difficult to tilt, gently tap the cage and inner race with a brass punch and hammer

11.4g Pry the balls out of the cage, one at a time

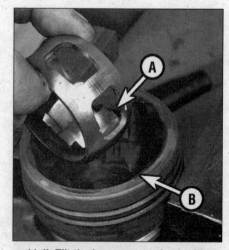

11.4h Tilt the inner race and cage 90-degrees, then align the windows in the cage (A) with the lands of the housing (B) and rotate the inner race up and out of the outer race

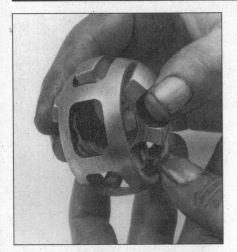

11.4i Align the inner race lands with the cage window and rotate the inner race out of the cage

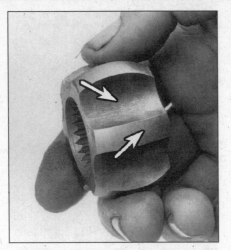

11.4j After cleaning the components with solvent, check the inner race lands and grooves for pitting and score marks

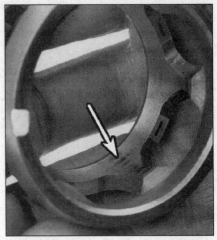

11.4k Check the cage for cracks, pitting and score marks - shiny spots are normal and don't affect operation

11.4l With the race and cage tilted at 90-degrees, lower the assembly into the housing

11.4m Rotate the assembly by gently tapping with a hammer and brass punch, then . . .

11.4n . . . press the balls into the cage windows, repeating until all of the balls are installed

11.4o On models so equipped, use needle-nose pliers to lower a new snap-ring into the groove . . .

11.4p . . . then seat it into the groove with snap-ring pliers

8

11.4q Apply grease through the splined hole, then insert a wooden dowel (with a diameter slightly less than that of the axle) through the splined hole and push down - the dowel will force the grease into the joint - repeat until the bearing is completely packed

11.4r Install the small clamp and the boot on the driveaxle and apply grease to the inside of the axle boot until . . .

11.4s . . . the level is up to the end of axle

11.4t Position the CV joint assembly on the driveaxle, aligning the splines, then use a soft-face hammer to drive the joint onto the driveaxle until the snap-ring is seated in the groove (models with a snap-ring) or until the inner race of the joint is the same distance from the paint mark (applied as in Step 11.4) as it was on removal

11.4u Seat the inner end of the boot in the groove and install the retaining clamp, then do the same on the other end of the boot - tighten boot clamps with the special tool (see illustration 11.3w)

Chapter 9 Brakes

Contents

Specifications

General

Brake fluid type	See Chapter 1

Disc brakes

Pad minimum thickness	See Chapter 1
Disc minimum thickness	Refer to the dimension cast into the disc
Disc runout (maximum)	0.003 inch
Disc thickness variation (maximum)	0.0005 inch

Rear drum brakes

Shoe lining minimum thickness	See Chapter 1
Drum maximum diameter	Refer to the dimension cast into the drum
Drum taper (maximum)	0.003 inch
Out-of-round (maximum)	0.002 inch

Torque specifications

	Ft-lbs (unless otherwise indicated)
Brake hose-to-caliper banjo bolt	33
Caliper mounting bolts	38
Wheel cylinder-to-backing plate bolts	180 in-lbs
Master cylinder-to-power brake booster nuts	20
Power brake booster retaining nuts	20

9

1   General information

All vehicles covered by this manual are equipped with hydraulically operated front and rear brake systems. The front brakes are disc type, and the rear brakes are drum type. All brakes are self-adjusting.

The hydraulic system consists of two separate circuits. The master cylinder has separate reservoirs for the two circuits; in the event of a leak or failure in one hydraulic circuit, the other circuit will remain operative. A visual warning of circuit failure, air in the system, or other pressure differential conditions in the brake system is given by a warning light activated by a failure warning switch in the master cylinder.

The parking brake mechanically operates the rear brakes only. It's activated by a pull-handle in the center console between the front seats.

The power brake booster, located in the engine compartment on the firewall, uses engine manifold vacuum and atmospheric pressure to provide assistance to the hydraulically operated brakes.

After completing any operation involving the disassembly of any part of the brake system, always test drive the vehicle to check for proper braking performance before resuming normal driving. Test the brakes while driving on a clean, dry, flat surface. Conditions other than these can lead to inaccurate test results. Test the brakes at various speeds with both light and heavy pedal pressure. The vehicle should stop evenly without pulling to one side or the other. Avoid locking the brakes because this slides the tires and diminishes braking efficiency and control.

Tires, vehicle load and front end alignment are factors which also affect braking performance.

2   Anti-lock Brake System (ABS) and Enhanced Traction System (ETS) - general information

Anti-lock Brake System (ABS)

Refer to illustrations 2.2, 2.4a and 2.4b

A four-wheel Anti-lock Brake System (ABS) maintains vehicle maneuverability, directional stability, and optimum deceleration under severe braking conditions on most road surfaces. It does so by monitoring the rotational speed of the wheels and controlling the brake line pressure to the wheels during braking. This prevents the wheels from locking up on slippery roads or during hard braking.

On 1995 through 1999 models, the hydraulic control unit/motor pack assembly **(see illustration)**, which is located in the left front corner of the engine compartment, is bolted to the left side of the master cylinder. The control unit/motor pack controls hydraulic pressure to the front calipers and rear wheel cylinders by modulating hydraulic pressure to prevent wheel lock-up.

On 1995 through 1999 models, the Electronic Brake Control Module (EBCM) is located above the left (driver's side) kick panel. The

2.2 The hydraulic control unit/motor pack assembly is an integral part of the master cylinder assembly (1995 through 1999)

EBCM monitors the ABS system and controls the anti-lock valve solenoids. It accepts and processes information received from the brake switch and wheel speed sensors to control the hydraulic line pressure and avoid wheel lock-up. It also monitors the system and stores fault codes which indicates specific problems.

The 2000 model varies from previous models with the Electronic Brake Control Module/Brake Pressure Modulator Valve (EBCM/BPMV) being a separate unit that is not an integral part of the master cylinder - it is located to the left side of the master cylinder. This system operates the ABS system in the same manner as the previous two different assemblies.

Each wheel speed sensor assembly consists of a variable reluctance sensor and a "toothed ring," with an air gap between them. The front wheel speed sensors **(see illustration)** are mounted on the inner side of each steering knuckle; the toothed rings are installed on the outer CV joints. The rear sensors **(see illustration)** are installed in the rear brake backing plates; the toothed rings are installed on the back side of each hub and bearing unit. The air gap between the sensors and the toothed rings is not adjustable, and the sensors themselves are not rebuildable. If a rear sensor or ring malfunctions, the entire hub and bearing assembly must be replaced. The front sensor can be replaced separately.

A wheel speed sensor measures wheel speed by monitoring the rotation of the toothed ring. As the teeth of the ring move through the magnetic field of the sensor, an AC voltage is generated. This signal frequency increases or decreases in proportion to the speed of the wheel. The EBCM monitors these three signals for changes in wheel speed; if it detects the sudden deceleration of a wheel, i.e. wheel lockup, the EBCM activates the ABS system.

2.4a The front wheel speed sensors are mounted on the inner side of the steering knuckles, in close proximity to the toothed rings on the outer CV joints

2.4b The rear wheel speed sensors are mounted in the rear drum brake backing plates; the toothed rings (not visible in this photo) are installed on the inner side of the flange for each hub and bearing unit

3.3 Depress the piston into the caliper with a large C-clamp

The ABS system has self-diagnostic capabilities. Each time the vehicle is started, the EBCM runs a self-test. The red BRAKE warning light should come on briefly then go out. The EBCM also monitors the ABS system continuously during vehicle operation. If the ABS INOP light on the dash comes on while you're driving, there is a fault somewhere in the ABS system and the ABS system may be inoperative, but the brakes should still function in their non-ABS mode. Take the vehicle to a dealer service department or other qualified repair shop immediately and have the ABS serviced.

If the ABS INOP light *flashes*, however, there is a more serious fault in the ABS system which may have affected the regular braking system. Pull over immediately and have the vehicle towed to a dealer or other qualified repair shop for service.

Although a special electronic tester is necessary to properly diagnose the system, the home mechanic can perform a few preliminary checks before taking the vehicle to a dealer service department or repair shop which is equipped with this tester.

a) Make sure the brake calipers are in good condition.
b) Check the electrical connector at the controller.
c) Check the fuses.
d) Follow the wiring harness to the speed sensors and brake light switch and make sure all connections are secure and the wiring isn't damaged.

If the above preliminary checks don't rectify the problem, the vehicle should be diagnosed by a dealer service department or other qualified repair shop.

Enhanced Traction System (ETS)

The Enhanced Traction System (ETS) limits wheel slip during acceleration by cutting fuel to certain cylinders, by retarding the ignition, by upshifting the transaxle, or by some combination of the three. The EBCM monitors wheel speed through the ABS wheel speed sensors. When one or more wheels begins to spin faster than the others, the EBCM calculates the reduction in torque needed to stop wheelspin and restore traction. Depending on the torque reduction necessary, the calculated torque reduction is some combination of fuel cutoff, spark retard, and/or upshifting the transaxle. This recommendation is relayed from the EBCM to the Power Train Control Module (PCM), which manipulates fuel cutoff and spark retard to produce the requisite torque. If the wheels still spin, the PCM requests an upshift to the next higher gear.

3   Disc brake pads - replacement

Refer to illustrations 3.3 and 3.4a through 3.4j
Warning: *Disc brake pads must be replaced on both front wheels at the same time - never replace the pads on only one wheel. Also, the dust created by the brake system may contain asbestos, which is harmful to your health. Never blow it out with compressed air and don't inhale any of it. An approved filtering mask should be worn when working on the brakes. Do not, under any circumstances, use petroleum-based solvents to clean brake parts. Use brake system cleaner only!*

1   Loosen the front wheel lug nuts, raise the front of the vehicle and support it securely on jackstands. Apply the parking brake. Remove the front wheels.

2   Remove about two-thirds of the fluid from the master cylinder reservoir and discard it. **Caution:** *Do not spill brake fluid on painted surfaces.* Position a drain pan under the brake assembly and clean the caliper and surrounding area with brake system cleaner.

3   Position a large C-clamp over the caliper and squeeze the piston back into in its bore **(see illustration)** to provide room for the new brake pads. As the piston is depressed to the bottom of its caliper bore, the fluid in the master cylinder will rise. Make sure it doesn't overflow. If necessary, siphon off some of the fluid.

4   Work on one brake assembly at a time. Follow the accompanying photos, beginning with **illustration 3.4a**. Be sure to stay in order and read the caption under each illustration.

5   While the pads are removed, inspect the caliper for seal leaks (evidenced by moisture around the piston boot) and for any damage to the piston dust boots. If excessive moisture is evident, rebuild the caliper (see Section 4). When you're done, make sure you tighten the caliper bolts to the torque listed in this Chapter's Specifications.

3.4a Before getting started, wash the front brake assembly with brake cleaner to protect yourself from brake dust, which might contain asbestos, a known carcinogen

3.4b To detach the front caliper from the steering knuckle, remove these two bolts (upper and lower arrows); don't remove the banjo bolt (center arrow) for the brake hose unless you intend to overhaul the caliper

3.4c While you're replacing the pads, hang the caliper with a piece of wire

9

3.4d Remove the outer brake pad

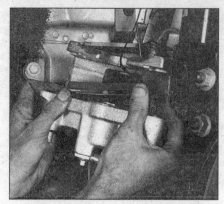

3.4e Remove the inner brake pad

3.4f Pull the two caliper guide pins out of their sleeves, clean them, inspect them for wear, replacing as necessary . . .

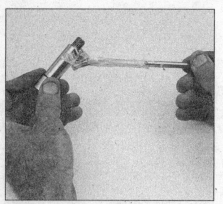

3.4g . . . then lubricate them with silicone grease and install them in the caliper

3.4h Apply anti-squeal compound to the backing plates of the new pads

3.4i Install the new inner pad

6   Install the brake pads on the opposite wheel, then install the wheels and lower the vehicle. Tighten the lug nuts to the torque listed in the Chapter 1 Specifications.

7   Pump the brakes several times to seat the pads against the disc, then check the fluid level and add some, if necessary (see Chapter 1).

8   Check the operation of the brakes before driving the vehicle in traffic. Try to avoid heavy brake applications until the brakes have been applied lightly several times to seat the pads.

---

4   Disc brake caliper - removal, overhaul and installation

Warning: *The dust created by the brake system may contain asbestos, which is harmful to your health. Never blow it out with compressed air and don't inhale any of it. An approved filtering mask should be worn when working on the brakes. Do not, under any circumstances, use petroleum-based solvents to clean brake parts. Use brake system cleaner only!*

Note: *If an overhaul is indicated (usually because of fluid leaks, a stuck piston or broken bleeder screw) explore all options before beginning this procedure. New and factory rebuilt calipers are available on an exchange basis, which makes this job quite easy. If you decide to rebuild the calipers, make sure rebuild kits are available before proceeding. Always rebuild or replace the calipers in pairs - never rebuild just one of them.*

Removal

1   Remove the cover from the brake fluid reservoir and siphon off two-thirds of the fluid into a container and discard it.

2   Loosen the wheel lug nuts, raise the front or rear of the vehicle and place it securely on jackstands. Remove the front or rear wheel.

3   Reinstall one wheel lug nut (flat side toward the disc) to hold the disc in place. Don't remove both calipers at the same time. Instead,

remove only one caliper at a time so you can use the other assembled unit for reference.

4   Push the caliper piston back into its bore with a C-clamp **(see illustration 3.3)**. As the piston is depressed to the bottom of the caliper bore, the fluid in the master cylinder will rise. Make sure that it does not overflow.

5   Remove the banjo fitting bolt holding the brake hose, then remove and discard the sealing washers found on either side of the banjo fitting. Always use new sealing washers when reinstalling the brake hose.

6   To prevent brake fluid leakage and contamination, plug the openings in the caliper and brake hose. **Note:** *If you're just removing the*

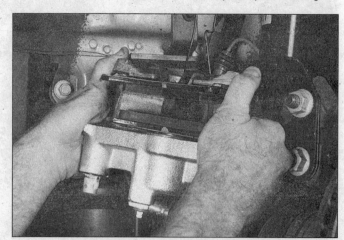

3.4j Install the new outer pad, then install the caliper and tighten the caliper guide pins to the torque listed in this Chapter's Specifications

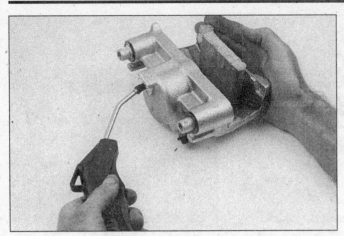

4.10 Using a block of wood as a cushion, ease the piston out of the caliper bore with compressed air

caliper for access to other components, don't disconnect the hose.

7   Remove the caliper (see Section 3) and separate the caliper from the disc.

8   Remove the brake pads from the caliper (see Section 3).

Overhaul

Refer to illustrations 4.10, 4.11, 4.12, 4.16, 4.17, 4.18 and 4.19

Note: *Purchase a brake caliper overhaul kit for your particular vehicle before beginning this procedure.*

9   Clean the exterior of the brake caliper with brake system cleaner (never use gasoline, kerosene or any petroleum-based solvents), then place the caliper on a clean workbench.

10   Place a wooden block or shop rag in the caliper as a cushion, then use compressed air to remove the piston from the caliper **(see illustration)**. Use only enough air pressure to ease the piston out of the bore. If the piston is blown out, even with the cushion in place, it may be damaged. **Warning:** *Never place your fingers in front of the piston in an attempt to catch or protect it when applying compressed air - serious injury could occur.*

11   Carefully pry the dust boot out of the caliper bore **(see illustration)**.

12   Using a wooden or plastic tool, remove the piston seal from the groove in the caliper bore **(see illustration)**. Metal tools may cause bore damage.

13   Remove the caliper bleeder valve, then remove and discard the sleeves and bushings from the caliper ears. Also discard all rubber parts.

14   Clean the remaining parts with brake fluid or brake system cleaner. Allow them to drain and then shake them vigorously to remove as much fluid as possible.

15   Carefully examine the piston for nicks and burrs and loss of plating. If surface defects are present, parts must be replaced. Check the caliper bore in a similar way, but light polishing with crocus cloth is permissible to remove light corrosion and stains. Discard the mounting bolts if they are corroded or damaged.

16   When assembling, lubricate the piston bore and seal with clean brake fluid; position the seal in the caliper bore groove. Make sure the seal seats properly and isn't twisted **(see illustration)**.

17   Lubricate the piston with clean brake fluid, then install a new boot in the piston groove **(see illustration)**.

18   Insert the piston squarely into the caliper bore, then apply force to bottom the piston in the bore **(see illustration)**.

19   Position the dust boot in the caliper counterbore, then use a seal driver to drive it into position **(see illustration)**. If you don't have a seal driver, carefully tap the boot into place with a blunt punch, contacting

4.11 Pry the dust boot out of the caliper bore with a screwdriver - make sure you don't nick or gouge anything

4.12 To remove the seal from the caliper bore, use a plastic or wooden tool, such as a pencil

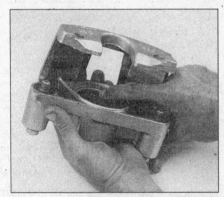

4.16 Lubricate the piston bore and seal with clean brake fluid before installing the new seal in the caliper bore groove

4.17 Install a new dust boot in the piston groove

4.18 Lubricate the piston with clean brake fluid, insert it squarely in the bore and carefully push it into the caliper

4.19 Use a hammer and seal driver to seat the boot in the caliper

9

5.5 Check the runout of the brake disc
with a dial indicator

5.6a The minimum allowable thickness of
the disc is cast into the hub area

5.6b Use a micrometer to check the
thickness of the disc and compare this
measurement to the minimum allowable
thickness cast into the disc

only the outer circumference (casing) of the boot. Make sure that the boot is evenly installed below the caliper face.

20 Install the bleeder valve.

21 Installation is the reverse of removal. Always use new sealing washers when connecting the brake hose. Tighten the caliper guide pins or mounting bolts to the torque listed in this Chapter's Specifications. Be sure to fill the master cylinder and bleed the brakes (see Section 10).

5 Brake disc - inspection, removal and installation

Refer to illustrations 5.5, 5.6a and 5.6b

1 Loosen the wheel lug nuts, raise the vehicle and place it securely on jackstands. Remove the wheel.

2 Remove the caliper assembly (see Section 4). It is not necessary to disconnect the brake hose. After removing the caliper mounting bolts, hang the caliper out of the way on a piece of wire. Never hang the caliper by the brake hose because damage to the hose will occur.

3 Inspect the disc surfaces. Light scoring or grooving is normal, but deep grooves or severe erosion is not. If pulsating has been noticed during application of the brakes, suspect disc runout.

4 Reinstall the wheel lug nuts - flat side toward the disc - to hold the disc in a flat, vertical plane during the following inspection. You may have to place washers under the lug nuts in order for them to tighten

onto the disc.

5 Attach a dial indicator to the caliper mounting bracket, turn the disc and note the amount of runout (see illustration). Check both inner and outer surfaces. If the runout is more than the specified allowable maximum, the disc must be removed from the vehicle and taken to an automotive machine shop for resurfacing.

6 The minimum thickness of the disc is cast into the inside surface of the hub part of the disc (see illustration). Using a micrometer, measure the thickness of the disc (see illustration). If it is less than the specified minimum, replace the disc. Also measure the disc thickness at several points to determine variations in the surface. Any variation over 0.0005-inch may cause pedal pulsations during brake application. If this condition exists and the disc thickness is not below the minimum, the disc can be removed and taken to an automotive machine shop for resurfacing.

7 If the disc needs to be removed for repair or replacement, it can be pulled off after the lug nuts are removed.

6 Drum brake shoes - replacement

Refer to illustrations 6.4a through 6.4x and 6.5

Warning: *Drum brake shoes must be replaced on both rear wheels at the same time - never replace the shoes on only one wheel. Also, the dust created by the brake system contains asbestos, which is harmful to your health. Never blow it out with compressed air and don't inhale any of it. An approved filtering mask should be worn when working on the brakes. Do not, under any circumstances, use petroleum-based solvents to clean brake parts. Use brake cleaner only!*

Caution: *Whenever the brake shoes are replaced, the return and hold-down springs should also be replaced. Due to the continuous heating/cooling cycle the springs are subjected to, they lose tension over a period of time and may allow the shoes to drag on the drum and wear at a much faster rate than normal.*

Note: *All four rear shoes must be replaced at the same time, but to avoid mixing up parts, work on only one brake assembly at a time.*

1 Put the shift lever in Park (automatic transaxle) or Reverse (manual transaxle). Release the parking brake.

2 Loosen the wheel lug nuts, raise the rear of the vehicle and support it securely on jackstands. Block the front wheels to keep the vehicle from rolling off the jackstands.

3 Remove the rear wheels.

4 Remove the brake drum. **Note:** *If the brake drum is stuck, make sure the parking brake is completely released, then apply some penetrating oil to the hub-to-brake drum joint. Allow the oil to soak in, then try to pull the drum off. If the drum still won't come off, the brake shoes*

6.4a Before removing anything, clean the brake assembly with
brake system cleaner - DO NOT use compressed air to
blow the dust out of the brake assembly!

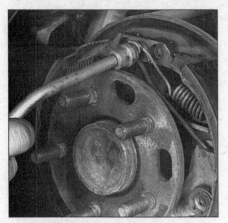

6.4b Remove the return springs with a brake spring tool

6.4c Remove the hold-down springs and pins by pushing the retainer in with and turning it 90-degrees - the tool shown here is available at most auto parts stores

6.4d Lift up on the actuator lever and remove the actuating link from the anchor pin pivot along with the actuator lever and return spring (arrows)

6.4e Spread the shoes apart at the top and remove the parking brake strut

6.4f With the shoe assembly spread to clear the hub flange, lift it away from the backing plate

6.4g Disconnect the parking brake lever from the cable and remove the shoe assembly

will have to be retracted. This is done by slackening the parking brake cables (see Section 12), then tapping off the brake drum with a rubber mallet. To replace the brake shoes, refer to the accompanying photographs, beginning with **illustration 6.4a**. Be sure to follow the sequence step-by-step and read each caption.

5   Before reinstalling the drum, check it for cracks, score marks, deep scratches and hard spots, which will appear as blue discolored

areas. If the hard spots can't be removed with fine emery cloth or if any of the other conditions listed above exist, the drum must be taken to an automotive machine shop to have it resurfaced. **Note:** *The drums should be resurfaced, regardless of the surface appearance, to impart a smooth finish and ensure a perfectly round drum (which will eliminate brake pedal pulsations related to out-of-round drums). At the very least, if you don't have the drums resurfaced, remove the glaze from*

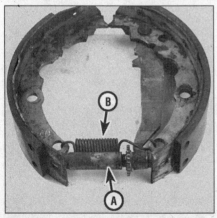

6.4h Remove the adjusting screw (A) and spring (B) from the shoe assembly - be sure to note the how they're positioned

6.4i Remove the parking brake lever by prying off the C-clip

6.4j Install the parking brake lever on the new brake shoe and press the C-clip into place with needle-nose pliers

9

6.4k Lubricate the contact surfaces of the backing plate with high-temperature brake grease

6.4l Lubricate the adjuster screw threads and socket end prior to installation

6.4m Connect the parking brake lever to the cable

6.4n Spread the brake assembly apart sufficiently to clear the hub flange and raise it into position

6.4o Install the parking brake strut and spring

6.4p Make sure the parking brake strut is positioned in the shoes properly (arrows)

the surface with sandpaper or emery cloth using a swirling motion. If the drum won't "clean up" before the maximum (discard) diameter is reached in the machining operation, install a new one. The maximum diameter is cast into each brake drum **(see illustration)**.

6   Install the brake drum. If the shoes are still interfering with the drum, retract them (see Step 4 in Section 12). When the drum is turned, make sure the shoes don't drag on the drum.

7   Mount the wheel, install the lug nuts, then lower the vehicle.

8   Make a number of forward and reverse stops to adjust the brakes until satisfactory pedal feel is obtained. After adjustment, make sure that both wheels turn freely.

7   Wheel cylinder - removal, overhaul and installation

Refer to illustrations 7.2 and 7.5

Note: *Obtain a wheel cylinder rebuild kit before beginning this procedure.*

1   Remove the brake shoes (see Section 6).

2   Remove the brake line fitting from the rear of the wheel cylinder **(see illustration)**. Use a flare-nut wrench, if available, to prevent rounding off the corners of the fitting. Cap the brake line to prevent contamination and excessive fluid loss.

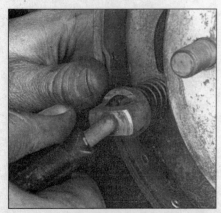

6.4q Install the primary brake shoe hold-down pin and spring

6.4r Attach the actuator link and lever to the secondary brake shoe

6.4s Install the actuator lever return spring

6.4t Install the secondary brake shoe hold-down pin and spring

6.4u Install the return springs

6.4v Center the brake shoe assembly so the drum will slide over it

6.4w Turn the adjusting screw so the drum fits snugly over the shoes, then back-off the adjustment until the drum can be turned without the shoes dragging

6.4x Remove the glaze from the drum with sandpaper or emery cloth, using a swirling motion

6.5 The maximum allowable diameter is cast into the brake drum

3   Remove the wheel cylinder retaining bolts from the rear of the backing plate.

4   Remove the cylinder and place it on a clean workbench.

5   Remove the bleeder valve, seals, pistons, boots and spring assembly from the cylinder body **(see illustration)**.

6   Clean the wheel cylinder with brake fluid or brake system cleaner. Do not, under any circumstances, use petroleum-based solvents to clean brake parts.

7   Use filtered, unlubricated compressed air to remove excess fluid from the wheel cylinder and to blow out the passages.

8   Check the cylinder bore for corrosion and scoring. Crocus cloth may be used to remove light corrosion and stains, but the cylinder must be replaced with a new one if the defects cannot be removed easily, or if the bore is scored.

9   Lubricate the new seals with clean brake fluid.

10   Assemble the brake cylinder, making sure the lips of the seals

7.2 Unscrew the brake line fitting (upper arrow) from the rear of the wheel cylinder with a flare-nut wrench to prevent rounding off the corners of the nut, then remove the wheel cylinder retaining bolts (lower arrows)

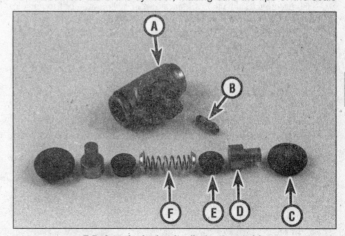

7.5 A typical wheel cylinder assembly

A Wheel cylinder body	D Piston
B Bleeder screw	E Seal
C Boot	F Spring

9

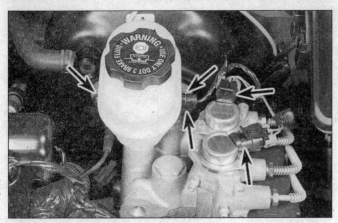

8.3 Unplug the electrical connectors for the fluid level warning switch (center arrow) and the ABS solenoids (two right arrows); to detach the master cylinder from the power brake booster, remove the two mounting nuts (upper arrows) (1995 through 1999)

face inward, and the boots are properly seated.

11   Place the wheel cylinder in position on the backing plate. Thread the brake line fitting into the cylinder, being careful not to cross-thread it. Don't tighten the fitting yet.

12   Install the wheel cylinder retaining bolts and tighten them to the torque listed in this Chapter's Specifications. Now tighten the brake line fitting securely.

13   Bleed the brake system (see Section 10).

8   Master cylinder - removal and installation

Caution: On 1995 through 1999 models, *the master cylinder and ABS hydraulic unit/motor pack can be detached from the power brake booster as a single assembly, but the master cylinder itself must not be detached from the hydraulic unit/motor pack. If the master cylinder must be unbolted from the hydraulic unit/motor pack, it must be done at a dealer service department, after the dealer has used a Tech-1 scan tool to relieve tension on the gears inside the ABS unit. This procedure can only be performed with the master cylinder/hydraulic unit/motor pack assembly operational. If the master cylinder must be replaced (unbolted from the hydraulic unit/motor pack), have it done by a dealer service department or other qualified repair shop.*

Removal

Refer to illustrations 8.3 and 8.4

1   Completely cover the front fender and cowling area of the vehicle; brake fluid can ruin painted surfaces if it is spilled.

2   Remove the brake fluid from the master cylinder reservoir and discard it. **Caution:** *Do not spill brake fluid on painted surfaces.* Place a

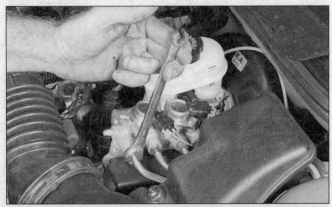

8.4 Unscrew the four brake line fittings with a flare-nut wrench (1995 through 1999)

drain pan or bundle of rags or newspapers under the master cylinder assembly and clean the area around the brake line fittings with brake system cleaner.

3   On 1995 through 1999 models, unplug the electrical connectors for the fluid warning switch and the ABS solenoids **(see illustration)**. On 2000 models, unplug the electrical connector for the fluid warning switch.

4   Disconnect the brake line fittings **(see illustration)**. Use a flare-nut wrench, if available, to prevent rounding off the corners of the fittings. Rags or newspapers should be placed under the master cylinder to soak up the fluid that will drain out.

5   Remove the two master cylinder mounting nuts **(see illustration 8.3)**, and remove the master cylinder from the vehicle. Do not to bend the hydraulic lines running to the combination valve. Plug the ends of both of the lines immediately to prevent air from entering the ABS modulator and to protect the system from moisture and dirt.

Installation

6   Anytime the master cylinder is removed, the brake hydraulic system must be bled. It's much easier to bleed the rest of the system quickly and effectively if you "bench bleed" the master cylinder before installing it on the vehicle. Bench bleed the master cylinder as follows.

7   Insert threaded plugs of the correct size into the four brake line outlet holes (1995-1999 models), or two brake line outlet holes (2000 models). Fill the reservoir with brake fluid (the master cylinder should be supported in such a manner that brake fluid will not spill out during the bench bleeding procedure).

8   Loosen one plug at a time and push the piston assembly into the bore to force air from the master cylinder. To prevent air from being drawn back into the cylinder, the appropriate plug must be tightened before allowing the piston to return to its original position.

9   Stroke the piston three or four times for each outlet to assure that all air has been expelled.

10   Refill the master cylinder reservoirs and install the diaphragm and cap assembly. **Note:** *The reservoir should only be filled to the top of the reservoir divider to prevent overflowing when the cap and diaphragm are installed.*

11   The remainder of installation is the reverse of removal. Make sure you bleed the rest of the system when you're done (see Section 10).

9   Brake hoses and lines - inspection and replacement

Inspection

1   About every six months, loosen the wheel lug nuts, raise the vehicle, place it securely on jackstands, remove the wheels, and inspect the flexible hoses which connect the steel brake lines with the front and rear brake assemblies. Look for cracks, chafing, leaks, blisters and any other damage. These are important and vulnerable parts of the brake system, so your inspection should be thorough. You'll need a flashlight and mirror to do the job right. If a hose exhibits any of the above conditions, replace it as follows.

Brake hose

Refer to illustrations 9.2 and 9.3

2   Using an open-end wrench on the hose fitting to keep the hose from turning, disconnect the brake line from the hose fitting, being careful not to bend the frame bracket or brake line **(see illustration)**.

3   Use pliers to remove the U-clip from the female fitting at the bracket **(see illustration)**, then remove the hose from the bracket.

4   At the caliper end of the hose, remove the banjo bolt from the fitting block, then remove the hose and the sealing washers on either side of the fitting block.

5   When installing the hose, always use new sealing washers on either side of the fitting block and lubricate all bolt threads with clean brake fluid before installing them.

6   With the fitting flange engaged with the caliper locating ledge, attach the hose to the caliper and tighten it to the torque listed in this Chapter's Specifications.

7   Without twisting the hose, install the female fitting in the hose

9.2 Using an open-end wrench on the brake hose, loosen the brake line from the hose fitting with a flare-nut wrench, being careful not to bend the frame bracket or brake line

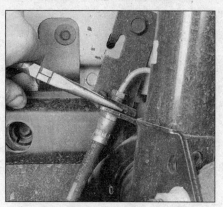

9.3 Use pliers to remove the U-clip from the female fitting at the bracket, then remove the hose from the bracket

10.21 When bleeding the brakes, a hose is connected to the bleeder screw and then submerged in brake fluid - air will be seen as bubbles in the tube and container (all air must be expelled before moving to the next brake or component)

bracket (it will fit the bracket in only one position).

8   Install the U-clip retaining the female fitting to the frame bracket.

9   Attach the brake line to the hose fitting and tighten it securely.

10   When the brake hose installation is complete, there should be no kinks in the hose. Also make sure that the hose does not contact any part of the suspension. If you're replacing a front brake hose, verify this by turning the wheels to the extreme left and right positions. If the hose contacts anything, disconnect it and correct the installation as necessary.

11   Fill the master cylinder reservoir and bleed the system (see Section 10).

Steel brake lines

12   When replacing brake lines, be sure to buy the correct replacement parts. Don't use copper or any other tubing for brake lines.

13   Prefabricated brake lines, with the ends already flared and fittings installed, are available at auto parts stores and dealer service departments. If necessary, carefully bend the line to the proper shape. A tube bender is recommended for this. **Caution:** *Don't crimp or damage the line.*

14   When installing the new line, make sure it's securely supported in the brackets with plenty of clearance between moving or hot components.

15   After installation, check the master cylinder fluid level and add fluid as necessary. Bleed the brake system as outlined in the next Section and test the brakes carefully before driving the vehicle in traffic.

10 Brake hydraulic system - bleeding

Refer to illustration 10.21

1   Bleeding the hydraulic system is necessary to remove air whenever it is introduced into the brake system.

2   The manufacturer specifies that it will be necessary to bleed the *entire* system whenever air is allowed into *any* part of the system.

3   Before beginning the bleeding procedure, the ABS pistons in the hydraulic unit must be returned to the top of their travel. To do this, start the engine and let it run for 10 seconds. Watch the ABS light - it should turn off after three seconds or so. **Warning:** *If the light doesn't go off, a GM Tech 1 scan tool must be connected to diagnose the problem. Have the vehicle towed to a dealer service department or other repair shop equipped with the necessary tool.*

4   If the light turned off like it is supposed to, turn the ignition off and repeat the procedure. If the light goes off again after three seconds, the brakes can now be bled.

5   Have an assistant on hand, as well as a supply of new brake fluid, an empty clear container, a length of clear plastic tubing to fit over the bleeder valve and a wrench to open and close the bleeder valve.

6   Remove the cap from the brake fluid reservoir and add fluid, if

necessary (see Chapter 1). Don't allow the fluid level to drop too low during this procedure - check it frequently. Reinstall the cap.

7   Bleed the master cylinder as follows:

Models without bleeder valves on the hydraulic modulator

8   Disconnect the two forward brake lines from the master cylinder with a flare-nut wrench **(see illustration 8.4)**.

9   Fill the reservoir with brake fluid until fluid begins to flow from the two forward outlets.

10   Reconnect the forward lines to the master cylinder and tighten the tube nut fittings.

11   Slowly depress and hold down the brake pedal.

12   Loosen the fittings again and allow any air in the fluid to be purged, then retighten the fittings.

13   Slowly release the pedal and wait 15 seconds.

14   Repeat this procedure until all air is removed from the two front outlets.

15   Disconnect the two rear tube nut fittings and repeat this procedure until all air is expelled from those ports as well. Proceed to Step 17.

Models with bleeder valves on the hydraulic modulator (1995 through 1999)

16   Attach the bleeder hose to the rearmost bleeder valve on the modulator assembly. With the other end of the hose in a jar partially filled with clean brake fluid, open the valve slowly and have an assistant depress the brake pedal. Keep the pedal down until fluid flows. Close the valve and release the brake pedal. Repeat this until no air is evident as the fluid enters the jar, then repeat the procedure for the forward bleeder valve. After bleeding the hydraulic modulator assembly, bleed the rest of the braking systems as described below.

All models

17   When all air has been removed from the master cylinder, slowly depress the brake pedal and verify that the pedal feels firm. If it does, you're done. If it doesn't, proceed to the next step and bleed the rest of the system.

18   Check the fluid level again, adding fluid as necessary. Do not use old brake fluid. It contains moisture which will allow the fluid to boil, rendering the brakes useless. When bleeding, make sure the fluid coming out of the bleeder is not only free of bubbles, but clean also.

19   Raise the vehicle and support it securely on jackstands.

20   Beginning at the right rear brake, loosen the bleeder screw slightly, then tighten it to a point where it's snug but can still be loosened quickly and easily.

21   Place one end of the tubing over the bleeder screw and submerge the other end in brake fluid in the container **(see illustration)**.

9

11.7 To detach the power brake booster from the firewall, remove these two lower nuts (arrows) and the two upper nuts (not visible in this photo)

11.8 The booster pushrod is attached to a pin at the top of the brake pedal; the pushrod is retained by a small spring clip which can be pried off with a screwdriver

22   Open the bleeder screw and have your assistant slowly depress the brake pedal and hold the pedal firmly depressed. Watch for air bubbles to exit the submerged end of the tube. When the fluid flow slows, tighten the screw, then have your assistant slowly release the pedal. Wait five seconds before proceeding.

23   Repeat Step 22 until no more air is seen leaving the tube, then tighten the bleeder screw and proceed to the left rear brake, right front brake and left front brake, in that order. Be sure to check the fluid in the master cylinder reservoir frequently.

24   After bleeding the right rear wheel brake, repeat Steps 20 through 23 on the left rear brake, right front brake and the left front brake, in that order.

25   Lower the vehicle and check the fluid level in the brake fluid reservoir, adding fluid as necessary.

26   Check the operation of the brakes. The pedal should feel solid when depressed, with no sponginess. If necessary, repeat the entire process. **Warning:** *Do not operate the vehicle if you are in doubt about the effectiveness of the brake system.*

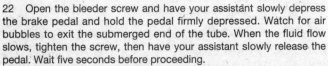

11   Power brake booster - check, removal and installation

Operating check

1   Depress the pedal and start the engine. If the pedal goes down slightly, operation is normal.

2   Depress the brake pedal several times with the engine running and make sure that there is no change in the pedal reserve distance.

Airtightness check

3   Start the engine and turn it off after one or two minutes. Depress the brake pedal several times slowly. If the pedal goes down farther the first time but gradually rises after the second or third depression, the booster is airtight.

4   Depress the brake pedal while the engine is running, then stop the engine with the pedal depressed. If there is no change in the pedal reserve travel after holding the pedal for 30 seconds, the booster is airtight.

Removal and installation

Refer to illustrations 11.7 and 11.8

Note: *Dismantling of the power brake unit requires special tools. If a problem develops, it is recommended that a new or factory-exchange unit be installed rather than trying to overhaul the original booster.*

5   Remove the mounting nuts which hold the master cylinder to the power brake unit (see Section 8). Move the master cylinder forward, but be careful not to bend or kink the lines leading to the master cylinder. If there is any strain on the lines, disconnect them at the master cylinder and plug the ends.

6   Disconnect the vacuum hose leading to the front of the power brake booster. Cover the end of the hose.

7   Inside the vehicle, loosen the four nuts that secure the booster to the firewall **(see illustrations)**. Do not remove these nuts at this time.

8   Disconnect the power brake pushrod from the brake pedal **(see illustration)**. Do not force the pushrod to the side when disconnecting it.

9   Now remove the four booster mounting nuts and carefully lift the unit out of the engine compartment.

10   When installing, loosely install the four mounting nuts, then connect the pushrod to the brake pedal. If the old retaining clip doesn't snap into place, replace it with a new clip. Tighten the booster retaining nuts to the torque listed in this Chapter's Specifications. Reconnect the vacuum hose. Install the master cylinder (see Section 8). If the brake lines were disconnected from the master cylinder, bleed the brake system to eliminate any air which has entered the system (see Section 10).

12.4 To adjust either rear brake assembly, push the actuator off the star wheel with a screwdriver (the actuator must be held off the star wheel to turn the wheel counterclockwise), then rotate the star wheel adjuster counterclockwise to retract the brake shoes, or clockwise to expand the shoes

13.8 Squeeze the tangs on the cable retainer and disengage the front parking brake cable from the bracket on the parking brake lever base

13.9 Pull on the front cable to rotate the reel until the pawl drops into the notch

12 Parking brake - adjustment

Refer to illustration 12.4

1   The vehicles covered by this manual are equipped with a self-adjusting parking brake lever assembly that automatically takes up slack in the cables as they stretch. There is no need to adjust the parking brake cables manually during their service life. However, whenever new rear brake shoes are installed on models with rear drum brakes, make sure the new shoes are properly adjusted; the automatic cable adjuster mechanism can't do its job if the rear shoes are out of adjustment. It's also a good idea to check cable adjustment whenever the brake shoes are replaced, or the brake drum is removed for some other reason.

2   Loosen the rear wheel lug nuts, raise the rear of the vehicle and support it securely on jackstands. Block the front wheels. Remove the rear wheels.

3   Remove the rear drums.

4   Using a screwdriver, lift the actuator off the adjuster wheel **(see illustration)**. Using a brake adjusting tool, or another screwdriver, turn the adjuster wheel until the shoes drag on the drum as the drum is turned. Since there are no adjustment holes in the brake backing plates on these vehicles, you'll have to expand the shoes a bit, then install the drum, expand the shoes a bit, install the drum, etc. until you feel the drum drag against the shoes. Then back off the adjuster wheel a little until the drum rotates freely.

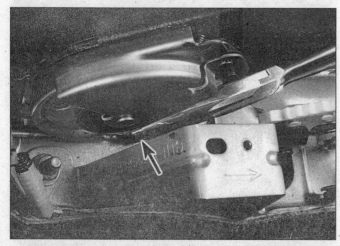

13.11 Using a pair of needle-nose pliers, disengage the plug (arrow) on the end of the front cable from the reel mechanism

5   Adjust the other rear brake the same way.

6   Apply the parking brake and verify that the shoes hold the drums. Release the parking brake and make sure that both rear wheels turn freely and that there is no brake drag in either direction.

7   Remove the jackstands and lower the vehicle.

13 Parking brake cables - replacement

Front cable

Refer to illustrations 13.8, 13.9 and 13.11

1   Raise the vehicle and place it securely on jackstands.

2   The rear part of the front cable is attached to the floorpan by a clip which is an integral part of the cable. Remove the clip retaining screw and detach the clip from the floorpan.

3   Pull the front cable toward the rear of the vehicle to put some slack in the cable and hold it there.

4   Disengage the front cable from the rear cable at the cable junction.

5   Lower the vehicle.

6   Remove the center console (see Chapter 11) and, on Pontiacs, the shift lever boot.

7   Fully release the parking brake lever.

8   Squeeze the tangs of the cable retainer together **(see illustration)** and disengage the cable from the parking brake lever bracket.

9   Pull on the cable to rotate the ratcheting reel until the pawl drops into the notch, locking the reel into place **(see illustration)**.

10   Slowly release the cable and allow the notch to catch the leg of the pawl spring.

11   Disengage the plug on the end of the cable from the reel assembly **(see illustration)**.

12   Remove the left rocker panel/door sill plate.

13   Remove the grommet and retainer from the floorpan.

14   The forward part of the front cable is attached to the floorpan by a retaining clip. Bend this clip open and disengage the cable.

15   Remove the front parking brake cable.

16   Installation is the reverse of removal.

Rear cable

Refer to illustrations 13.21 and 13.22

17   Fully release the parking brake lever.

18   Loosen the rear wheel lug nuts, raise the rear of the vehicle and place it securely on jackstands. Remove the rear wheel(s).

19   Remove the rear drum and disassemble the brake assembly (see Section 6).

20   Detach the cable from the rear shoe parking brake lever **(see illustration 6.4g)**.

9

13.21 Squeeze the retainer and pull the parking brake cable through the brake backing plate

13.22 Disengage the rear parking brake cable from the front cable junction

21   Squeeze the cable retainer with a pair of pliers **(see illustration)** and pull the cable through the backing plate.
22   Disengage the rear cable from the front cable at the cable junction **(see illustration)**.
23   Disconnect the cable from all clips and brackets.
24   Installation is the reverse of removal.

14   Brake light switch - check, adjustment and replacement

Check

Refer to illustration 14.1

1   The brake light switch **(see illustration)** is located on the brake pedal bracket. You'll need to remove the knee bolster (the trim panel beneath the steering column) to get to the switch and connector.
2   With the brake pedal in the fully released position, the switch plunger is pressed into the switch housing, and the brake light circuit is open. When the brake pedal is depressed, the plunger protrudes from the switch, which closes the circuit and sends current to the brake lights. On models with cruise control, the brake light switch is also a switch for the cruise control system when the system is operating. With the brake pedal in the released position, the circuit is closed; when the brake pedal is depressed, the brake light switch opens the circuit and cuts off current to the cruise control module, which turns off the cruise control system.

3   If the brake lights are inoperative, check the fuse (see Chapter 12).
4   If the fuse is okay, verify that voltage is available at the switch by backprobing the connector **(see illustration 14.1)**.
5   If there's no voltage to the switch, use a test light to find the open circuit condition between the fuse panel and the switch. If there is voltage to the switch, close the switch (depress the brake pedal) and verify that there's voltage on the other side of the switch.
6   If there's no voltage on the other side of the switch with the brake pedal depressed, replace the switch (see Step 7). If voltage is available, check for voltage at the brake lights. If no power is present at the brake light sockets, look for an open circuit condition between the switch and the brake lights.

Replacement

Refer to illustration 14.9

7   Remove the knee bolster.
8   Unplug the electrical connector(s) from the switch.
9   Remove the switch from the bracket **(see illustration)**.
10   The switch must be adjusted as it's installed (see below).

Adjustment

11   Depress the brake pedal, insert the switch into its bracket and push it in until it's fully seated.
12   Slowly pull the brake pedal to the rear until you no longer hear any "clicking" sounds. The switch should now be adjusted.
13   The remainder of installation is the reverse of removal.

14.1 A typical brake light switch

1   Voltage in (brake light circuit)
2   Voltage out (brake light circuit)
3   Voltage in (cruise control circuit)
4   Voltage out (cruise control circuit)

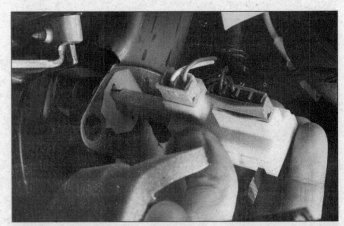

14.9 To remove the brake light switch, simply pull it straight out, then unplug the electrical connectors; to install the switch, depress the brake pedal, insert the switch into its bracket and push it in until it's fully seated, then slowly pull the brake pedal to the rear until you no longer hear any "clicking" sounds

Chapter 10 Suspension and steering systems

Contents

Specifications

Torque specifications

Front suspension

Ft-lbs (unless otherwise indicated)

Balljoint-to-steering knuckle nut	41 (up to 50, to install cotter pin)
Control arm	
Front bushing bolt	
1998 and earlier	89, plus an additional 180-degrees rotation
1999	79
Rear vertical bushing bolt	125
Hub and bearing assembly-to-steering knuckle bolts	70
Stabilizer bar	
Link nuts	156 in-lbs
Bushing clamp bolts	49
Strut assembly	
Strut-to-steering knuckle nuts	133
Strut-to-body nuts	18
Strut-to-body bolt	18
Strut damper shaft nut	52
Crossmember bolts	
1995 and 1996 models	96
1997 models	71, plus an additional 180-degrees rotation
1998 models	
Left rear outboard bolts	71, plus an additional 90-degrees rotation
Right rear outboard bolts	71, plus an additional 90-degrees rotation
Rear inboard bolts	71, plus an additional 90-degrees rotation
Front bolts	66, plus an additional 90-degrees rotation
1999 models	71

Rear suspension

Hub and bearing assembly retaining bolts	44
Rear axle beam trailing arm pivot bolts	
1995 through 1999	44
2000	88
Shock absorber and coil spring assembly	
Lower shock bolt/nut	52
Upper shock nut and bolts	
1995 through 1999	21
2000	18

Steering

Steering gear mounting bolts	89
U-joint coupling pinch bolt	30
Power steering pump mounting bolts	22
Steering wheel nut	30
Tie-rod end-to-steering knuckle nuts	84 in-lbs, plus an additional 210-degrees
Wheel lug nuts	See Chapter 1

1.1 Front suspension and steering components

1	Strut and spring assembly	4	Stabilizer link	7	Steering gear boot
2	Steering knuckle	5	Control arm	8	Tie-rod
3	Balljoint	6	Stabilizer bar	9	Tie-rod end

1   General information

Refer to illustrations 1.1 and 1.2

The front suspension **(see illustration)** is a MacPherson strut design. The upper ends of the struts are attached to the body; the lower ends of the struts are bolted to the steering knuckles. The lower ends of the steering knuckles are attached to the control arms by balljoints. The inner ends of the control arms are attached to the cross-member. A stabilizer bar reduces body lean during cornering. The stabilizer is attached to the crossmember by a pair of bushing clamps and to the control arms by link bolts.

The rear suspension **(see illustration)** is a semi-independent design which uses an axle beam with integral trailing arms and a pair of shock absorber/coil spring assemblies. The axle trailing arms are attached to the body. The upper ends of the shocks are attached to the vehicle body; the lower ends are bolted to the axle beam.

The rack-and-pinion steering gear, which is located behind the engine/transaxle assembly, is bolted to the suspension crossmember. The steering gear turns the steering knuckles via a pair of tie-rod assemblies, each of which consists of an inner tie-rod and a tie-rod end. The inner tie-rods are attached to the steering gear; the outer tie-rods, or tie-rod ends, are attached to the steering knuckles. All models are equipped with power steering.

Warning: *Whenever any of the suspension or steering fasteners are loosened or removed, they must be inspected and, if necessary, replaced with new ones of the same part number or of original equipment quality and design. Torque specifications must be followed for proper reassembly and component retention. Never attempt to heat or straighten any suspension or steering components. Instead, replace* any bent or damaged part with a new one.

Note 1: *These vehicles have a combination of standard and metric fasteners on the various suspension and steering components, so it would be a good idea to have both types of tools available when beginning work.*

Note 2: *On models equipped with a Delco Loc II or Theftlock audio system, be sure the lockout feature is turned off and make sure you have the correct activation code before performing any procedure which requires disconnecting the battery.*

2   Stabilizer bar and bushings - removal and installation

Removal

Refer to illustrations 2.2, 2.3 and 2.4

1   Loosen the lug nuts on both front wheels, raise the front of the vehicle and support it securely on jackstands. Apply the parking brake and block the rear wheels to keep the vehicle from rolling off the jackstands. Remove the front wheels.

2   Remove the nuts from the upper ends of the link bolts that connect the stabilizer bar to the control arms **(see illustration)**. Note the order in which the bushings, spacers and washers are installed on the links; they must be installed in exactly the same order in which they're removed.

3   Support the rear of the crossmember with a jack, then remove the rear center and outer crossmember retaining bolts **(see illustration)**. Lower the crossmember about three inches.

4   Remove the stabilizer bar bushing clamp bolts and bushing

1.2 Rear suspension components

| 1 | Shock absorber and coil spring assembly | 2 | Rear axle beam assembly | 3 | Rear axle pivot bolts |

2.2 To disconnect the stabilizer bar from the link bolts that connect it to the control arms, remove the link bolt nut (arrow) from each end; make sure that you note the order in which the bushings, spacers and washers on the link are installed

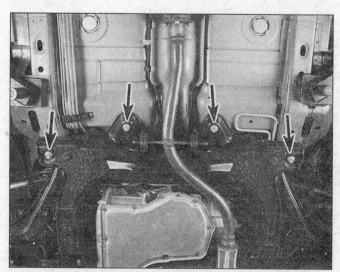

2.3 To lower the rear part of the crossmember, remove these four bolts (arrows); be sure to put a sturdy floor jack under the crossmember before unbolting it

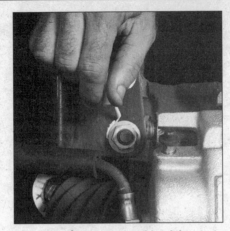

2.4 To detach the stabilizer bar from the crossmember, remove the bushing clamp bolt (arrow) and clamp from each bushing

3.2 To mark the relationship of the strut to the steering knuckle, paint or scribe lines around the nuts

3.3 To remove the strut-to-knuckle bolts, drive them out with a hammer and brass punch

clamps from the upper side of the crossmember **(see illustration)**.

5   Remove the stabilizer bar and bushings through the wheel well.

6   Inspect the stabilizer bushings for wear and damage and replace them if necessary. To ease installation of the new bushings, spray the inside and outside of the bushings with a silicone-based lubricant. Do not use petroleum-based lubricants on rubber parts. Inspect the link bushings, spacers and washers for wear and replace as necessary.

Installation

7   Install the stabilizer bar bushings and clamps, guide the stabilizer through the wheel well, over the crossmember and into position.

8   Center the stabilizer bar, install the bushing clamp bolts and hand tighten them.

9   Install the spacers, bushings and washers on the links in the same order in which they were removed. Tighten the link nuts to the torque listed in this Chapter's Specifications.

10   Tighten the bushing clamp bolts to the torque listed in this Chapter's Specifications.

11   Raise the crossmember, install the crossmember retaining bolts and tighten them to the torque listed in this Chapter's Specifications.

12   Install the wheels and lower the vehicle. Tighten the lug nuts to the torque listed in the Chapter 1 Specifications.

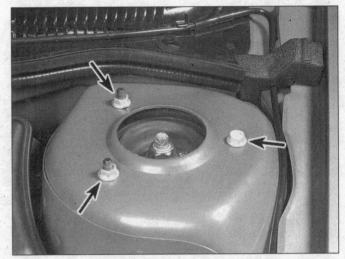

3.5 Remove the strut upper mounting nuts and bolt (arrows) while supporting the strut assembly

3   Strut and coil spring assembly (front) - removal, inspection and installation

Removal

Refer to illustrations 3.2, 3.3 and 3.5

1   Loosen the wheel lug nuts, raise the front of the vehicle and support it securely on jackstands. Apply the parking brake and block the rear wheels to keep the vehicle from rolling off the jackstands. Remove the wheel.

2   Mark the strut-to-steering knuckle relationship by making a line around the strut-to-steering knuckle nuts **(see illustration)**.

3   Remove the nuts from the strut-to-knuckle bolts and knock the bolts out with a brass punch and a hammer **(see illustration)**.

4   Separate the strut from the steering knuckle. Be careful not to overextend the inner CV joint or stretch the brake hose. If necessary, support the control arm with a jack.

5   Support the strut and spring assembly with one hand and remove the upper strut mounting nuts and bolt **(see illustration)**. Remove the strut and spring assembly.

Inspection

6   Check the strut body for leaking fluid, dents, cracks and other obvious damage which would warrant repair or replacement.

7   Check the coil spring for chips and cracks in the spring coating (this will cause premature spring failure due to corrosion). Inspect the spring seat for hardening, cracks and general deterioration.

8   If wear or damage is evident, replace the strut and/or coil spring as necessary (see Section 4).

Installation

9   Guide the strut assembly up into the fenderwell and insert the upper mounting studs through the holes in the shock tower. Once the studs protrude from the shock tower, install the nuts and bolt so the strut won't fall back through. This may require an assistant, since the strut is quite heavy and awkward.

10   Slide the steering knuckle into the strut flange and insert the two bolts. Install the nuts, align the marks you made prior to disassembly and tighten the nuts to the torque listed in this Chapter's Specifications.

11   Install the wheel, lower the vehicle and tighten the wheel lug nuts to the torque listed in the Chapter 1 Specifications.

12   Tighten the upper mounting nuts and bolt to the torque listed in this Chapter's Specifications.

13   Drive the vehicle to a dealer service department or an alignment shop to have the front wheel alignment checked and, if necessary, adjusted (this is only necessary if the strut has been modified for camber adjustment).

4.3a Mark the relationship of the coil spring to the upper spring seat and insulator and to the strut mount . . .

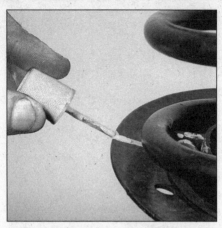

4.3b . . . and to the lower seat of the strut

4.4 Following the tool manufacturer's instructions, install the spring compressor on the spring and compress it sufficiently to relieve all pressure from the upper spring seat

4.5 Using a wrench on the damper shaft to prevent it from turning, loosen the damper shaft nut

4.6a Remove the washer . . .

4   Strut/shock absorber or coil spring - replacement

1   If the struts/shock absorbers or coil springs exhibit the telltale signs of wear (leaking fluid, loss of damping capability, chipped, sagging or cracked coil springs) explore all options before beginning any work. The strut/shock absorber assemblies are not serviceable and must be replaced if a problem develops. However, strut assemblies complete with springs may be available on an exchange basis, which eliminates much time and work. Whichever route you choose to take, check on the cost and availability of parts before disassembling your vehicle. **Warning:** *Disassembling a strut is potentially dangerous and utmost attention must be directed to the job, or serious injury may result. Use only a high-quality spring compressor and carefully follow the manufacturer's instructions furnished with the tool. After removing the coil spring from the strut assembly, set it aside in a safe, isolated area.*

Disassembly

Refer to illustrations 4.3a, 4.3b, 4.4, 4.5, 4.6a, 4.6b, 4.7, 4.8 and 4.9

2   Remove the strut and spring assembly (see Section 3) or shock absorber/coil spring assembly (see Section 9). Mount the strut assembly in a vise. Line the vise jaws with wood or rags to prevent damage to the unit and don't tighten the vise excessively.

3   Mark the relationship of the coil spring to the upper insulator and mount and to the lower seat **(see illustrations)**.

4   Following the tool manufacturer's instructions, install the spring compressor (which can be obtained at most auto parts stores or at equipment rental yards) on the spring and compress it sufficiently to relieve all pressure from the upper spring seat **(see illustration)**. This

4.6b . . . and the strut mount; inspect the bearing in the mount for smooth operation and the rubber portion of the mount for cracking and general deterioration, if the bearing doesn't turn smoothly, or if there's any separation of the rubber, replace the mount

10

can be verified by wiggling the spring.

5   Loosen the damper shaft nut **(see illustration)**.

6   Remove the washer and strut mount **(see illustrations)**. Inspect the bearing in the strut mount for smooth operation. If it doesn't turn smoothly, replace the strut mount. Check the rubber portion of the strut mount for cracking and general deterioration. If there is any separation of the rubber, replace it.

7   Remove the upper spring seat and insulator from the damper

4.7 Remove the upper spring seat and insulator from the damper shaft; inspect the insulator for cracking and hardness and, if necessary, replace it

4.8 Remove the rubber jounce bumper and dust shield from the damper shaft

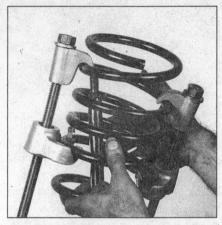

4.9 Carefully lift the compressed spring from the assembly and set it in a safe place; Do NOT *place your head near the end of the spring!*

4.12 Place the coil spring onto the lower insulator, with the end of the spring butted against the spring stop on the insulator

4.15a Install the strut mount . . .

4.15b . . . and washer

shaft **(see illustration)**. Check the insulator for cracking and hardness; replace it if necessary.

8   Remove the rubber jounce bumper and dust shield from the damper shaft **(see illustration)**.

9   Carefully lift the compressed spring from the assembly **(see illustration)** and set it in a safe place. **Warning:** *Never place your head near the end of the spring!*

10   Check the lower insulator for wear, cracking and hardness and replace it if necessary.

Reassembly

Refer to illustrations 4.12, 4.15a, 4.15b and 4.16

11   If the lower insulator is being replaced, set it into position with the dropped portion seated in the lowest part of the seat. Extend the damper rod to its full length and install the rubber bumper.

12   Place the coil spring onto the lower insulator, with the end of the spring butted against the spring stop on the insulator **(see illustration)**.

13   Install the dust shield and rubber jounce bumper.

14   Install the upper insulator and spring seat.

15   Install the strut mount and washer **(see illustrations)**.

16   Install the nut and tighten it securely **(see illustration)**.

17   Install the strut/spring assembly (see Section 3) or shock absorber/coil spring assembly (see Section 9).

18   Repeat this entire procedure for the other strut or shock

4.16 Hold the damper shaft from turning, then tighten the nut securely (unless you have a special socket with a "window" in it for the back-up wrench, you won't be able to torque the nut until the strut assembly is installed in the vehicle and the vehicle is on the ground)

absorber/coil spring assembly.

19   After the vehicle has been lowered to the ground, tighten the damper shaft nuts to the torque listed in this Chapter's Specifications.

5.3 Remove the cotter pin and loosen - but don't remove - the nut (arrow) from the balljoint stud

5.4 Pop the balljoint out of the steering knuckle by striking the knuckle ballstud boss sharply with a hammer as shown

5.5 Once the ballstud is loose, separate the control arm from the steering knuckle by prying the ballstud out of the hole with a large prybar as shown

5.6a The front control arm pivot bolt (arrow) is accessed from the front of the suspension crossmember

5.6b To separate the rear part of the control arm from the suspension crossmember, remove the vertical bushing bolt (arrow)

5 Control arm - removal and installation

Removal

Refer to illustrations 5.3, 5.4, 5.5, 5.6a and 5.6b

1   Loosen the wheel lug nuts, raise the front of the vehicle and support it securely on jackstands. Apply the parking brake and block the rear wheels to keep the vehicle from rolling off the jackstands. Remove the wheel(s).

2   Disconnect the stabilizer bar from the control arm being removed (see Section 2). (If only one control arm is being removed, disconnect only that end of the stabilizer bar; if both control arms are being removed, disconnect both ends.)

3   Remove the cotter pin and loosen the balljoint stud-to-steering knuckle nut **(see illustration)**.

4   Using a hammer, strike the ballstud boss on the steering knuckle **(see illustration)**. until the ballstud pops loose from the knuckle. If the ballstud is frozen in the knuckle, you may have to use a special balljoint separator or a picklefork tool. Keep in mind that a picklefork will damage or destroy the boot, so it should be used only as a last resort.

5   Remove the nut from the ballstud. Using a large prybar positioned between the control arm and steering knuckle, separate the ballstud from the knuckle **(see illustration)**. **Caution:** *When removing the balljoint from the knuckle, be careful not to overextend the inner CV joint or it may be damaged.*

6   Remove the front control arm pivot bolt and the rear vertical bushing bolt **(see illustrations)** and detach the control arm.

7   The control arm bushings are replaceable, but special tools and expertise are necessary to do the job. Carefully inspect the bushings for hardening, excessive wear and cracks. If they appear to be worn or deteriorated, take the control arm to a dealer service department or other repair shop.

Installation

8   Position the control arm in the suspension crossmember and install the front pivot bolt and the rear vertical bushing bolt. Do not tighten them completely at this time.

9   Insert the balljoint stud into the steering knuckle boss, install the castellated nut and tighten it to the torque listed in this Chapter's Specifications. If necessary, tighten the nut a little more (up to, but not beyond, the specified maximum listed in the Specifications) if the cotter pin hole doesn't line up with an opening on the nut. Install a new cotter pin.

10   Install the stabilizer bar-to-control arm link bolt, bushings, spacers and washers (see Section 2) and tighten the link nut to the torque listed in this Chapter's Specifications.

11   Install the wheel and lower the vehicle. Tighten the lug nuts to the torque listed in the Chapter 1 Specifications.

12   With the weight of the vehicle on the suspension, tighten the control arm pivot bolt and the rear vertical bushing bolt to the torque listed in this Chapter's Specifications. **Note:** *You can raise the suspension with a floor jack positioned under the balljoint to simulate normal ride height.* **Caution:** *If the bolts aren't tightened with the weight of the vehicle on the suspension, control arm bushing damage may occur.*

10

6.3a Check for movement between the balljoint and steering knuckle when prying up

6.3b With the prybar positioned between the steering knuckle boss and the balljoint, pry down and check for play in the balljoint - if there's any play, replace the balljoint

7.6 A No. 55 Torx bit is required to remove the hub bolts - DO NOT use an Allen wrench or the bolts will be damaged

6   Balljoints - check and replacement

Check

Refer to illustration 6.3

1   Raise the front of the vehicle and support it securely on jack-stands. Apply the parking brake and block the rear wheels to keep the vehicle from rolling off the jackstands.

2   Visually inspect the rubber dust boot for damage, deterioration and leaking grease. If the boot is damaged, deteriorated or leaking, replace the balljoint.

3   Place a large prybar under the balljoint and resting on the wheel, then try to pry the balljoint up while feeling for movement between the balljoint and steering knuckle **(see illustration)**. Now, pry between the control arm and the steering knuckle and try to lever the control arm down while feeling for movement between the balljoint and steering knuckle **(see illustration)**. If any movement is evident in either check, the balljoint is worn out.

4   Have an assistant grasp the tire at the top and bottom and move the top of the tire in-and-out. Touch the balljoint stud castellated nut. If any looseness is felt, suspect a worn out balljoint stud or a widened hole in the steering knuckle boss. If the latter problem exists, the steering knuckle should be replaced as well as the balljoint.

5   Separate the control arm from the steering knuckle (Section 5). Using your fingers (don't use pliers), try to twist the stud in the socket. If the stud turns, replace the balljoint.

Replacement

6   Loosen the wheel lug nuts, raise the front of the vehicle and support it securely on jackstands. Apply the parking brake and block the rear wheels to keep the vehicle from rolling off the jackstands. Remove the wheel.

7   Separate the control arm from the steering knuckle (see Section 5). Temporarily insert the balljoint stud back into the steering knuckle (loosely). This will ease balljoint removal after Step 9 has been performed, as well as hold the assembly stationary while drilling out the rivets.

8   Using a 1/8-inch drill bit, drill a pilot hole into the center of each balljoint-to-control arm rivet. Be careful not to damage the CV joint boot in the process.

9   Using a 1/2-inch drill bit, drill the head off each rivet. Work slowly and carefully to avoid deforming the holes in the control arm.

10   Loosen (but don't remove) the stabilizer bar-to-control arm link nut. Pull the control arm and balljoint down to remove the balljoint stud from the steering knuckle, then dislodge the balljoint from the control arm.

11   Position the new balljoint on the control arm and install the bolts (supplied in the balljoint kit) from the top of the control arm. Tighten the

bolts to the torque specified in the new balljoint instruction sheet.

12   Insert the balljoint into the steering knuckle, install the castellated nut, tighten it to the torque listed in this Chapter's Specifications and install a new cotter pin. It may be necessary to tighten the nut some to align the cotter pin hole with an opening in the nut, which is acceptable (up to, but not beyond, the maximum torque, listed in this Chapter's Specifications). Never loosen the castellated nut to allow cotter pin insertion.

13   Tighten the stabilizer bar-to-control arm link nut to the torque listed in this Chapter's Specifications.

14   Install the wheel, lower the vehicle and tighten the lug nuts to the torque listed in the Chapter 1 Specifications.

7   Hub and bearing assembly (front) - removal and installation

Refer to illustrations 7.6 and 7.7

Warning: *Dust created by the brake system may contain asbestos, which is harmful to your health. Never blow it out with compressed air and don't inhale any of it. Do not, under any circumstances, use petroleum-based solvents to clean brake parts. Use brake cleaner or denatured alcohol only.*

Note: *The hub and bearing assembly is a sealed unit. If worn or damaged, it must be replaced.*

1   Loosen the wheel lug nuts, raise the front of the vehicle and support it securely on jackstands. Apply the parking brake and block the rear wheels to keep the vehicle from rolling off the jackstands. Remove the wheel.

2   Disconnect the stabilizer bar from the control arm (see Section 2).

3   Remove the balljoint-to-steering knuckle nut and separate the control arm from the knuckle (see Section 5).

4   Remove the caliper from the steering knuckle and hang it out of the way with a piece of wire (see Chapter 9).

5   Pull the disc off the hub and remove the driveaxle (see Chapter 8 if necessary).

6   Remove the hub retaining bolts from the steering knuckle **(see illustration)**. Remove the disc shield.

7   Wiggle the hub and bearing assembly back-and-forth and pull it out of the steering knuckle **(see illustration)**.

8   Clean the mating surfaces on the steering knuckle, bearing flange and knuckle bore.

9   Install a new O-ring around the rear of the hub assembly and push it up against the bearing flange. Lubricate the outside diameter of the bearing and the seal lips with high-temperature grease and insert the hub and bearing into the steering knuckle. Position the disc shield and install the three bolts. Tighten them to the torque listed in this Chapter's Specifications.

7.7 Pull the hub and bearing assembly and the disc shield out of the steering knuckle

9.2 Remove this single nut (arrow) from the trunk space (the other two bolts, the threads of which are visible in this photo, are removed from underneath after the vehicle is raised)

9.4 To disconnect the lower end of the shock absorber from the rear axle beam, remove this nut and bolt (arrows)

10   Install the driveaxle (see Chapter 8).

11   Attach the control arm to the steering knuckle (see Section 5).

12   Reconnect the stabilizer bar to the control arm (see Section 2).

13   Install the brake disc and caliper (see Chapter 9).

14   Install the hub nut and tighten it securely to seat the driveaxle in the hub. Prevent the axle from turning by inserting a screwdriver through the caliper and into a disc cooling vane.

15   Install the wheel, lower the vehicle and tighten the lug nuts to the torque listed in this Chapter's Specifications.

16   Tighten the hub nut to the torque listed in the Chapter 8 Specifications.

---

8   Steering knuckle and hub - removal and installation

Warning: *Dust created by the brake system may contain asbestos, which is harmful to your health. Never blow it out with compressed air and don't inhale any of it. Do not, under any circumstances, use petroleum-based solvents to clean brake parts. Use brake cleaner or denatured alcohol only.*

Removal

1   Loosen the wheel lug nuts, raise the front of the vehicle and support it securely on jackstands. Apply the parking brake and block the rear wheels to keep the vehicle from rolling off the jackstands. Remove the wheel.

2   Remove the hub nut. Insert a screwdriver through the caliper and into a disc cooling vane to prevent the driveaxle from turning.

3   Remove the caliper and suspend it out of the way with a piece of wire. Lift the disc off the hub.

4   Mark the position of the two strut-to-knuckle nuts and remove them **(see illustration 3.2)**. Don't drive out the bolts at this time.

5   Separate the control arm balljoint from the steering knuckle (see Section 5).

6   Attach a puller to the hub flange and push the driveaxle out of the hub (see Chapter 8). Hang the driveaxle with a piece of wire to prevent damage to the inner CV joint.

7   Support the knuckle and drive out the two strut-to-knuckle bolts with a hammer and brass punch. Separate the steering knuckle from the strut.

Installation

8   Position the knuckle in the strut and insert the two splined bolts. Tap the bolts into place and install the nuts, but don't tighten them at this time.

9   Install the driveaxle in the hub.

10   Connect the control arm to the steering knuckle and tighten the castellated nut to the torque listed in this Chapter's Specifications.

Install a new cotter pin.

11   Align the strut-to-knuckle nuts with the previously applied marks and tighten them to the torque listed in this Chapter's Specifications.

12   Install the brake disc and caliper.

13   Tighten the hub nut securely to seat the driveaxle in the hub.

14   Install the wheel, lower the vehicle and tighten the lug nuts to the torque listed in the Chapter 1 Specifications.

15   Tighten the hub nut to the torque listed in the Chapter 8 Specifications.

---

9   Shock absorber and coil spring assembly (rear) - removal and installation

Refer to illustrations 9.2, 9.4 and 9.5

Caution: *Don't remove both shock absorbers at the same time. They limit the downward travel of the rear suspension and damage to the brake hoses and lines may occur if the suspension is allowed to hang.*

1   Open the trunk and peel back the side trim panel to expose the upper shock mount.

2   Remove the mount *nut* **(see illustration)**. (The two *bolts* you see must be removed from the wheel well side, after the vehicle is raised and the wheel is removed.)

3   Loosen the wheel lug nuts, raise the rear of the vehicle and support it securely on jackstands. Block the front wheels to keep the vehicle from rolling off the jackstands. Remove the rear wheels.

4   Support the trailing arm with a jack and remove the lower shock absorber mounting nut and bolt **(see illustration)**.

5   Remove the two upper mount bolts **(see illustration)** and detach

9.5 To detach the upper end of the shock absorber and coil spring assembly from the vehicle body, remove these two bolts (arrows)

10

10.3a Remove the four hub and bearing assembly bolts with a no. 55 Torx bit

10.3b Put a wrench on these nuts (arrows) as you unscrew each Torx bolt

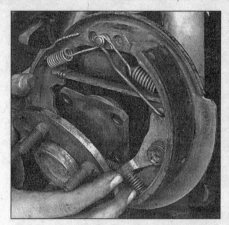

10.4a Angle the hub and bearing assembly out through the brake assembly

10.4b Temporarily reinstall two bolts (arrows) to retain the brake assembly to the trailing arm, rather than let it hang by the brake line

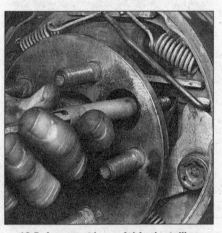

10.5 A magnet is useful for installing the bolts

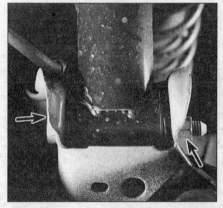

11.7 To disconnect the rear axle beam from the vehicle, remove the pivot bolt and nut (arrows) from the forward end of each trailing arm

the shock absorber and coil spring from the vehicle. Check the shock absorber for dents and leaking fluid. Check the coil spring for chips and cracks.

6   Inspect the upper mount for cracks, hardening, separation and other damage.

7   If any of these conditions are found, replace the shock absorber or coil spring as necessary (see Section 4 for the disassembly and reassembly procedure - it's similar to a front strut).

8   Installation is the reverse of the removal procedure.

10   Hub and bearing assembly (rear) - removal and installation

Warning: *Dust created by the brake system may contain asbestos, which is harmful to your health. Never blow it out with compressed air and don't inhale any of it. Do not, under any circumstances, use petroleum-based solvents to clean brake parts. Use brake cleaner or denatured alcohol only.*

Removal

Refer to illustrations 10.3a, 10.3b, 10.4a, 10.4b and 10.5

Note: *The rear hub and bearing assembly is a sealed unit; if it's worn or damaged, it must be replaced.*

1   Loosen the wheel lug nuts, raise the rear of the vehicle and support it securely on jackstands. Block the front wheels to keep the vehicle from rolling off the jackstands. Remove the wheel.

2   Pull the brake drum off the hub. If difficulty is encountered, refer to Chapter 9 for the removal procedure.

3   Using a no. 55 Torx bit on the bolts and a wrench on the nuts (on the backside of the backing plate), remove the four hub-to-trailing arm bolts, accessible by turning the hub flange so the circular cutout exposes each bolt **(see illustrations)**. Save the upper rear bolt for last, because there isn't much clearance between the bolt and the parking brake strut.

4   Remove the hub and bearing assembly, maneuvering it out through the brake assembly. Reinstall two bolts through the brake backing plate into the trailing arm to avoid hanging the brake assembly by the hydraulic line **(see illustrations)**.

Installation

5   Position the hub and bearing assembly on the trailing arm and align the holes in the backing plate. Install the bolts, beginning with the upper rear bolt. A magnet is useful for guiding the bolts through the hub flange and into position **(see illustration)**. After all four bolts have been installed, tighten them to the torque listed in this Chapter's Specifications.

6   Install the brake drum and wheel. Lower the vehicle and tighten the wheel lug nuts to the torque listed in the Chapter 1 Specifications.

11   Rear axle beam - removal, inspection and installation

Warning: *Dust created by the brake system may contain asbestos, which is harmful to your health. Never blow it out with compressed air and don't inhale any of it. Do not, under any circumstances, use petroleum-based solvents to clean brake parts. Use brake cleaner or denatured alcohol only.*

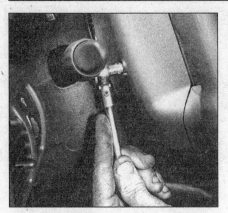

12.3 Remove the airbag module retaining screws (the other screw is in the same location on the other side of the steering column)

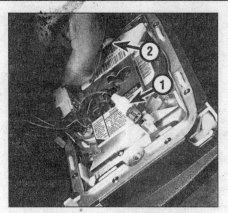

12.4 Unplug the electrical connectors for the airbag module (the big yellow connector), the horn lead (1) and the horn ground (2)

12.6 There should already be alignment marks on the steering wheel and steering shaft; if not, make alignment marks on the steering wheel hub and shaft at 12 o'clock

Removal

Refer to illustration 11.7

1   Loosen the wheel lug nuts, raise the rear of the vehicle and place it securely on jackstands. Block the front wheels and remove the rear wheels.

2   Remove the rear brake drums (see Chapter 9).

3   Disconnect the brake lines from the wheel cylinders and disconnect the brake hoses from the brake lines (see Chapter 9). Plug the brake lines to prevent moisture and contamination from entering the brake system.

4   Detach the brake backing plates from the axle beam and suspend them from the coil springs with pieces of wire. It isn't necessary to remove the brake shoes or the parking brake cable from the backing plate.

5   Support the axle beam with a floor jack. Place a block of wood between the axle and the jack head to protect the axle.

6   Disconnect the lower ends of the shock absorbers from the axle beam (see Section 9).

7   Remove the pivot bolts from the forward ends of the axle beam trailing arms **(see illustration)**.

8   Remove the axle beam assembly.

Inspection

9   Inspect the trailing arm bushings for cracks, deformation and wear. If they're damaged or worn out, take the axle beam assembly to a dealer service department or an automotive machine shop to have the old ones pressed out and new ones pressed in.

Installation

10   Installation is the reverse of removal. Be sure to tighten all fasteners to the torque listed in this Chapter's Specifications, but make sure to tighten the pivot bolts after the suspension is at normal ride height.

11   Lower the vehicle and tighten the wheel lug nuts to the torque listed in the Chapter 1 Specifications.

12   Bleed the brakes (see Chapter 9).

12   Steering wheel - removal and installation

Refer to illustrations 12.3, 12.4, 12.6 and 12.7

Warning: *These models have airbags. Always turn the steering wheel to the straight ahead position, place the ignition switch in the Lock position, remove the Air Bag fuse and unplug the yellow Connector Position Assurance (CPA) connectors at the base of the steering column and under the right side of the instrument panel before working in the vicinity of the impact sensors, steering column or instrument panel to avoid the possibility of accidental deployment of the airbag, which could cause personal injury (see Chapter 12).*

12.7 Use a steering wheel puller to separate the steering wheel from the shaft - DO NOT attempt to remove the wheel with a hammer!

Caution: *On models equipped with a Delco Loc II or Theftlock audio system, be sure the lockout feature is turned off before performing any procedure which requires disconnecting the battery.*

1   Disconnect the cable from the negative battery terminal.

2   Disable the airbag system (see Chapter 12 and **Warning** above).

3   Remove the airbag module retaining screws **(see illustration)**.

4   Remove the airbag module and unplug the electrical connectors **(see illustration)**. **Warning:** *Carry the airbag module with the trim side facing away from you. Set the airbag module in an isolated area with the trim side facing up.*

5   Remove the steering wheel retaining nut.

6   The steering wheel and the steering shaft should already be marked **(see illustration)**. If they're not, mark their relationship to one another before removing the steering wheel.

7   Remove the steering wheel with a puller **(see illustration)**. **Warning:** *Don't allow the steering shaft to turn with the steering wheel removed. If for some reason the steering shaft does turn, refer to Section 8 in Chapter 12 for the airbag coil centering procedure.*

8   Installation is the reverse of removal. Be sure to tighten the steering wheel nut to the torque listed in this Chapter's Specifications.

9   To enable the airbag system, refer to Chapter 12.

13   Tie-rod ends - removal and installation

Refer to illustrations 13.2a, 13.2b and 13.3

1   Loosen the wheel lug nuts, raise the vehicle and place it securely

10

13.2a Before removing the tie-rod end, loosen the jam nut . . .

13.2b . . . and mark the position of the tie-rod end on the inner tie-rod

on jackstands. Remove the wheel.

2   Loosen the tie-rod end jam nut **(see illustration)** and mark the position of the tie-rod end on the threaded portion of the tie-rod **(see illustration)**.

3   Remove the cotter pin and loosen (but do not remove) the castle nut from the tie-rod end balljoint stud, then install a small puller **(see illustration)** and break loose the tie-rod end from the steering knuckle. Remove the nut and detach the tie-rod end.

4   Unscrew the old tie-rod end and install the new one. Make sure the new tie-rod end is aligned with the mark you made on the threads of the tie-rod.

5   Installation is the reverse of removal. Be sure to tighten the tie-rod end balljoint nut to the torque listed in this Chapter's Specifications. Tighten the jam nut securely.

13.3 To separate the tie-rod end from the steering knuckle, loosen - but don't remove - the ballstud nut, then install a balljoint removal tool (shown) or a small puller to pop the ballstud out of the knuckle (DO NOT pound on the stud!)

14 Steering gear boots - replacement

Refer to illustrations 14.4a and 14.4b

1   If a steering gear boot is torn, dirt and moisture can damage the steering gear. Replace it.

2   Loosen the wheel lug nuts, raise the vehicle and place it securely on jackstands. Remove the front wheels.

3   Disconnect the tie-rod ends from the steering knuckles and remove them from the tie-rods (see Section 13). Also remove the jam nuts.

4   Remove the boot clamps **(see illustration)** and slide the boots off the tie-rods.

5   Installation is the reverse of removal. Be sure to use new clamps on the boots.

15 Steering gear - removal and installation

Refer to illustrations 15.7, 15.8 and 5.11

Warning 1: *These models have airbags. Always turn the steering wheel to the straight ahead position, place the ignition switch in the Lock position, remove the Air Bag fuse and unplug the yellow Connector Position Assurance (CPA) connectors at the base of the steering column and under the right side of the instrument panel before working in the vicinity of the impact sensors, steering column or instrument panel to avoid the possibility of accidental deployment of the airbag, which could cause personal injury (see Chapter 12).*

Warning 2: *Make sure the steering shaft is not turned while the steering gear is removed or you could damage the airbag system. To pre-*

14.4a Squeeze the outer boot clamp with a pair of pliers and slide it down the tie-rod

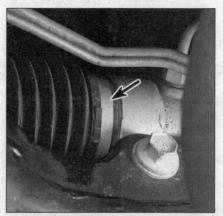

14.4b Cut off the inner boot clamps (arrow) with diagonal cutters and slide off the old boots

15.7 Remove the steering gear mounting bolts (arrows)

15.8 Unscrew these tube nuts (arrows) and disconnect the power steering pressure and return lines from the steering gear

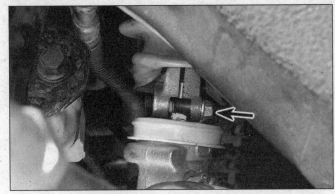

15.11 Mark the relationship of the lower U-joint coupling to the steering gear input shaft, then remove the pinch bolt (arrow)

vent the shaft from turning, turn the ignition key to the lock position before beginning work or run the seat belt through the steering wheel and clip the seat belt into place.

1   Park the vehicle with the wheels pointing straight ahead.
2   Remove the left side under-dash panel.
3   Follow the steering column down towards the firewall and locate the intermediate shaft upper pinch bolt. Mark the relationship of the intermediate shaft to the steering shaft, then remove the bolt.
4   Loosen the wheel lug nuts, raise the vehicle and support it securely on jackstands.
5   Remove the front wheels.
6   Disconnect the tie-rod ends from the steering knuckles (see Section 13).
7   Remove the steering gear mounting bolts **(see illustration)**.
8   Place a drain pan under the left (driver's side) of the steering gear, then disconnect the pressure and return hoses attached to the power steering gear assembly **(see illustration)**. Note: *If available, use a flare-nut wrench to prevent rounding-off the fittings.* Plug the ends of the disconnected hoses and the holes in the power steering housing to prevent contamination.
9   Remove the plastic shroud protecting the lower intermediate shaft U-joint coupling.
10   Mark the relationship of the lower intermediate shaft U-joint coupling to the steering gear.
11   Remove the pinch bolt from the intermediate shaft-to-steering gear U-joint **(see illustration)**.
12   Remove the intermediate shaft assembly.
13   Support the suspension crossmember with a floor jack.
14   Remove the two rear suspension crossmember retaining bolts and the four other crossmember bolts, then lower the crossmember slightly to provide sufficient clearance for removal of the steering gear.
15   Remove the steering gear through the left wheel well.
16   Installation is the reverse of removal. Be sure to tighten all fasteners

to the torque listed in this Chapter's Specifications.
17   Top up the power steering pump reservoir when you're done and bleed the power steering system (see Section 17).

16   Power steering pump - removal and installation

Removal

Refer to illustrations 16.4, 16.6 and 16.7

1   Disconnect the cable from the negative battery terminal. **Caution:** *If the stereo in your vehicle is equipped with an anti-theft system, make sure you have the correct activation code before disconnecting the battery.*
2   Remove the drivebelt on 2.2L engines (see Chapter 1). Remove the air intake duct from the engine on 2.3L and 2.4L engines (see Chapter 4).
3   Disconnect the inlet and outlet hoses from the power steering pump.
4   Unbolt the power steering pump from the bracket on 2.2L engines **(see illustration)** or from the rear of the camshaft housing on 2.3L and 2.4L engines.
5   Remove the pump.
6   On 2.2L engines, remove the pulley from the power steering pump with a suitable pulley removal tool **(see illustration)**. Pulley removal and installation tools are available at most auto parts stores.
7   Install the pulley on the new pump using a special installation tool **(see illustration)**. **Caution:** *Do not use a press to install the pulley.* The pulley should be installed so the face of the pulley is flush with the end of the pump shaft.
8   Installation is the reverse of removal. Tighten the pump bolts and the fittings securely.
9   Prime the pump by turning the pulley in the reverse direction to that of normal rotation (counterclockwise as viewed from the front) until air bubbles cease to emerge from the fluid when observed

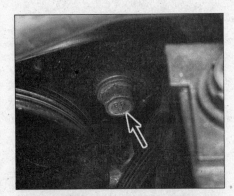

16.4 To remove the power steering pump, on 2.2L engines remove this bolt (arrow) and the other two bolts, which are accessed through the holes in the pulley

16.6 A typical power steering pump pulley removal tool

16.7 A typical power steering pump pulley installation tool

10

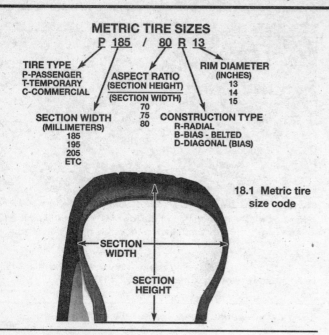

METRIC TIRE SIZES

P 185 / 80 R 13

TIRE TYPE
P-PASSENGER
T-TEMPORARY
C-COMMERCIAL

ASPECT RATIO
(SECTION HEIGHT)
(SECTION WIDTH)
70
75
80

RIM DIAMETER
(INCHES)
13
14
15

SECTION WIDTH
(MILLIMETERS)
185
195
205
ETC

CONSTRUCTION TYPE
R-RADIAL
B-BIAS - BELTED
D-DIAGONAL (BIAS)

18.1 Metric tire size code

SECTION WIDTH

SECTION HEIGHT

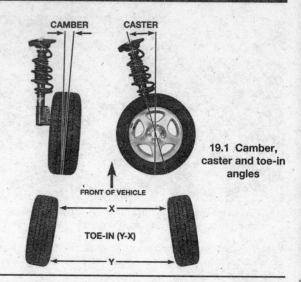

19.1 Camber, caster and toe-in angles

CAMBER   CASTER

FRONT OF VEHICLE

X

TOE-IN (Y-X)

Y

through the reservoir filler cap.
10   Install the drivebelt (see Chapter 1).
11   Bleed the power steering system (see Section 17).

17   Power steering system - bleeding

1   This is not a routine operation and normally will only be required when the system has been dismantled and reassembled.
2   Fill the reservoir to the correct level with fluid of the recommended type and allow it to remain undisturbed for at least two minutes.
3   Start the engine and run it for two or three seconds only. Check the reservoir and add more fluid as necessary.
4   Repeat the operations described in the preceding paragraph until the fluid level remains constant.
5   Raise the front of the vehicle until the wheels are clear of the ground.
6   Start the engine and increase the speed to about 1500 rpm. Now turn the steering wheel gently from stop-to-stop. Check the reservoir fluid level.
7   Lower the vehicle to the ground and, with the engine still running, move the vehicle forward sufficiently to obtain full right lock followed by full left lock. Recheck the fluid level. If the fluid in the reservoir is extremely foamy, allow the vehicle to stand for a few minutes with the engine switched off and then repeat the previous operations. At the same time, check the belt tightness and check for a bent or loose pulley. Check also to make sure the power steering hoses are not touching any other part of the vehicle, especially sheet metal or the exhaust manifold.
8   The procedures above will normally remedy an extreme foam condition and/or an objectionably noisy pump (low fluid level and/or air in the power steering fluid are the leading causes of this condition). If, however, either or both conditions persist after a few trials, the power steering system will have to be thoroughly checked. Do not drive the vehicle until the condition(s) have been remedied.

18   Wheels and tires - general information

Refer to illustration 18.1

All vehicles covered by this manual are equipped with metric-sized fiberglass or steel-belted radial tires **(see illustration)**. Use of other size or type of tires may affect the ride and handling of the vehicle. Don't mix different types of tires, such as radials and bias belted, on the same vehicle as handling may be seriously affected. It's recom-

mended that tires be replaced in pairs on the same axle, but if only one tire is being replaced, be sure it's the same size, structure and tread design as the other. Because tire pressure has a substantial effect on handling and wear, the pressure on all tires should be checked at least once a month or before any extended trips (see Chapter 1).

Wheels must be replaced if they are bent, dented, leak air, have elongated bolt holes, are heavily rusted, out of vertical symmetry or if the lug nuts won't stay tight. Wheel repairs that use welding or peening are not recommended.

Tire and wheel balance is important to the overall handling, braking and performance of the vehicle. Unbalanced wheels can adversely affect handling and ride characteristics as well as tire life. Whenever a tire is installed on a wheel, the tire and wheel should be balanced by a shop with the proper equipment.

19   Front end alignment - general information

Refer to illustration 19.1

A front end alignment refers to the adjustments made to the front wheels so they are in proper angular relationship to the suspension and the ground **(see illustration)**. Front wheels that are out of proper alignment not only affect steering control, but also increase tire wear. Camber and toe-in are the only angles that can be adjusted on the vehicles covered by this manual, but caster should also be measured to determine if any suspension parts are bent.

Getting the proper front wheel alignment is a very exacting process, one in which complicated and expensive machines are necessary to perform the job properly. Because of this, you should have a technician with the proper equipment perform these tasks. We will, however, use this space to give you a basic idea of what is involved with front end alignment so you can better understand the process and deal intelligently with the shop that does the work.

Camber is the tilting of the front wheels from vertical when viewed from the front of the vehicle. On the vehicles covered in this manual, camber can only be adjusted by elongating the lower strut-to-knuckle hole.

Caster is the tilting of the top of the front steering axis from the vertical: a tilt toward the rear is positive caster and a tilt toward the front is negative caster. Caster is not adjustable on these vehicles.

Toe-in is the turning in of the front wheels. The purpose of a toe specification is to ensure parallel rolling of the front wheels. In a vehicle with zero toe-in, the distance between the front edges of the wheels will be the same as the distance between the rear edges of the wheels. The actual amount of toe-in is normally only a fraction of an inch. Toe-in adjustment is controlled by the position of the tie-rod end on the tie-rod. Incorrect toe-in will cause the tires to wear improperly by making them scrub against the road surface.

Chapter 11 Body

Contents

1 General information

Warning: *The models covered by this manual are equipped with airbags. Always disable the airbag system (see Chapter 12) before working in the vicinity of the impact sensors, steering column or instrument panel. Failure to follow these procedures may cause accidental deployment of the airbag, which could cause personal injury. The airbag circuits are easily identified by yellow insulation covering the entire wiring harness. Do not use electrical test equipment on any of these wires or tamper with them in any way.*

Caution: *If the vehicle is equipped with a Delco Loc II or Theftlock audio system, make sure you have the correct activation code before disconnecting the battery.*

The models covered by this manual are designed with a "unibody" type construction. The major body components, floor pan and front and rear frame side rails are welded together to create a rigid structure which supports the remaining body components, drivetrain, front and rear suspension components and other mechanical components.

Certain body components are particularly vulnerable to accident damage and can be unbolted and repaired or replaced. Among these parts are the body moldings, front fenders, doors, bumpers, the hood, trunk lid and all glass.

Only general body maintenance practices and body panel repair procedures within the scope of the do-it-yourselfer are included in this Chapter.

2 Body - maintenance

1   The condition of your vehicle's body is very important, because the resale value depends a great deal on it. It's much more difficult to repair a neglected or damaged body than it is to repair mechanical components. The hidden areas of the body, such as the wheel wells, the frame and the engine compartment, are equally important, although they don't require as frequent attention as the rest of the body.

2   Once a year, or every 12,000 miles, it's a good idea to have the underside of the body steam cleaned. All traces of dirt and oil will be removed and the area can then be inspected carefully for rust, damaged brake lines, frayed electrical wires, damaged cables and other problems. The front suspension components should be greased after completion of this job.

3   At the same time, clean the engine and the engine compartment with a steam cleaner or water-soluble degreaser.

4   The wheel wells should be given close attention, since undercoating can peel away and stones and dirt thrown up by the tires can cause the paint to chip and flake, allowing rust to set in. If rust is found, clean down to the bare metal and apply an anti-rust paint.

5   The body should be washed about once a week. Wet the vehicle thoroughly to soften the dirt, then wash it down with a soft sponge and plenty of clean soapy water. If the surplus dirt is not washed off very carefully, it can wear down the paint.

6   Spots of tar or asphalt thrown up from the road should be removed with a cloth soaked in solvent.

11

7   Once every six months, wax the body and chrome trim. If a chrome cleaner is used to remove rust from any of the vehicle's plated parts, remember that the cleaner also removes part of the chrome, so use it sparingly.

3   Vinyl trim - maintenance

Don't clean vinyl trim with detergents, caustic soap or petroleum-based cleaners. Plain soap and water works just fine, with a soft brush to clean dirt that may be ingrained. Wash the vinyl as frequently as the rest of the vehicle. After cleaning, application of a high-quality rubber and vinyl protectant will help prevent oxidation and cracks. The protectant can also be applied to weatherstripping, vacuum lines and rubber hoses, which often fail as a result of chemical degradation, and to the tires.

4   Upholstery and carpets - maintenance

1   Every three months remove the floor mats and clean the interior of the vehicle (more frequently if necessary). Use a stiff whisk broom to brush the carpeting and loosen dirt and dust, then vacuum the upholstery and carpets thoroughly, especially along seams and crevices.
2   Dirt and stains can be removed from carpeting with basic household or automotive carpet shampoos available in spray cans. Follow the directions and vacuum again, then use a stiff brush to bring back the "nap" of the carpet.
3   Most interiors have cloth or vinyl upholstery, either of which can be cleaned and maintained with a number of material-specific cleaners or shampoos available in auto supply stores. Follow the directions on the product for usage, and always spot-test any upholstery cleaner on an inconspicuous area (bottom edge of a back seat cushion) to ensure that it doesn't cause a color shift in the material.
4   After cleaning, vinyl upholstery should be treated with a protectant. **Note:** *Make sure the protectant container indicates the product can be used on seats - some products may make a seat too slippery.* **Caution:** *Do not use protectant on vinyl-covered steering wheels.*
5   Leather upholstery requires special care. It should be cleaned regularly with saddlesoap or leather cleaner. Never use alcohol, gasoline, nail polish remover or thinner to clean leather upholstery.
6   After cleaning, regularly treat leather upholstery with a leather conditioner, rubbed in with a soft cotton cloth. Never use car wax on leather upholstery.
7   In areas where the interior of the vehicle is subject to bright sunlight, cover leather seating areas of the seats with a sheet if the vehicle is to be left out for any length of time.

5   Body repair - minor damage

Flexible plastic body panels (front and rear bumper fascia)

The following repair procedures are for minor scratches and gouges. Repair of more serious damage should be left to a dealer service department or qualified auto body shop. Below is a list of the equipment and materials necessary to perform the following repair procedures on plastic body panels. Although a specific brand of material may be mentioned, it should be noted that equivalent products from other manufacturers may be used instead.

Wax, grease and silicone removing solvent
Cloth-backed body tape
Sanding discs
Drill motor with three-inch disc holder
Hand sanding block
Rubber squeegees
Sandpaper
Non-porous mixing palette
Wood paddle or putty knife
Curved tooth body file
GM part no. 12345744 Compoxy Repair Material, or equivalent (for flexible panels)

1   Remove the damaged panel, if necessary or desirable. In most cases, repairs can be carried out with the panel installed.
2   Clean the area(s) to be repaired with a wax, grease and silicone removing solvent applied with a water-dampened cloth.
3   If the damage is structural, that is, if it extends through the panel, clean the backside of the panel area to be repaired as well. Wipe dry.
4   Sand the rear surface about 1-1/2 inches beyond the break.
5   Cut two pieces of fiberglass cloth large enough to overlap the break by about 1-1/2 inches. Cut only to the required length.
6   Mix the adhesive from the repair kit according to the instructions included with the kit, and apply a layer of the mixture approximately 1/8-inch thick on the backside of the panel. Overlap the break by at least 1-1/2 inches.
7   Apply one piece of fiberglass cloth to the adhesive and cover the cloth with additional adhesive. Apply a second piece of fiberglass cloth to the adhesive and immediately cover the cloth with additional adhesive insufficient quantity to fill the weave.
8   Allow the repair to cure for 20 to 30 minutes at 60-degrees to 80-degrees F.
9   If necessary, trim the excess repair material at the edge.
10   Remove all of the paint film over and around the area(s) to be repaired. The repair material should not overlap the painted surface.
11   With a drill motor and a sanding disc (or a rotary file), cut a "V" along the break line approximately 1/2-inch wide. Remove all dust and loose particles from the repair area.
12   Mix and apply the repair material. Apply a light coat first over the damaged area; then continue applying material until it reaches a level slightly higher than the surrounding finish.
13   Cure the mixture for 20 to 30 minutes at 60-degrees to 80-degrees F.
14   Roughly establish the contour of the area being repaired with a body file. If low areas or pits remain, mix and apply additional adhesive.
15   Block sand the damaged area with sandpaper to establish the actual contour of the surrounding surface.
16   If desired, the repaired area can be temporarily protected with several light coats of primer. Because of the special paints and techniques required for flexible body panels, it is recommended that the vehicle be taken to a paint shop for completion of the body repair.

Steel body panels
See photo sequence

Repair of minor scratches

17   If the scratch is superficial and does not penetrate to the metal of the body, repair is very simple. Lightly rub the scratched area with a fine rubbing compound to remove loose paint and built-up wax. Rinse the area with clean water.
18   Apply touch-up paint to the scratch, using a small brush. Continue to apply thin layers of paint until the surface of the paint in the scratch is level with the surrounding paint. Allow the new paint at least two weeks to harden, then blend it into the surrounding paint by rubbing with a very fine rubbing compound. Finally, apply a coat of wax to the scratch area.
19   If the scratch has penetrated the paint and exposed the metal of the body, causing the metal to rust, a different repair technique is required. Remove all loose rust from the bottom of the scratch with a pocket knife, then apply rust inhibiting paint to prevent the formation of rust in the future. Using a rubber or nylon applicator, coat the scratched area with glaze-type filler. If required, the filler can be mixed with thinner to provide a very thin paste, which is ideal for filling narrow scratches. Before the glaze filler in the scratch hardens, wrap a piece of smooth cotton cloth around the tip of a finger. Dip the cloth in thinner and then quickly wipe it along the surface of the scratch. This will ensure that the surface of the filler is slightly hollow. The scratch can now be painted over as described earlier in this section.

Repair of dents

20 When repairing dents, the first job is to pull the dent out until the affected area is as close as possible to its original shape. There is no point in trying to restore the original shape completely as the metal in the damaged area will have stretched on impact and cannot be restored to its original contours. It is better to bring the level of the dent up to a point which is about 1/8-inch below the level of the surrounding metal. In cases where the dent is very shallow, it is not worth trying to pull it out at all.

21 If the back side of the dent is accessible, it can be hammered out gently from behind using a soft-face hammer. While doing this, hold a block of wood firmly against the opposite side of the metal to absorb the hammer blows and prevent the metal from being stretched.

22 If the dent is in a section of the body which has double layers, or some other factor makes it inaccessible from behind, a different technique is required. Drill several small holes through the metal inside the damaged area, particularly in the deeper sections. Screw long, self-tapping screws into the holes just enough for them to get a good grip in the metal. Now the dent can be pulled out by pulling on the protruding heads of the screws with locking pliers.

23 The next stage of repair is the removal of paint from the damaged area and from an inch or so of the surrounding metal. This is done with a wire brush or sanding disk in a drill motor, although it can be done just as effectively by hand with sandpaper. To complete the preparation for filling, score the surface of the bare metal with a screwdriver or the tang of a file, or drill small holes in the affected area. This will provide a good grip for the filler material. To complete the repair, see the subsection on filling and painting later in this Section.

Repair of rust holes or gashes

24 Remove all paint from the affected area and from an inch or so of the surrounding metal using a sanding disk or wire brush mounted in a drill motor. If these are not available, a few sheets of sandpaper will do the job just as effectively.

25 With the paint removed, you will be able to determine the severity of the corrosion and decide whether to replace the whole panel, if possible, or repair the affected area. New body panels are not as expensive as most people think and it is often quicker to install a new panel than to repair large areas of rust.

26 Remove all trim pieces from the affected area except those which will act as a guide to the original shape of the damaged body, such as headlight shells, etc. Using metal snips or a hacksaw blade, remove all loose metal and any other metal that is badly affected by rust. Hammer the edges of the hole in to create a slight depression for the filler material.

27 Wire brush the affected area to remove the powdery rust from the surface of the metal. If the back of the rusted area is accessible, treat it with rust inhibiting paint.

28 Before filling is done, block the hole in some way. This can be done with sheet metal riveted or screwed into place, or by stuffing the hole with wire mesh.

29 Once the hole is blocked off, the affected area can be filled and painted. See the following subsection on filling and painting.

Filling and painting

30 Many types of body fillers are available, but generally speaking, body repair kits which contain filler paste and a tube of resin hardener are best for this type of repair work. A wide, flexible plastic or nylon applicator will be necessary for imparting a smooth and contoured finish to the surface of the filler material. Mix up a small amount of filler on a clean piece of wood or cardboard (use the hardener sparingly). Follow the manufacturer's instructions on the package, otherwise the filler will set incorrectly.

31 Using the applicator, apply the filler paste to the prepared area. Draw the applicator across the surface of the filler to achieve the desired contour and to level the filler surface. As soon as a contour that approximates the original one is achieved, stop working the paste. If you continue, the paste will begin to stick to the applicator. Continue to add thin layers of paste at 20-minute intervals until the level of the filler is just above the surrounding metal.

32 Once the filler has hardened, the excess can be removed with a body file. From then on, progressively finer grades of sandpaper should be used, starting with a 180-grit paper and finishing with 600-grit wet-or-dry paper. Always wrap the sandpaper around a flat rubber or wooden block, otherwise the surface of the filler will not be completely flat. During the sanding of the filler surface, the wet-or-dry paper should be periodically rinsed in water. This will ensure that a very smooth finish is produced in the final stage.

33 At this point, the repair area should be surrounded by a ring of bare metal, which in turn should be encircled by the finely feathered edge of good paint. Rinse the repair area with clean water until all of the dust produced by the sanding operation is gone.

34 Spray the entire area with a light coat of primer. This will reveal any imperfections in the surface of the filler. Repair the imperfections with fresh filler paste or glaze filler and once more smooth the surface with sandpaper. Repeat this spray-and-repair procedure until you are satisfied that the surface of the filler and the feathered edge of the paint are perfect. Rinse the area with clean water and allow it to dry completely.

35 The repair area is now ready for painting. Spray painting must be carried out in a warm, dry, windless and dust free atmosphere. These conditions can be created if you have access to a large indoor work area, but if you are forced to work in the open, you will have to pick the day very carefully. If you are working indoors, dousing the floor in the work area with water will help settle the dust which would otherwise be in the air. If the repair area is confined to one body panel, mask off the surrounding panels. This will help minimize the effects of a slight mismatch in paint color. Trim pieces such as chrome strips, door handles, etc., will also need to be masked off or removed. Use masking tape and several thickness of newspaper for the masking operations.

36 Before spraying, shake the paint can thoroughly, then spray a test area until the spray painting technique is mastered. Cover the repair area with a thick coat of primer. The thickness should be built up using several thin layers of primer rather than one thick one. Using 600-grit wet-or-dry sandpaper, rub down the surface of the primer until it is very smooth. While doing this, the work area should be thoroughly rinsed with water and the wet-or-dry sandpaper periodically rinsed as well. Allow the primer to dry before spraying additional coats.

37 Spray on the top coat, again building up the thickness by using several thin layers of paint. Begin spraying in the center of the repair area and then, using a circular motion, work out until the whole repair area and about two inches of the surrounding original paint is covered. Remove all masking material 10 to 15 minutes after spraying on the final coat of paint. Allow the new paint at least two weeks to harden, then use a very fine rubbing compound to blend the edges of the new paint into the existing paint. Finally, apply a coat of wax.

6 Body repair - major damage

1 Major damage must be repaired by an auto body/frame repair shop with the necessary welding and hydraulic straightening equipment.

2 If the damage has been serious, it is vital that the structure be checked for proper alignment or the vehicle's handling characteristics may be adversely affected. Other problems, such as excessive tire wear and wear in the driveline and steering may occur.

3 Due to the fact that all of the major body components (hood, fenders, etc.) are separate and replaceable units, any seriously damaged components should be replaced rather than repaired. Sometimes these components can be found in a wrecking yard that specializes in used vehicle components, often at considerable savings over the cost of new parts.

7 Hinges and locks - maintenance

Once every 3000 miles, or every three months, the hinges and latch assemblies on the doors, hood and trunk should be given a few drops of light oil or lock lubricant. The door latch strikers should also

11

These photos illustrate a method of repairing simple dents. They are intended to supplement *Body repair - minor damage* in this Chapter and should not be used as the sole instructions for body repair on these vehicles.

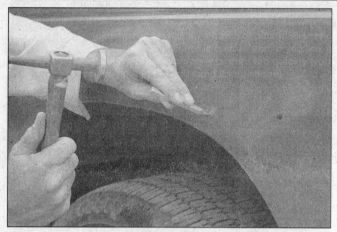

1   If you can't access the backside of the body panel to hammer out the dent, pull it out with a slide-hammer-type dent puller. In the deepest portion of the dent or along the crease line, drill or punch hole(s) at least one inch apart . . .

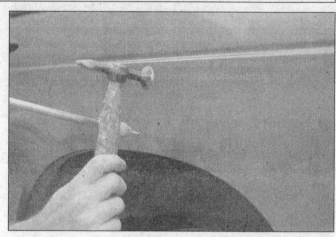

2   . . . then screw the slide-hammer into the hole and operate it. Tap with a hammer near the edge of the dent to help 'pop' the metal back to its original shape. When you're finished, the dent area should be close to its original contour and about 1/8-inch below the surface of the surrounding metal

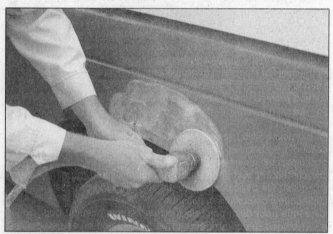

3   Using coarse-grit sandpaper, remove the paint down to the bare metal. Hand sanding works fine, but the disc sander shown here makes the job faster. Use finer (about 320-grit) sandpaper to feather-edge the paint at least one inch around the dent area

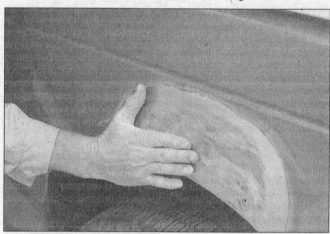

4   When the paint is removed, touch will probably be more helpful than sight for telling if the metal is straight. Hammer down the high spots or raise the low spots as necessary. Clean the repair area with wax/silicone remover

5   Following label instructions, mix up a batch of plastic filler and hardener. The ratio of filler to hardener is critical, and, if you mix it incorrectly, it will either not cure properly or cure too quickly (you won't have time to file and sand it into shape)

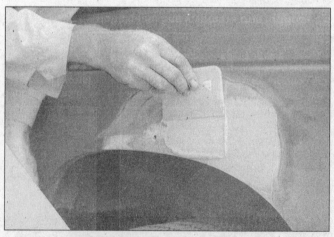

6   Working quickly so the filler doesn't harden, use a plastic applicator to press the body filler firmly into the metal, assuring it bonds completely. Work the filler until it matches the original contour and is slightly above the surrounding metal

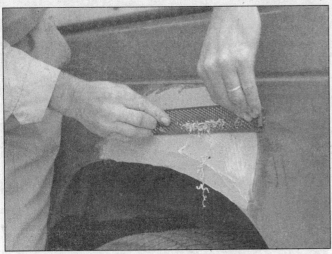

7   Let the filler harden until you can just dent it with your fingernail. Use a body file or Surform tool (shown here) to rough-shape the filler

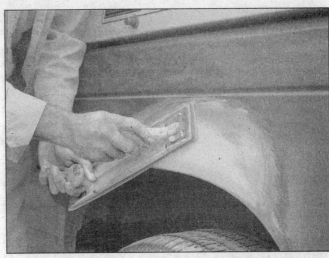

8   Use coarse-grit sandpaper and a sanding board or block to work the filler down until it's smooth and even. Work down to finer grits of sandpaper - always using a board or block - ending up with 360 or 400 grit

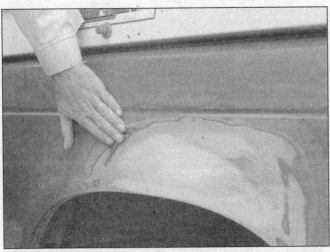

9   You shouldn't be able to feel any ridge at the transition from the filler to the bare metal or from the bare metal to the old paint. As soon as the repair is flat and uniform, remove the dust and mask off the adjacent panels or trim pieces

10   Apply several layers of primer to the area. Don't spray the primer on too heavy, so it sags or runs, and make sure each coat is dry before you spray on the next one. A professional-type spray gun is being used here, but aerosol spray primer is available inexpensively from auto parts stores

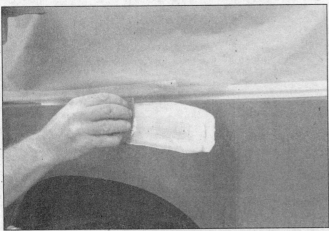

11   The primer will help reveal imperfections or scratches. Fill these with glazing compound. Follow the label instructions and sand it with 360 or 400-grit sandpaper until it's smooth. Repeat the glazing, sanding and respraying until the primer reveals a perfectly smooth surface

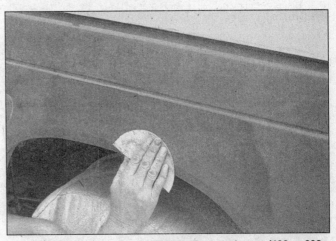

12   Finish sand the primer with very fine sandpaper (400 or 600-grit) to remove the primer overspray. Clean the area with water and allow it to dry. Use a tack rag to remove any dust, then apply the finish coat. Don't attempt to rub out or wax the repair area until the paint has dried completely (at least two weeks)

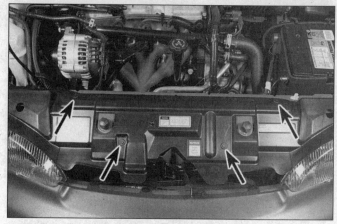

9.2 Mark the relationship of the hinges to the hood as shown, then, with the help of an assistant to hold the hood, remove the retaining bolts (arrows) from each hinge plate and lift off the hood

9.10a To remove the hood latch cover, release the push-in retainers (arrows) (1996 model shown, other model years similar)

be lubricated with a thin coat of grease to reduce wear and ensure free movement. Lubricate the door and trunk locks with spray-on graphite lubricant.

8   Windshield and fixed glass - replacement

Replacement of the windshield and fixed glass requires the use of special fast setting adhesive/caulk materials. These operations should be left to a dealer or a shop specializing in glass work.

9   Hood - removal, installation and adjustment

Removal and installation

Refer to illustration 9.2
Note: *The hood is somewhat awkward to remove and install; at least two people should perform this procedure.*
1   Open the hood, then place blankets or pads over the fenders and cowl area of the body. This will protect the body and paint as the hood is lifted off.
2   Make marks or scribe a line around the hood hinge to ensure proper alignment during installation **(see illustration)**.
3   Disconnect any cables or wires that will interfere with removal.
4   With an assistant supporting the weight of the hood, remove the

hinge-to-hood bolts and lift off the hood.
6   Installation is the reverse of removal. Align the hinge bolts with the marks made in step 2.

Adjustment

Refer to illustrations 9.10a, 9.10b, 9.10c, 9.10d and 9.11
7   Fore-and-aft and side-to-side adjustment of the hood is done by moving the hinge plate slot after loosening the bolts or nuts.
8   Scribe a line around the entire hinge plate so you can determine the amount of movement.
9   Loosen the bolts or nuts and move the hood into correct alignment. Move it only a little at a time. Tighten the hinge bolts and carefully lower the hood to check the position.
10   If necessary after installation, the entire hood latch assembly can be adjusted up-and-down as well as from side-to-side on the radiator support so the hood closes securely and flush with the fenders. First, remove the hood latch cover **(see illustrations)**. To make the adjustment, scribe a line or mark around the hood latch to provide a reference point **(see illustration)**, then loosen the latch retaining bolts **(see illustration)** and reposition the latch assembly, as necessary. Following adjustment, retighten the mounting bolts.
11   Finally, adjust the hood bumpers **(see illustration)** on the radiator support so the hood, when closed, is flush with the fenders.
12   The hood latch assembly, as well as the hinges, should be periodically lubricated with white, lithium-base grease to prevent binding and wear.

9.10b To release a push-in retainer, simply pry up the center part with a tool like this one or a flat-blade screwdriver

9.10c Mark the relationship of the hood latch to the crossmember to provide a visual indicator of the adjustment

9.10d To adjust the position of the closed hood, loosen the hood latch bolts (arrows) and move the hood latch a little at a time, alternately closing the hood and noting its position in relation to the gaps between the hood and the fenders

9.11 Screw the hood bumpers in or out to adjust the hood flush with the fenders (left bumper shown, right bumper identical)

10.2a To disconnect the cable from the hood latch mechanism, pry the cable ferrule out of the latch assembly . . .

10.2b . . . and disengage the cable end plug from its slot in the latch

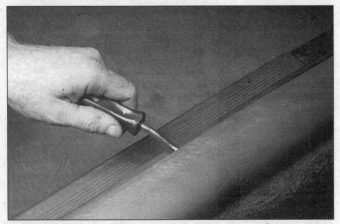

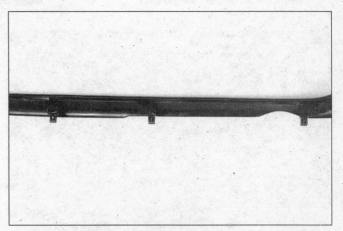

10.6a Gently pry the carpet retainer loose with a suitable tool

10.6b There are three locating pins under the front door portion of the carpet retainer on sedan models (shown) and two more pins under the rear door portion. On coupes there are four pins under the front door part

10 Hood latch and release cable - removal and installation

Warning: *The models covered by this manual are equipped with airbags. Always disable the airbag system (see Chapter 12) before working in the vicinity of the impact sensors, steering column or instrument panel. Failure to follow these procedures may cause accidental deployment of the airbag, which could cause personal injury. The airbag circuits are easily identified by yellow insulation covering the entire wiring harness. Do not use electrical test equipment on any of these wires or tamper with them in any way.*

Hood latch

Refer to illustrations 10.2a and 10.2b

1   Scribe a line around the latch to aid alignment when installing, then remove the retaining bolts to the radiator support **(see illustration 9.10d)**. Remove the latch.
2   Disconnect the hood release cable by disengaging the cable from the latch assembly **(see illustrations)**.
3   Installation is the reverse of removal. Adjust the latch so the hood engages securely when closed and the hood bumpers are slightly compressed (see Section 9).

Release cable

Refer to illustrations 10.6a, 10.6b and 10.7

4   Disconnect the release cable from the hood latch assembly as described in Step 2.
5   Unclip the release cable from the engine wiring harness. Attach a piece of wire to the cable.

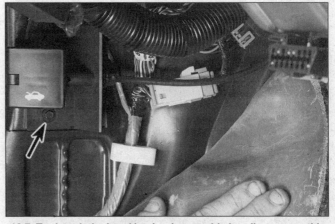

10.7 To detach the hood latch release cable handle, remove this retaining screw (arrow)

6   Working in the passenger compartment, remove the kick panel/carpet retainer **(see illustrations)**, then peel back the carpet to expose the hood latch release cable and handle.
7   Remove the hood latch release handle retaining screw **(see illustration)**.
8   Trace the cable forward to the grommet where the cable goes through the firewall and pry the grommet out of the firewall. Pull the

11

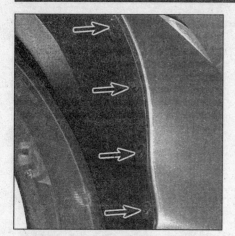 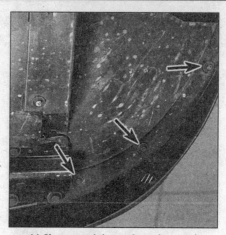 

11.2a To detach the front bumper fascia from the splash shield, remove these four sheet metal screws (arrows) in each wheel well . . .

11.2b . . . and these three (arrows) underneath (right wheel well shown, left wheel well identical)

11.3 To detach the front bumper fascia from the fenders, remove these two bolts (arrows) from each fender

 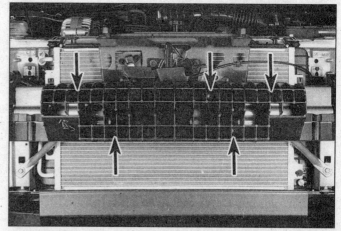

11.4 To detach the front bumper fascia from the impact bar, remove these four push-in retainers; to release each retainer, simply pry it up with a screwdriver (Chevrolet model shown, Pontiac models similar)

11.7 To detach the front energy absorber from the impact bar, drill out the nine pop rivets (five shown here, other four - located near ends of absorber - not visible in this photo); to locate the pop rivets, look for the semi-circular cutouts in the leading edge of the square above and below each rivet hole (non-Z24 model shown; Z24 models use six push-in retainers instead)

handle and cable rearward into the passenger compartment.

8   Disconnect the guide wire from the old cable and fasten it to the new cable.

9   With the new cable attached to the wire, pull the wire back through the firewall until the new cable reaches the latch assembly. Make sure that the grommet is properly seated on both sides of the hole in the firewall. Push on the grommet with your fingers from the passenger compartment side to seat the grommet in the firewall correctly.

10   Install the carpet and kick panel/carpet retainer.

11   Install the hood latch release handle retaining screw and tighten it securely.

12   The remainder of installation is the reverse of removal.

11   Bumpers - removal and installation

Warning: *The models covered by this manual are equipped with airbags. Always disable the airbag system (see Chapter 12) before working in the vicinity of the impact sensors, steering column or instrument panel. Failure to follow these procedures may cause accidental deployment of the airbag, which could cause personal injury. The airbag circuits are easily identified by yellow insulation covering the entire wiring harness. Do not use electrical test equipment on any of*

these wires or tamper with them in any way.

Caution: *If the vehicle is equipped with a Delco Loc II or Theftlock audio system, make sure you have the correct activation code before disconnecting the battery.*

Front bumper

Refer to illustrations 11.2a, 11.2b, 11.3, 11.4 and 11.7

1   Raise the front of the vehicle and support it securely on jackstands.

2   Disconnect the bumper fascia from the left and right wheel well splash shields **(see illustrations)**.

3   Detach the bumper fascia from the fenders **(see illustration)**. **Note:** *Chevrolet models use two bolts to fasten the front bumper fascia to each front fender, while Pontiac models use two nuts and one screw to fasten the front bumper fascia to the fender.*

4   Disconnect the bumper fascia from the front impact bar **(see illustration)**.

5   Before removing the bumper fascia, unplug the electrical connectors for the turn signal and parking lights (see Chapter 12).

6   Remove the bumper fascia.

7   To remove the front bumper energy absorber **(see illustration)** on most vehicles, drill out all nine pop rivets (Z24 models use six push-in retainers instead). To unlock a push-in retainer, pull the head out; to

11.11 Inside the rear compartment, there are four rear bumper fascia retaining screws (arrows), two per side (left side screws shown)

11.12 There are four more rear bumper fascia retaining screws (arrows) - two per side - inside the rear wheel wells

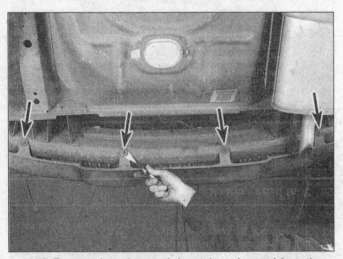

11.13 Remove these four push-in retainers (arrows) from the underside of the rear bumper fascia, then remove the fascia

11.15 To detach the rear bumper energy absorber, drill out the rivets (Pontiac Coupe and all Chevrolet models except the Z24) or remove the push-in retainers (Pontiac sedans and Chevrolet Z24 models); the number of rivets may vary - look for semicircular cutouts in the upper and lower edges of certain boxes, indicating a rivet in that box

lock it into place, push the head down.

8   To remove the front bumper impact bar, simply remove the four impact bar retaining nuts (two at each end).

9   Installation is the reverse of removal. On all vehicles except Z24 models, you'll have to use new pop rivets.

Rear bumper

Refer to illustrations 11.11, 11.12, 11.13, 11.15 and 11.16

10   Raise the rear of the vehicle and place it securely on jackstands.

11   Working inside the rear compartment, remove the upper storage net retainers for access to the four rear fascia retaining screws - two per side **(see illustration)** - located inside the compartment. Remove these four screws.

12   Working inside the rear wheel wells, remove the four rear fascia retaining screws - two per wheel well **(see illustration)**.

13   Working underneath the rear fascia, remove the four push-in retainers **(see illustration)**.

14   Remove the rear bumper fascia.

15   To remove the rear bumper energy absorber **(see illustration)**, drill out the rivets (Pontiac Coupe and all Chevrolet models except the Z24) or remove the push-in retainers (Pontiac sedans and Chevrolet Z24 models).

16   To remove the rear bumper impact bar, remove the six retaining nuts (three at each end of the bar) from the stud plates **(see illustration)**.

17   Installation is the reverse of removal.

11.16 To detach the rear impact bar from the stud plates, remove six nuts (arrows) - three per side

11

12.4a To remove the wheel housing, release these three push-in retainers (arrows) in the upper part of the housing . . .

12 Fender (front) - removal and installation

Refer to illustrations 12.4a, 12.4b, 12.4c, 12.5, 12.8, 12.9a, 12.9b and 12.10

1   Remove the hood (see Section 9).
2   Loosen the front wheel lug nuts. Raise the vehicle, support it securely on jackstands and remove the front wheels.
3   Remove the splash shield (see illustrations 11.2a and 11.2b).

4   Remove the wheel housing and the fender flare (see illustrations).
5   Remove the lower fender bracket bolts and the lower fender bracket (see illustration).
6   Remove the front side marker light (see Chapter 12).
7   Detach the front bumper fascia from the fender (see illustration 11.3).
8   Remove the three fender insulator push-in retainers (see illustration) and remove the insulator.
9   Remove the two center fender bracket bolts and the upper fender bolt (see illustrations) and the center fender bracket.
10   Remove the two lower hood hinge bolts (see illustration) and the lower hood hinge. Remove the two upper fender mounting bolts.
11   Remove the lower fender mounting bolts (see illustration 12.4c).
12   Detach the fender. It's a good idea to have an assistant support the fender while it's being moved away from the vehicle to prevent damage to the surrounding body panels.
13   Installation is the reverse of removal.

13 Trunk lid - removal, installation and adjustment

Note: *The trunk lid is heavy and somewhat awkward to remove and install - at least two people should perform this procedure.*

Removal and installation

Refer to illustration 13.3

1   Open the trunk lid and cover the edges of the trunk compartment with pads or cloths to protect the painted surfaces when the lid is removed.

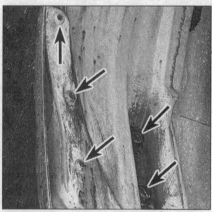

12.4b . . . and these two (right arrows), plus three more (not visible in this photo); to remove the fender flare, remove these three screws (left arrows) . . .

12.4c . . . and these two (right arrows); the two bolts (arrows) to the left of the fender flare are the lower rear fender mounting bolts

12.5 To detach the lower front part of the fender, remove these two bolts (arrows) and remove the lower fender bracket

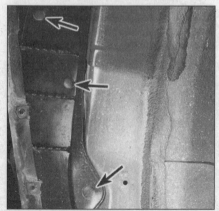

12.8 To remove the fender insulator, remove these three push-in retainers (arrows)

12.9a To detach the rear part of the fender, remove the center fender bracket bolts (arrows) . . .

12.9b . . . and the upper bolt (arrow)

12.10 Remove the lower hood hinge bolts (left arrows) and remove the lower hood hinge; to detach the upper part of the fender, remove the two bolts on the right (arrows)

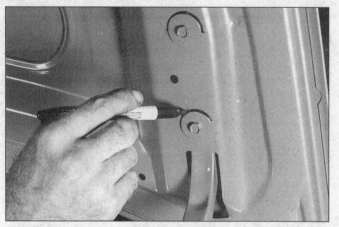

13.3 Before removing the four (two per side) trunk lid hinge bolts, be sure to mark the relationship of the hinge to the trunk lid to ensure correct alignment when the lid is installed

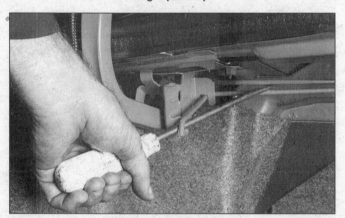

13.12 To readjust the tension on the trunk lid hinges, lever the ends of the torque rods up or down, one notch at a time, with a screwdriver as shown (raising the torque rods a notch makes it easier to raise the lid, and harder to lower it; moving the rods down a notch makes it harder to raise the lid, and easier to lower it)

2   Disconnect any cables or wire harness connectors attached to the trunk lid that would interfere with removal.
3   Use a felt-tip marker or scribe to make alignment marks around the trunk lid hinge **(see illustration)**.
4   While an assistant supports the lid, remove the hinge bolts from both sides and lift the trunk lid off the vehicle.
5   Installation is the reverse of removal. **Note:** *When reinstalling the trunk lid, align the hinge with the marks made during removal.*

Adjustment

Refer to illustration 13.12

6   Fore-and-aft and side-to-side adjustment of the trunk lid is done by moving the hood in relation to the hinge plate after loosening the bolts or nuts.
7   Scribe a line around the entire hinge plate as described earlier in this section so you can judge the amount of movement.
8   Loosen the bolts or nuts and move the trunk lid into correct alignment. Move it only a little at a time. Tighten the hinge bolts or nuts and carefully lower the trunk lid to check the alignment.
9   If necessary after installation, the entire trunk lid latch assembly can be adjusted up and down as well as from side to side on the trunk lid so the lid closes securely and is flush with the rear quarter panels. To do this, scribe a line around the trunk lid latch mounting bolts to provide a reference point. Then loosen the bolts and reposition the latch assembly as necessary. Following adjustment, retighten the mounting bolts.
10   Adjust the bumpers on the trunk lid, so that the trunk lid is flush

14.2 Mark the relationship of the trunk lid latch to the trunk lid and remove the bolts (arrows)

with the rear quarter panels when closed.
11   The trunk lid latch assembly, as well as the hinges, should be periodically lubricated with white lithium-base grease to prevent sticking and wear.
12   The effort required to raise or lower the trunk lid can also be adjusted, by moving the position of the ends of a pair of torque rods installed between the two trunk lid hinges **(see illustration)**. Each rod can be adjusted up or down in one of three positions. To increase the effort required to raise the trunk lid, or to decrease the effort required to lower the lid, relocate the rods down a notch. To decrease the effort needed to raise the lid, or to increase the effort needed to lower the lid, raise the rods a notch. To move a rod up or down, use a short section of pipe of a suitable diameter over the end of the rod; don't try to move a rod with your hand - it's too stiff.

14   Trunk lid latch, striker and lock cylinder - removal and installation

Caution: *If the vehicle is equipped with a Delco Loc II or Theftlock audio system, make sure you have the correct activation code before disconnecting the battery.*

Latch

Refer to illustration 14.2 and 14.3

1   On models with a remote release system, remove the locking clip, flip up the door on the latch assembly and disengage the release cable from the latch assembly.
2   Mark the relationship of the latch to the trunk lid **(see illustration)**.

11

14.3 To detach the lock release box from the latch assembly, release this push-in retainer (arrow)

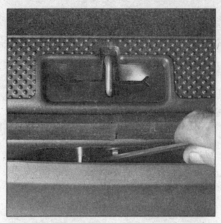

14.6a The striker retaining nuts and reinforcement plate can be accessed through the gap between the rear bumper fascia and the body

14.6b You'll have to remove this sill plate to remove the striker

3   If you're replacing the latch or the lock cylinder, not just unbolting it from the trunk lid to swap it over to a new trunk lid, release the push-in retainer **(see illustration)** that secures the lock release box to the latch and detach the lock release box from the latch assembly. If you're repairing or replacing the trunk lid, rather than the latch and lock cylinder, it's not necessary to detach the lock release box from the latch assembly; simply remove both of them as a single assembly.

4   Remove the latch retaining bolts and remove the latch.

5   Installation is the reverse of removal. Be sure to align the latch carefully with the marks you made prior to removal.

Striker

Refer to illustrations 14.6a and 14.6b

6   Using a wrench inserted down into the space between the bumper fascia and the body, remove the two striker retaining nuts **(see illustration)**, remove the striker reinforcement plate, detach the sill plate **(see illustration)** and remove the striker.

7   Installation is the reverse of removal, but don't tighten the striker retaining nuts until you have closed the trunk lid and verified that the trunk latch and the striker are properly aligned.

Lock cylinder

Refer to illustration 14.10

8   Remove the trunk lid latch (see Steps 1 through 4).

9   Detach the lock release box from the latch assembly **(see illustration 14.3)**.

10   To detach the lock cylinder from the trunk lid on a Chevrolet model, pry off the retainer **(see illustration)**. To detach the lock cylinder on a Pontiac model, drill out the two rivets.

11   Installation is the reverse of removal. Don't forget the gasket between the lock cylinder and the trunk lid. On Pontiac models, install the new lock cylinder with new rivets, if available, or use nuts and bolts.

15   Door trim panel - removal and installation

Refer to illustrations 15.5a, 15.5b, 15.6a, 15.6b, 15.6c, 15.7, 15.8a and 15.8b

1   On models with power door locks and/or power windows, disconnect the cable from the negative terminal of the battery. **Caution:** *If the vehicle is equipped with a Delco Loc II or Theftlock audio system, make sure you have the correct activation code before disconnecting the battery.*

2   On Pontiac models, pull the inside door handle to the open position and carefully pry the upper edge of the handle bezel out of the door trim panel with a flat-blade screwdriver. Lift the bezel upward and remove it from the trim panel. On front doors, flip it over and unplug the

14.10 To detach the lock cylinder from the trunk lid on a Chevrolet (shown), pry off this retainer with a screwdriver; to detach the lock cylinder on a Pontiac model, drill out the two rivets

15.5a Use a window regulator handle removal tool to pop the handle retaining clip loose, then remove the handle

15.5b If you don't have a window regulator handle removal tool, work a clean shop rag up between the handle and the door trim panel from underneath as shown and pull up on the rag to pop the handle retaining clip loose, then remove the handle

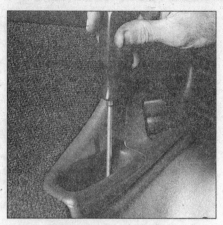

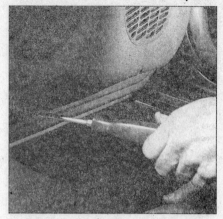

15.6a Remove the trim panel retaining screws inside this recess in the arm rest

15.6b Remove all trim panel retaining screws from the rear (shown) and the front edges of the trim panel

15.6c Remove all trim panel retaining screws from the lower edge of the trim panel (Chevrolet model shown, Pontiac models similar)

electrical connector from the power door lock switch. Remove the bezel from the trim panel.

3   On Chevrolet models with power door locks, pry the power door lock switch out of the door trim panel with a flat-blade screwdriver, unplug the electrical connector from the switch and remove the switch.

4   On models with power windows, pry the power window switch out of the trim panel with a flat-blade screwdriver, unplug the electrical connector from the switch and remove the switch.

5   On models with manually-operated windows, remove the window regulator handle **(see illustrations)**.

6   Remove the trim panel screws **(see illustrations)**.

7   Once all of the screws are removed, detach the trim panel from the door, disconnect any electrical connectors and remove the trim panel from the vehicle by lifting it up and away from the door **(see illustration)**.

8   If you're planning to repair or replace anything inside the door itself, you'll have to remove the water deflector between the trim panel and the door **(see illustration)**. To remove part of the deflector, simply peel it loose from the door **(see illustration)**. If you need to remove the entire deflector, drill out the two rivets which secure the armrest hanger to the door and remove the hanger.

9   After you're done servicing the door component(s), reattach the deflector to the door. Use extra sealant if necessary. Make sure that the deflector is securely attached to the door all the way around its

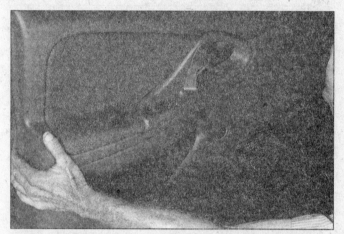

15.7 To detach the trim panel from the door, carefully lift it up, then away, from the door (Chevrolet model shown, Pontiac models similar)

perimeter. If you removed the armrest hanger, reattach it with a pair of suitable sheet metal screws or small bolts and nuts.

10   Installation is otherwise the reverse of removal.

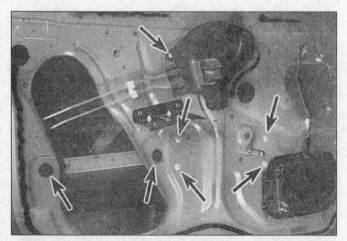

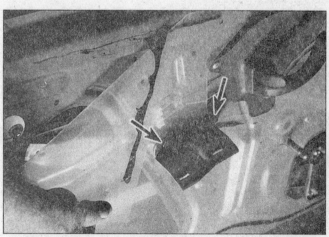

15.8a The water deflector is the plastic liner between the trim panel and the door. Locations of rivets and screws for the window regulator and door handle are indicated by arrows

15.8b To get at the inner door components, simply peel off the deflector in the area in which you need to work; if you need to remove the entire deflector, drill out the two rivets (arrows) for the armrest hanger and remove the hanger

11

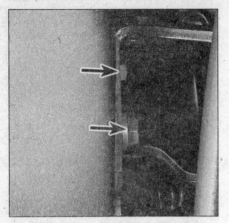

16.7a To detach the door from the vehicle, remove the upper hinge bolts (arrows) . . .

16.7b . . . and the lower hinge bolts (arrow). There are two bolts, but only one is visible

16.11 To adjust the door lock striker, loosen these Torx bolts (arrows) and move the striker up, down or sideways as necessary

16 Door - removal, installation and adjustment

Removal and installation

Refer to illustrations 16.7a and 16.7b

Note: *The door is heavy and somewhat awkward to remove and install - at least two people should perform this procedure. This procedure applies to both front and rear doors.*

1   Raise the window completely and disconnect the negative cable from the battery if equipped with power windows. **Caution:** *If the vehicle is equipped with a Delco Loc II or Theftlock audio system, make sure you have the correct activation code before disconnecting the battery.*

2   Open the door all the way and support it on jacks or blocks covered with rags to prevent damaging the paint.

3   Remove the door trim panel and enough of the water deflector to access all electrical wiring harnesses (see Section 15).

4   Unplug all electrical connectors, and detach all ground wires and harness retaining clips from the door. It's a good idea to label all connectors to aid the reassembly process.

5   Working on the door side, detach the rubber conduit between the body and the door. Pull the wiring harness through the conduit hole and remove the wiring from the door.

6   Mark around the door hinges with a pen or a scribe to facilitate realignment during reassembly.

7   Have an assistant hold the door, remove the hinge-to-door bolts **(see illustrations)** and lift the door off.

8   Installation is the reverse of removal.

Adjustment

Refer to illustration 16.11

9   Having proper door to body alignment is a critical part of a well functioning door assembly. First check the door hinge pins and bushings for excessive play. If the door can be lifted (1/16-inch or more) without the car body lifting with it the hinge pins and bushings should be replaced.

10   Door-to-body alignment adjustments are made by loosening the hinge-to-body bolts or hinge-to-door bolts and moving the door. Proper body alignment is achieved when the top of the doors are parallel with the roof section, the front door is flush with the fender, the rear door is flush with the rear quarter panel and the bottom of the doors are aligned with the lower rocker panel. If these goals can't be reached by adjusting the hinge-to-body or hinge-to-door bolts, body alignment shims may have to be purchased and inserted behind the hinges to achieve correct alignment.

11   To adjust the door closed position, verify that the door latch is contacting the center of the striker. If it isn't, loosen the striker bolts **(see illustration)** and adjust the striker as necessary (up, down or sideways) to provide optimal engagement with the latch mechanism. Tighten the striker bolts securely once the striker is adjusted.

17 Door handles, lock cylinder and latch - removal and installation

1   Remove the door trim panel (see Section 15) and peel away the water deflector in the vicinity of the component you're planning to remove.

Inside handle

2   Drill out the rivet securing the inside handle **(see illustration 15.8a).** Slide the handle to the rear to disengage its locating tabs from the door frame.

3   Pull out the handle, disengage the actuating rods from the backside of the handle and remove the handle. Note which rod is connected to which lever on the handle. The rods must be reattached exactly the same way when the handle is installed.

4   Installation is the reverse of removal. If you don't have a pop rivet installer, reattach the handle assembly to the door with a suitable sheet metal screw or a nut and bolt.

Outside handle

5   Lift up the outside handle and drill out the two handle assembly rivets.

6   Pull out the handle assembly, disengage the actuator rod from the lever on the backside of the handle assembly and remove the handle. Note how the rod is connected to the handle lever; it must be reattached exactly the same way when the handle is installed.

7   Installation is the reverse of removal. If you don't have a pop rivet installer, reattach the handle assembly to the door with nuts and bolts.

Lock cylinder

Refer to illustration 17.8

8   Pry off the lock cylinder retainer clip **(see illustration)** that secures the lock cylinder to the door.

9   Disengage the actuator rod from the lock cylinder lever and remove the lock cylinder.

10   Installation is the reverse of removal.

Latch

Refer to illustration 17.11

11   Remove the screws securing the latch to the door **(see illustration).**

12   Working through the large access hole, position the latch as necessary to disengage the actuator rods from the outside door handle and lock cylinder and the inside handle. Note how these rods are con-

17.8 To detach the lock cylinder from the door, insert a large screwdriver through the small hole above the lock cylinder and pry off this retainer

17.11 To detach the latch assembly from the door, remove these three bolts (arrows)

18.3a To detach the window from the regulator assembly, remove this nut (arrow) . . .

nected to the latch assembly. They must be installed exactly the same way during reassembly. Remove the latch assembly.

13   Installation is the reverse of removal.

18   Door window glass - removal and installation

Refer to illustrations 18.3a and 18.3b

1   Remove the door trim panel and the plastic water deflector (see Section 15).
2   Lower the window glass all the way down into the door.
3   Remove the regulator-to-window retaining nuts **(see illustrations)**.
4   Drill out the rivets which secure the sashes to the window **(see illustration 18.3b)**. Pull off the sashes and sash spacers and remove the window from the door by pulling it up and out.
5   Installation is the reverse of removal.

19   Window regulator - removal and installation

1   On models with power windows, detach the negative battery cable. **Caution:** *If the vehicle is equipped with a Delco Loc II or Theft-*

lock audio system, make sure you have the correct activation code before disconnecting the battery.
2   Remove the door trim panel and the plastic water deflector (see Section 15).
3   Remove the window glass assembly (see Section 18).
4   On models with power windows, unplug the electrical connector from the window regulator motor.
5   On front doors, remove the window regulator bolts **(see illustration 15.8a)**.
6   Drill out the rivets that secure the window regulator to the door frame **(see illustration 15.8a)**.
7   Remove the regulator assembly.
8   Installation is the reverse of removal.

20   Outside mirrors - removal and installation

Refer to illustrations 20.2a, 20.2b and 20.3
Caution: *If the vehicle is equipped with a Delco Loc II or Theftlock audio system, make sure you have the correct activation code before disconnecting the battery.*
1   On models with power mirrors, disconnect the negative cable from the battery.
2   Remove the mirror trim panel **(see illustrations)**. On models with

18.3b . . . and this one (arrow); the rivets holding together the sashes, sash spacers and window must be drilled out before the window can be removed from the door (upper arrow points to the rivet for the rear sash; rivet for front sash not visible in previous photo)

20.2a To get to the outside mirror retaining nuts, remove this Phillips screw . . .

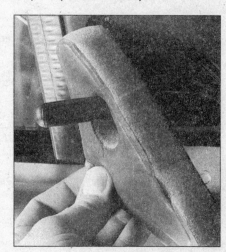

20.2b . . . and remove the triangular trim panel

11

20.3 To detach the outside mirror from the door, remove these retaining nuts (arrows)

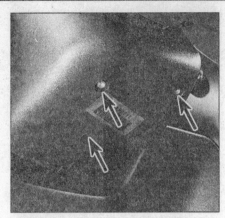

21.1 To detach the steering column covers from the steering column, remove these three screws (arrows) from the lower cover

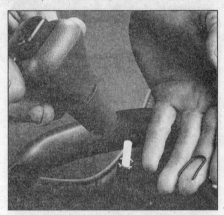

22.3 To detach the knob from the shift lever, pry out this retainer and pull the knob straight up

22.4a Pry the gear position indicator trim panel straight up to disengage the retaining clips . . .

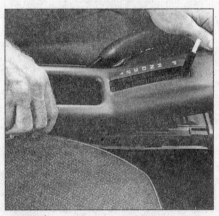

22.4b . . . then remove the trim panel

22.5a To detach the center console from the floor, remove this screw (arrow) from the front storage well . . .

power mirrors, unplug the electrical connector from the power mirror switch.
3   Remove the three mirror retaining nuts **(see illustration)** and detach the mirror from the vehicle.
4   Installation is the reverse of removal.

21   Steering column covers - removal and installation

Refer to illustration 21.1
1   Remove the three screws from the lower steering column cover half **(see illustration)** and remove the steering column covers.
2   Installation is the reverse of removal.

22   Center console - removal and installation

Refer to illustrations 22.3, 22.4a, 22.4b, 22.5a, 22.5b, 22.5c and 22.5d
Warning: *The models covered by this manual are equipped with airbags. Always disable the airbag system before working in the vicinity of the impact sensors, steering column or instrument panel to avoid the possibility of accidental deployment of the airbag(s), which could cause personal injury (see Chapter 12). The yellow wires and connectors routed through the console are for this system. Do not use electrical test equipment on these yellow wires or tamper with them in any way while working around the console.*
1   Disconnect the negative cable from the battery. **Caution:** *If the vehicle is equipped with a Delco Loc II or Theftlock audio system, make sure you have the correct activation code before disconnecting*

the battery.
2   Apply the parking brake lever and place the shift lever in the Neutral position.
3   On vehicles equipped with an automatic transaxle pry out the shift lever knob retaining clip and remove the knob **(see illustration)**. On vehicles equipped with a manual transaxle unscrew the shift lever knob and pry out the shift lever boot.
4   On vehicles equipped with an automatic transaxle pry out and remove the gear position indicator trim bezel **(see illustrations)**.
5   Remove the console retaining screws, raise up the console **(see illustrations)**, unplug any electrical connectors and remove the console from the vehicle.
6   Installation is the reverse of removal.

23   Dashboard trim panels - removal and installation

Warning: *The models covered by this manual are equipped with airbags. Always disable the air bag system before working in the vicinity of the impact sensors, steering column or instrument panel to avoid the possibility of accidental deployment of the airbag(s), which could cause personal injury (see Chapter 12). The yellow wires and connectors routed through the instrument panel are for this system. Do not use electrical test equipment on these yellow wires or tamper with them in any way while working around the instrument panel.*
1   Disconnect the negative battery cable. **Caution:** *If the vehicle is equipped with a Delco Loc II or Theftlock audio system, make sure you have the correct activation code before disconnecting the battery.*
2   Disable the airbag system (see Chapter 12).

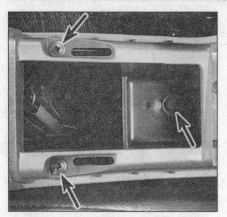

22.5b . . . remove these two screws (arrows) from the mid-console . . .

22.5c . . . remove these two screws from the rear storage well of the console . . .

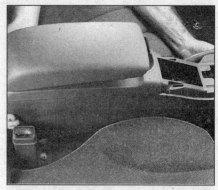

22.5d . . . then lift the console straight up and remove it (this Chevrolet console is typical; the number and location of retaining screws may vary slightly on different models)

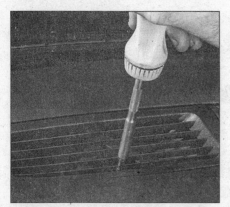

23.6a To detach the center defroster grille, remove this retaining screw . . .

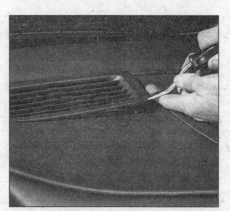

23.6b . . . then carefully pry the grille out of the instrument panel trim pad

23.9 To remove either end cap, simply pry it off as shown

Sound insulators

3   The sound insulators are the flat plastic trim panels underneath the left and right ends of the dash that must be removed to access everything located between the dash and the firewall (clutch pedal position switch, brake lights switch, cruise control switch, wire harnesses, etc.).

4   To remove either sound insulator, remove the screws along the trailing edge of the insulator (three screws for the left insulator, two for the right) then tilt the trailing edge of the insulator down, unplug any electrical connectors, then pull it to the rear to remove it.

5   Installation is the reverse of removal.

Defroster grilles and vent grilles

Refer to illustrations 23.6a and 23.6b

6   To remove the center defroster grille on Chevrolet models, remove the retaining screw, then gently pry up the upper grille **(see illustrations)**. To remove the center defroster grille on Pontiac models, simply pry the grille out of the dash using a small flat bladed screwdriver. Try not to scratch the plastic trim around the grille or the dash surface. To remove the smaller defroster grilles at the left and right ends of the instrument panel trim pad on Chevrolet models, simply pry them out.

7   To remove a left or right vent grille, remove the retaining screw and pry off the grille.

8   Installation is the reverse of removal.

Instrument panel end caps

Refer to illustration 23.9

9   To remove either end cap on a Chevrolet model, simply pry it off **(see illustration)**. To remove either end cap on a Pontiac model, remove the retaining screw.

23.12 Remove the instrument panel trim pad retaining screw (arrow) in the recess for the center defroster grille (Pontiacs have two screws in this area)

10   Installation is the reverse of removal. Make sure the grommets for the locating pins are in good shape, then align the locating pins on the back of the end cap with their respective grommets. If any of the grommets are cracked or torn, replace them (or the end cap may not stay on).

Instrument panel trim pad

Refer to illustrations 23.12, 23.14a, 23.14b, 23.16a, 23.16b and 23.17

11   Remove the center defroster grille (see Step 6).

12   Remove the retaining screw(s) in the center defroster grille opening **(see illustration)**. Chevrolet models have one screw; Pontiacs have two.

13   On Pontiac models, remove the valance retaining screws and the

11

23.14a Remove this instrument panel trim pad retaining screw (arrow) from the left end . . .

23.14b . . . and this screw (arrow) from the right end

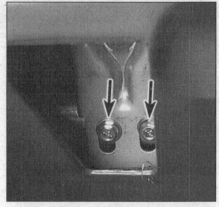

23.16a Inside the glovebox area, remove these two bolts (arrows) from the passenger airbag module bracket . . .

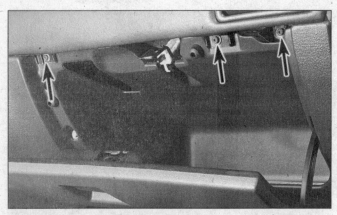

23.16b . . . and remove these three instrument panel trim pad retaining screws (arrows)

23.17 Lift the instrument panel trim pad straight up to pop the three locating pins above the instrument cluster loose from their respective holes in the instrument cluster trim plate

23.20a To detach the instrument cluster trim plate from the instrument panel, remove this screw (arrow) . . .

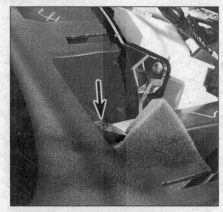

23.20b . . .this screw (arrow) . . .

23.20c . . . and this one (arrow)

valance (the valance is the trim piece between the instrument panel trim pad and the windshield).

14   Remove the instrument panel end caps (see Step 9), then remove the screws located under the end caps **(see illustrations)**.

16   Open the glovebox and remove the retaining bolts from the passenger side airbag module bracket **(see illustration)**, then remove the instrument panel trim pad retaining screws **(see illustration)**.

17   Carefully inspect the perimeter of the trim pad and remove any remaining retaining screws (Pontiacs have more screws along the upper front edge of the trim pad), then grasp the panel securely and detach it from the instrument panel **(see illustration)**.

18   Installation is the reverse of removal.

Instrument panel cluster trim plate (Chevrolet models)

Refer to illustrations 23.20a, 23.20b, 23.20c, 23.21 and 23.22

19   Remove the instrument panel trim pad (see steps 11 through 17).

20   Remove the instrument cluster trim plate retaining screws **(see illustrations)**.

23.21 Work a prying tool around the perimeter of the instrument cluster trim plate in the area surrounding the heater/air conditioning controls and the radio, gently prying the trim plate loose

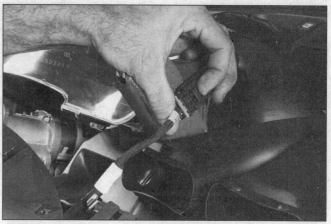

23.22 Pull the instrument cluster trim plate toward you and unplug the electrical connector for the trim plate wire harness (it's not necessary to unplug the individual connectors for the dimmer switch or for the cigarette lighter unless you're replacing those components)

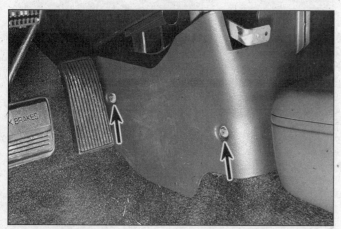
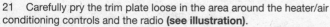

23.28 To remove the heater core/air conditioning evaporator trim panels, remove these two screws (arrows) from the left panel and the other two screws from the right panel . . .

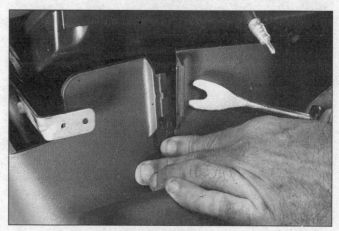

23.29 . . . then disengage the locking tab and slot in the middle and pull the two trim panels apart

21 Carefully pry the trim plate loose in the area around the heater/air conditioning controls and the radio **(see illustration)**.
22 Lift up the trim plate and pull it toward you, then unplug the electrical connector for the trim plate wiring harness **(see illustration)**. Remove the trim plate.
23 Installation is the reverse of removal.

Accessory trim plate (Pontiac models)

24 Remove the accessory trim plate-to-instrument panel retaining screws.
25 Open the glove box door and remove the rest of the accessory trim plate retaining screws.
26 Carefully pry the accessory trim plate to the rear to detach it from the instrument panel.
27 Installation is the reverse of removal.

Heater core/air conditioning evaporator trim panels

Refer to illustrations 23.28 and 23.29

28 Remove the four trim panel retaining screws **(see illustration)**.
29 Disengage the locking tab and slot in the middle and pull the trim panels apart **(see illustration)**.
30 Installation is the reverse of removal.

24 Instrument panel - removal and installation

Refer to illustrations 24.7, 24.8a, 24.8b, 24.8c, 24.8d and 24.8e
Warning: *The models covered by this manual are equipped with airbags. Always disable the air bag system before working in the vicinity of the impact sensors, steering column or instrument panel to avoid the possibility of accidental deployment of the airbag(s), which could cause personal injury (see Chapter 12). The yellow wires and connectors routed through the instrument panel are for this system. Do not use electrical test equipment on these yellow wires or tamper with them in any way while working around the instrument panel.*
1 Disconnect the negative battery cable. **Caution:** *If the vehicle is equipped with a Delco Loc II or Theftlock audio system, make sure you have the correct activation code before disconnecting the battery.*
2 Disable the airbag system (see Chapter 12).
3 Remove the sound insulators, defroster grilles, end caps and instrument panel trim pad, and instrument cluster trim plate (Chevrolet models) or accessory trim plate (Pontiac models) (see Section 23).
4 Remove the heater and air conditioning control assembly (see Chapter 3).
5 Remove the radio (see Chapter 12). Remove the steering wheel (see Chapter 10), the combination switch and the windshield wiper switch (see Chapter 12) from the steering column.
6 Remove any air distribution duct retaining screws. (The air distri-

11

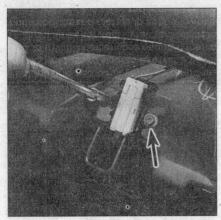

24.7 To detach the glovebox light, remove these two screws

24.8a To detach the left end of the instrument panel, remove this screw (upper arrow) and nut (lower arrow)

24.8b To detach the center part of the instrument panel, remove these three screws (arrows)

24.8c To detach the upper right part of the instrument panel, remove these two screws (arrows)

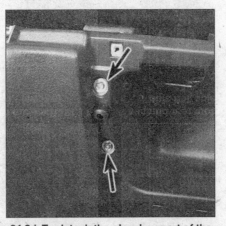

24.8d To detach the glovebox part of the instrument panel, remove these two screws (arrows)

24.8e To detach the right end of the instrument panel remove this nut (arrow)

bution duct must be detached from the instrument panel but it's not really necessary to remove it to remove the instrument panel.)
7   Detach the glovebox light **(see illustration)**.
8   Remove the instrument panel retaining screws **(see illustrations)**.
9   Disengage the instrument panel assembly from the body by lifting it up and back slightly.
10   Reach behind the instrument panel and unplug any electrical connectors that may still be connected.
11   Installation is the reverse of removal.

25 Seats - removal and installation

Front seat

1   If you just want to repair or replace the seat itself, or you want to remove the seat to create more room to work under the dash, unbolt the seat from the adjuster assembly. Locate the four bolts - one at each corner - securing the seat to the adjuster assembly (the bolt heads face down). Remove all four bolts.
2   Tilt the seat up, unplug any electrical connectors underneath and remove the seat.
3   If you want to replace the seat adjuster assembly, or you need to remove the adjuster to remove or replace the carpet, unbolt the adjuster from the floor. The seat adjuster assembly is bolted to the floor with four bolts.
4   Installation is the reverse of removal. Be sure to tighten all bolts securely.

Rear seat

5   Unlatch the rear seat back and tilt it forward.
6   Unhook the outer side belt retractor from the rear seat back.
7   Remove the spacer from right key pin.
8   Slide the seat back to the right and disengage the left key pin from the left key slot.
9   Slide the seat back to the left and disengage the right key slot from the right key slot.
10   Remove the seat back.
11   Press the seat cushion retainer to the rear with a screwdriver.
12   Lift the seat cushion up and remove it.
13   Installation is the reverse of removal.

26 Seat belts - check

1   Check the seat belts, buckles, latch plates and guide loops for any obvious damage or signs of wear.
2   Make sure the seat belt reminder light comes on when the key is turned on.
3   The seat belts are designed to lock up during a sudden stop or impact, yet allow free movement during normal driving. The retractors should hold the belt against your chest while driving and rewind the belt when the buckle is unlatched.
4   If any of the above checks reveal problems with the seat-belt system, replace parts as necessary.

Chapter 12   Chassis electrical system

Contents

1   General information

The electrical system is a 12-volt, negative ground type. Power for the lights and all electrical accessories is supplied by a lead/acid-type battery which is charged by the alternator.

This Chapter covers repair and service procedures for the various electrical components not associated with the engine. Information on the battery, alternator, ignition system and starter motor can be found in Chapter 5. It should be noted that when portions of the electrical system are serviced, the negative battery cable should be disconnected from the battery to prevent electrical shorts and/or fires.

Caution: *If the vehicle is equipped with a Delco Loc II or Theftlock audio system, make sure you have the correct activation code before disconnecting the battery.*

2   Electrical troubleshooting - general information

A typical electrical circuit consists of an electrical component, any switches, relays, motors, fuses, fusible links or circuit breakers related to that component and the wiring and electrical connectors that link the component to both the battery and the chassis. To help you pinpoint an electrical circuit problem, wiring diagrams are included at the end of this Chapter.

Before tackling any troublesome electrical circuit, first study the appropriate wiring diagrams to get a complete understanding of what makes up that individual circuit. Trouble spots, for instance, can often be narrowed down by noting if other components related to the circuit are operating properly. If several components or circuits fail at one time, chances are the problem is in a fuse or ground connection, because several circuits are often routed through the same fuse and ground connections.

Electrical problems usually stem from simple causes, such as loose or corroded connections, a blown fuse, a melted fusible link or a bad relay. Visually inspect the condition of all fuses, wires and connections in a problem circuit before troubleshooting it.

If testing instruments are going to be utilized, use the diagrams to plan ahead of time where you will make the necessary connections in order to accurately pinpoint the trouble spot.

The basic tools needed for electrical troubleshooting include a circuit tester or voltmeter (a 12-volt bulb with a set of test leads can also be used), a continuity tester, which includes a bulb, battery and set of test leads, and a jumper wire, preferably with a circuit breaker incorporated, which can be used to bypass electrical components. Before attempting to locate a problem with test instruments, use the wiring diagram(s) to decide where to make the connections.

Voltage checks

Voltage checks should be performed if a circuit is not functioning properly. Connect one lead of a circuit tester to either the negative battery terminal or a known good ground. Connect the other lead to an electrical connector in the circuit being tested, preferably nearest to the battery or fuse. If the bulb of the tester lights, voltage is present, which means that the part of the circuit between the electrical connector and the battery is problem free. Continue checking the rest of the circuit in the same fashion. When you reach a point at which no voltage is present, the problem lies between that point and the last test point with voltage. Most of the time the problem can be traced to a loose connection. **Note:** *Keep in mind that some circuits receive voltage only when the ignition key is in the Accessory or Run position.*

Finding a short

One method of finding shorts in a circuit is to remove the fuse and connect a test light or voltmeter in its place to the fuse terminals. There should be no voltage present in the circuit. Move the wiring harness from side-to-side while watching the test light. If the bulb goes on, there is a short to ground somewhere in that area, probably where the insulation has rubbed through. The same test can be performed on each component in the circuit, even a switch.

Ground check

Caution: *If the vehicle is equipped with a Delco Loc II or Theftlock audio system, make sure you have the correct activation code before disconnecting the battery.*

Perform a ground test to check whether a component is properly grounded. Disconnect the battery and connect one lead of a self-powered test light, known as a continuity tester, to a known good ground. Connect the other lead to the wire or ground connection being tested. If the bulb goes on, the ground is good. If the bulb does not go on, the ground is not good.

Continuity check

A continuity check is done to determine if there are any breaks in a circuit - if it is passing electricity properly. With the circuit off (no power in the circuit), a self-powered continuity tester can be used to check the circuit. Connect the test leads to both ends of the circuit (or to the "power" end and a good ground), and if the test light comes on the circuit is passing current properly. If the light doesn't come on, there is a break somewhere in the circuit. The same procedure can be used to test a switch, by connecting the continuity tester to the switch terminals. With the switch turned On, the test light should come on.

Finding an open circuit

When diagnosing for possible open circuits, it is often difficult to locate them by sight because oxidation or terminal misalignment are

3.1a The passenger compartment fuse block, on the left end of the instrument panel, is accessible after removing the cover; besides fuses, this block also houses at least one circuit breaker (arrow), such as the one shown here, which is an accessory circuit breaker (some passenger compartment fuse blocks also house circuit breakers for the power windows and other power accessories)

hidden by the electrical connectors. Merely wiggling an electrical connector on a sensor or in the wiring harness may correct the open circuit condition. Remember this when an open circuit is indicated when troubleshooting a circuit. Intermittent problems may also be caused by oxidized or loose connections.

Electrical troubleshooting is simple if you keep in mind that all electrical circuits are basically electricity running from the battery, through the wires, switches, relays, fuses and fusible links to each electrical component (light bulb, motor, etc.) and to ground, from which it is passed back to the battery. Any electrical problem is an interruption in the flow of electricity to and from the battery.

3   Fuses - general information

Refer to illustrations 3.1a, 3.1b, 3.1c and 3.3

1   The electrical circuits of the vehicle are protected by a combination of fuses, circuit breakers and fusible links. The two fuse blocks are located on the left end of the instrument panel and in the engine compartment **(see illustrations)**.

2   Each of the fuses is designed to protect a specific circuit, and the various circuits are identified on the fuse panel itself.

3   Miniaturized fuses are employed in the fuse block. These compact fuses, with blade terminal design, allow fingertip removal and replacement. If an electrical component fails, always check the fuse first. The easiest way to check fuses is with a test light. Check for power at the exposed terminal tips of each fuse. If power is present on one side of the fuse but not the other, the fuse is blown. A blown fuse

can also be confirmed by visually inspecting it **(see illustration)**.

4   Be sure to replace blown fuses with the correct type. Fuses of different ratings are physically interchangeable, but only fuses of the proper rating should be used. Replacing a fuse with one of a higher or lower value than specified is not recommended. Each electrical circuit needs a specific amount of protection. The amperage value of each fuse is molded into the fuse body.

If the replacement fuse immediately fails, don't replace it again until the cause of the problem is isolated and corrected. In most cases, the cause will be a short circuit in the wiring caused by a broken or deteriorated wire.

4   Fusible links - general information

Some circuits are protected by fusible links. Fusible links are circuit protection devices that are part of the wiring harness itself, that are designed to melt and open the circuit when a short causes excessive current flow. Fusible links on these models are on the right (passenger's side) of the engine compartment, on the inner wheelwell. Fusible links are used in circuits which are not ordinarily fused, such as the ignition circuit.

Although the fusible links appear to be a heavier gauge than the wire they are protecting, the appearance is due to the thick insulation. All fusible links are several wire gauges smaller than the wire they are designed to protect.

Fusible links cannot be repaired, but a new link of the same size wire can be put in its place. The procedure is as follows:

a) *Disconnect the negative cable from the battery.* **Caution:** *If the vehicle is equipped with a Delco Loc II or Theftlock audio system, make sure you have the correct activation code before disconnecting the battery.*

b) *Disconnect the fusible link from the wiring harness.*

c) *Cut the damaged fusible link out of the wiring just behind the electrical connector.*

d) *Strip the insulation back approximately 1/2-inch.*

e) *Position the electrical connector on the new fusible link and crimp it into place.*

f) *Use rosin core solder at each end of the new link to obtain a good solder joint.*

g) *Use plenty of electrical tape around the soldered joint. No wires should be exposed.*

h) *Connect the battery ground cable. Test the circuit for proper operation.*

5   Circuit breakers - general information

Circuit breakers protect components such as power windows, power door locks and headlights. On some models the circuit breaker resets itself automatically, so an electrical overload in a circuit breaker

3.1b To access the fuses in the engine compartment fuse block, remove this cover; notice that the function and amperage rating of each fuse location is clearly indicated on the cover

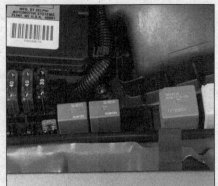

3.1c The engine compartment fuse block also contains several important relays; again, their functions are indicated on the cover

3.3 When a fuse blows, the element between the terminal melts - the fuse on the left is blown, the fuse on the right is good

protected system will cause the circuit to fail momentarily, then come back on. If the circuit doesn't come back on, check it immediately. Once the condition is corrected, the circuit breaker will resume its normal function. Some circuit breakers must be reset manually.

6   Relays - general information

General information

1   Several electrical accessories in the vehicle, such as the fuel injection system, horns, starter, and fog lamps use relays to transmit the electrical signal to the component. Relays use a low-current circuit (the control circuit) to open and close a high-current circuit (the power circuit). If the relay is defective, that component will not operate properly. The various relays are mounted in engine compartment **(see illustration 3.1c)** and several locations throughout the vehicle. If a faulty relay is suspected, it can be removed and tested using the procedure below or by a dealer service department or a repair shop. Defective relays must be replaced as a unit.

Testing

2   It's best to refer to the wiring diagram for the circuit to determine the proper hook-ups for the relay you're testing. However, if you're not able to determine the correct hook-up from the wiring diagrams, you may be able to determine the test hook-ups from the information that follows.

3   On most relays, two of the terminals are the relay's control circuit (they connect to the relay coil which, when energized, closes the large contacts to complete the circuit). The other terminals are the power circuit (they are connected together within the relay when the control-circuit coil is energized).

4   Most relays are marked as an aid to help you determine which terminals are the control circuit and which are the power circuit.

5   Connect a fused jumper wire between one of the two control circuit terminals and the positive battery terminal. Connect another jumper wire between the other control circuit terminal and ground. When the connections are made, the relay should click. On some relays, polarity may be critical, so, if the relay doesn't click, try swapping the jumper wires on the control circuit terminals.

6   With the jumper wires connected, check for continuity between the power circuit terminals as indicated by the markings on the relay.

7   If the relay fails any of the above tests, replace it.

7   Turn signal/hazard flasher - check and replacement

Warning: *The models covered by this manual are equipped with airbags. Always disable the airbag system before working in the vicinity of the impact sensors, steering column or instrument panel to avoid the possibility of accidental deployment of the airbag(s), which could cause personal injury (see Section 25). The yellow wires and connectors routed through the instrument panel are for this system. Do not use electrical test equipment on these yellow wires or tamper with them in any way while working under the instrument panel.*

Caution: *If the vehicle is equipped with a Delco Loc II or Theftlock audio system, make sure you have the correct activation code before disconnecting the battery.*

1   1995 models are equipped with separate turn signal and hazard flasher units while 1996 and later models are equipped with a combination turn signal/hazard flasher. The turn signal flasher on 1995 models is located at the left side of the dash, to the right of the steering column. The hazard flasher on 1995 models is plugged into the "convenience center," which is mounted at the top of the left kick panel. The convenience center also houses the alarm module, if equipped, and the horn relay. On 1996 and 1997 models the combination turn signal/hazard flasher is plugged into the convenience center which is mounted at the top of the left kick panel. On 1998 and later models the combination turn signal/hazard flasher is mounted to the brake pedal bracket above the steering column.

2   You can determine whether a flasher unit is operating correctly by verifying that it makes an audible clicking sound when it's energized.

3   If a turn signal light fails to blink on one side or the other, and the flasher unit does not make its characteristic clicking sound, a turn signal bulb is out.

3   If both turn signals fail to blink, the problem may be a blown fuse, a faulty flasher unit, a broken switch or a loose or open connection.

4   If the turn signal fuse has blown, look for a short in the wiring before installing a new fuse.

5   If the fuse is okay, verify that there's voltage to the flasher.

6   If there's voltage to the flasher, verify that there's voltage out of the flasher when it's energized. If there's no voltage to the flasher, check the combination switch (see Section 8)

7   If there's voltage out of the flasher, check the wiring between the switch and the turn signals (turn signal flasher) or between the switch and all the running lights (hazard flasher).

8   To access the flasher units on all models it will be necessary to remove the left sound insulator panel.

9   To replace the hazard flasher on 1995 models or the combination turn signal/hazard flasher on 1996 and 1997 models, simply unplug it from the convenience center at the top of the left kick panel.

10   To replace the turn signal flasher on 1995 models or the combination turn signal/hazard flasher on 1998 and later models, unplug the electrical connector and disengage the flasher from its holder on the brake pedal mounting bracket.

11   Make sure that the replacement unit is identical to the original. Compare the old one to the new one before installing it.

12   Installation is the reverse of removal.

8   Combination switch - removal and installation

Refer to illustrations 8.4 and 8.5

Warning: *The models covered by this manual are equipped with airbags. Always disable the airbag system before working in the vicinity of the impact sensors, steering column or instrument panel to avoid the possibility of accidental deployment of the airbag(s), which could cause personal injury (see Section 25). The yellow wires and connectors routed through the instrument panel are for this system. Do not use electrical test equipment on these yellow wires or tamper with them in any way while working under the instrument panel.*

Caution: *If the vehicle is equipped with a Delco Loc II or Theftlock audio system, make sure you have the correct activation code before disconnecting the battery.*

1   Detach the cable from the negative battery terminal and disable the airbag system (see Section 25).

2   Remove the steering wheel (see Chapter 10).

3   Remove the steering column covers (see Chapter 11).

4   Remove the combination switch mounting screws **(see illustration)**.

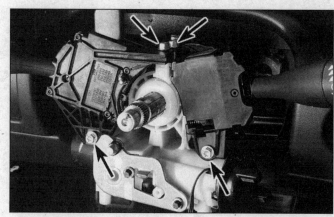

8.4 To remove the combination switch, remove the indicated screws (upper arrows and lower left arrow); to remove the windshield wiper/washer switch, remove the combination switch and remove the windshield wiper/washer switch mounting screw (lower right arrow)

12

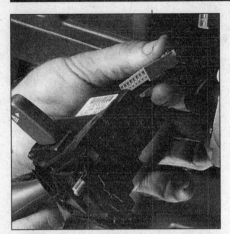

8.5 Detach the combination switch from the steering column housing assembly and unplug the electrical connector

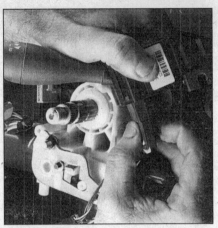

9.6 Detach the windshield wiper/washer switch from the steering column housing assembly and unplug the electrical connector

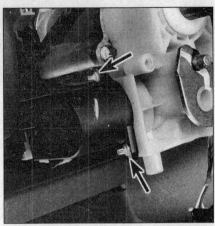

10.4 To remove the ignition switch from the steering column housing assembly, remove these two screws (arrows)

5   Detach the combination switch from the steering column housing assembly and unplug the electrical connector **(see illustration)**.
6   Installation is the reverse of removal.

9   Windshield wiper/washer switch - removal and installation

Refer to illustration 9.6

Warning: *The models covered by this manual are equipped with airbags. Always disable the airbag system before working in the vicinity of the impact sensors, steering column or instrument panel to avoid the possibility of accidental deployment of the airbag(s), which could cause personal injury (see Section 25). The yellow wires and connectors routed through the instrument panel are for this system. Do not use electrical test equipment on these yellow wires or tamper with them in any way while working under the instrument panel.*
Caution: *If the vehicle is equipped with a Delco Loc II or Theftlock audio system, make sure you have the correct activation code before disconnecting the battery.*

1   Detach the cable from the negative battery terminal and disable the airbag system (see Section 25).
2   Remove the steering wheel (see Chapter 10).
3   Remove the upper and lower steering column covers (see Chapter 11).
4   Remove the combination switch (see Section 8).

5   Remove the windshield wiper/washer switch mounting screws **(see illustration 8.4)**.
6   Detach the windshield wiper/washer switch from the steering column housing assembly and unplug the electrical connector **(see illustration)**.
7   Installation is the reverse of removal.

10   Ignition switch and key lock cylinder - removal and installation

Warning: *The models covered by this manual are equipped with airbags. Always disable the airbag system before working in the vicinity of the impact sensors, steering column or instrument panel to avoid the possibility of accidental deployment of the airbag(s), which could cause personal injury (see Section 25). The yellow wires and connectors routed through the instrument panel are for this system. Do not use electrical test equipment on these yellow wires or tamper with them in any way while working under the instrument panel.*
Caution: *If the vehicle is equipped with a Delco Loc II or Theftlock audio system, make sure you have the correct activation code before disconnecting the battery.*

1   Disable the airbag system (see Section 25).
2   Remove the tilt lever, if equipped.
3   Remove the upper and lower steering column covers (see Chapter 11).

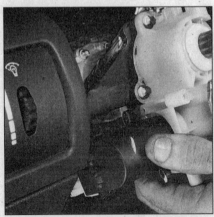

10.5a Move the ignition switch to the left to detach it from the steering column housing assembly . . .

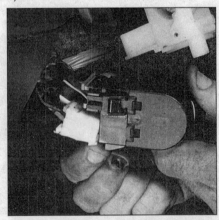

10.5b . . . then pull it back and down and unplug the electrical connectors

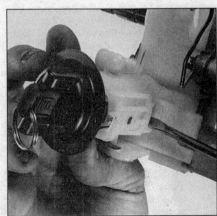

10.8 To remove the key lock cylinder, turn the ignition key to the Run position, depress the lock button with a screwdriver or other suitable tool, then pull out the lock cylinder

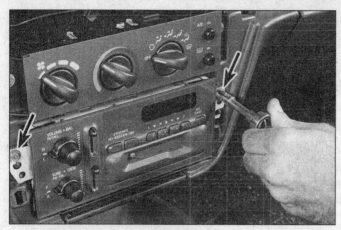

11.3 To detach the radio from the dash, remove these two screws (arrows)

11.4a Unplug the electrical connector(s) from the radio

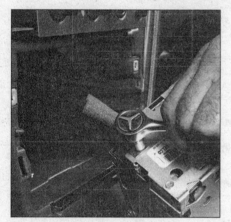

11.4b To detach the antenna lead from the radio, remove the hold-down screw . . .

11.4c . . . then pull out the antenna lead

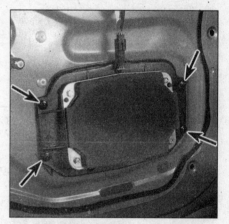

11.8 To remove the front speaker from the door, unplug the electrical connector and remove these four screws (arrows)

Ignition switch

Refer to illustrations 10.4, 10.5a and 10.5b

4   Remove the ignition switch retaining screws **(see illustration)**.
5   Detach the ignition switch from the steering column housing assembly and unplug the electrical connectors **(see illustrations)**.
6   Installation is the reverse of removal.

Key lock cylinder

Refer to illustration 10.8

7   Turn the key to the Run position.
8   Depress the lock button **(see illustration)** and remove the key lock cylinder.
9   To install the key lock cylinder, turn the key to the Run position, depress the lock button and, rotating the key counterclockwise about five degrees, gently push the lock cylinder into place.
10   Installation is otherwise the reverse of removal.

11 Radio and speakers - removal and installation

Warning: *The models covered by this manual are equipped with airbags. Always disable the airbag system before working in the vicinity of the impact sensors, steering column or instrument panel to avoid the possibility of accidental deployment of the airbag(s), which could cause personal injury (see Section 25). The yellow wires and connectors routed through the instrument panel are for this system. Do not use electrical test equipment on these yellow wires or tamper with them in any way while working under the instrument panel.*

Caution: *If the vehicle is equipped with a Delco Loc II or Theftlock audio system, make sure you have the correct activation code before disconnecting the battery.*

Radio

Refer to illustrations 11.3, 11.4a, 11.4b and 11.4c

1   Disable the airbag system (see Section 25).
2   Remove the instrument panel trim plate (Chevrolet models) or accessory trim plate (Pontiac models) (see Chapter 11).
3   Remove the radio retaining screws **(see illustration)**.
4   Pull out the radio and disconnect the electrical connections and antenna lead **(see illustrations)**.
5   Remove the radio from the instrument panel.
6   Installation is the reverse of removal.

Front speakers

Refer to illustration 11.8

7   Remove the front door trim panel (see Chapter 11).
8   Unplug the speaker electrical connector **(see illustration)** and remove the speaker retaining screws. Remove the speaker from the door.
9   Installation is the reverse of removal.

Rear speakers

10   Remove the rear seatback (see Chapter 11).
11   Remove one of the rear quarter trim panels.
12   Release the four plastic push-in retainers that secure the rear parcel shelf, disengage the rear seat belts from the slots in the parcel shelf and remove the parcel shelf.

12

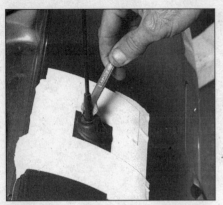

12.1 To detach the antenna mast from the bezel, loosen the nut at the bottom of the mast with a small wrench, back it off, and unscrew the mast

12.2a Peel back the carpet in the trunk area . . .

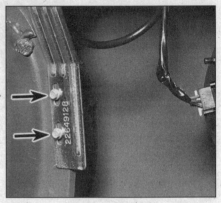

12.2b . . . then remove these two antenna bracket screws (arrows)

13   To disengage either rear speaker, push down on the tab on front of the speaker spacer, lift the spacer and pull the speaker forward. Unplug the electrical connector and remove the speaker from the vehicle.
14   Installation is the reverse of removal.

12   Antenna - removal and installation

Refer to illustrations 12.1, 12.2a, 12.2b and 12.3
Caution: *If the vehicle is equipped with a Delco Loc II or Theftlock audio system, make sure you have the correct activation code before disconnecting the battery.*
1   Use a small wrench to remove the antenna mast **(see illustration)**.
2   Open the trunk, peel back the trunk liner **(see illustration)** and remove the antenna base/bracket retaining screws **(see illustration)**.
3   Remove the antenna base/bracket from the outer bezel and the quarter panel then remove the antenna lead from the antenna base **(see illustration)**.
4   If the antenna lead is to be replaced detach the radio and disconnect the antenna lead from the backside of the radio (see Section 11). Then remove any retaining clips under the instrument panel securing the antenna lead.
5   To facilitate the removal of the antenna lead it will be necessary to remove the following interior trim panels: the right side kick panel, the right door sill plate and the right rear quarter trim panel. **Note:** *On 4-door models it will be necessary to remove the center pillar trim panel*

and the rear door sill plate.
6   Attach a piece of stiff wire to the end of the antenna lead in the trunk compartment.
7   Working in the interior of the vehicle pull the antenna lead from the trunk compartment into the passenger compartment.
8   Detach the wire from the end of the antenna lead and remove the antenna lead from the vehicle.
9   Fasten the end of the new antenna lead to the wire leading to the trunk compartment and pull the antenna lead into the trunk compartment.
10   The remainder of the installation is the reverse of removal.

13   Rear window defogger - check and repair

1   The rear window defogger consists of a number of horizontal elements baked onto the glass surface.
2   Small breaks in the element can be repaired without removing the rear window.

Check

Refer to illustrations 13.4, 13.5 and 13.7
3   Turn the ignition switch and defogger system switches to the ON position. Using a voltmeter, place the positive probe against the defogger grid positive terminal and the negative lead against the ground terminal. If the battery voltage is not indicated, check the fuse, defogger switch and related wiring.

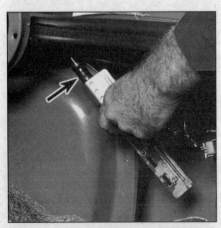

12.3 Remove the antenna base/bracket from the outer bezel and the quarter panel, then detach the antenna lead (arrow)

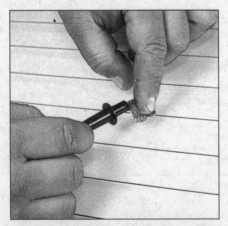

13.4 When measuring the voltage at the rear window defogger grid, wrap a piece of aluminum foil around the positive probe of the voltmeter and press the foil against the wire with your finger

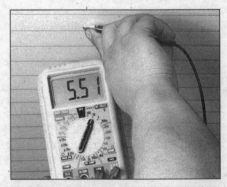

13.5 To determine if a heating element has broken, check the voltage at the center of each element - if the voltage is 6-volts, the element is unbroken - if the voltage is 12-volts, the element is broken between the center and the ground side - if there is no voltage, the element is broken between the center and the positive side

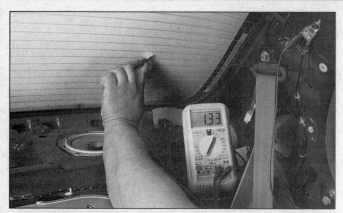

13.7 To find the break, place the voltmeter negative lead against the defogger ground terminal, place the voltmeter positive lead with the foil strip against the heating element at the positive terminal end and slide it toward the negative terminal end - the point at which the voltmeter reading changes abruptly is the point at which the element is broken

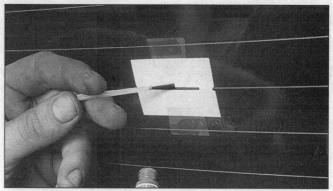

13.13 To use a defogger repair kit, apply masking tape to the inside of the window at the damaged area, then brush on the special conductive coating

4   When measuring voltage during the next two tests, wrap a piece of aluminum foil around the tip of the voltmeter positive probe and press the foil against the heating element with your finger **(see illustration)**.

5   Check the voltage at the center of each heating element **(see illustration)**. If the voltage is 6-volts, the element is okay (there is no break). If the voltage is 12-volts, the element is broken between the center of the element and the ground side. If the voltage is 0-volts the element is broken between the center of the element and positive side.

6   If none of the elements are broken, connect the negative lead to a good body ground. The reading should stay the same, if it doesn't the ground connection is bad.

7   To find the break, place the voltmeter negative lead against the defogger ground terminal. Place the voltmeter positive lead with the foil strip against the heating element at the positive terminal end and slide it toward the negative terminal end. The point at which the voltmeter deflects from several volts to zero is the point at which the heating element is broken **(see illustration)**.

Repair

Refer to illustration 13.13

8   Repair the break in the element using a repair kit specifically recommended for this purpose, such as Dupont paste No. 4817 (or equivalent). Included in this kit is plastic conductive epoxy.

9   Prior to repairing a break, turn off the system and allow it to cool off for a few minutes.

10   Lightly buff the element area with fine steel wool, then clean it

thoroughly with rubbing alcohol.

11   Use masking tape to mask off the area being repaired.

12   Thoroughly mix the epoxy, following the instructions provided with the repair kit.

13   Apply the epoxy material to the slit in the masking tape, overlapping the undamaged area about 3/4-inch on either end **(see illustration)**.

14   Allow the repair to cure for 24 hours before removing the tape and using the system.

14 Headlight bulb - replacement

Refer to illustrations 14.2, 14.3 and 14.4

Warning: *Halogen gas filled bulbs are under pressure and may shatter if the surface is scratched or the bulb is dropped. Wear eye protection and handle the bulbs carefully, grasping only the base whenever possible. Do not touch the surface of the bulb with your fingers because the oil from your skin could cause it to overheat and fail prematurely. If you do touch the bulb surface, clean it with rubbing alcohol.*

1   Remove the upper air intake splash shield.

2   Remove the headlight housing retaining bolts **(see illustration)** and pull the housing assembly out to access the bulb.

3   Working at the rear of the headlight housing, rotate the bulb holder counterclockwise to separate it from the housing while setting the housing aside **(see illustration)**.

4   Unplug the electrical connector from the bulb holder **(see illustration)**.

5   Without touching the glass with your bare fingers, plug the electrical connector into the bulb holder and install the bulb holder into the headlight housing.

14.2 To detach the headlight assembly from the radiator support, remove these two bolts (arrows)

14.3 To disengage the electrical connector and headlight bulb holder from the headlight assembly, turn the holder counterclockwise and pull it out of the housing

14.4 Unplug the electrical connector from the headlight bulb holder

12

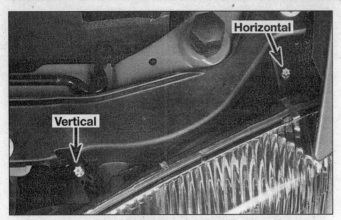

16.1a The headlight vertical and horizontal adjustment screws (arrows) are right behind the upper end of the headlight housing; you'll need a Torx-head tool for these screws (headlight trim piece removed for clarity)

6   The remainder of installation is the reverse of removal.
7   Make sure the headlights are correctly adjusted (see Section 16).

15 Headlight housing - removal and installation

Warning: *The models covered by this manual are equipped with airbags. Always disable the airbag system before working in the vicinity of the impact sensors, steering column or instrument panel to avoid the possibility of accidental deployment of the airbag(s), which could cause personal injury (see Section 25). The yellow wires and connectors routed through the instrument panel are for this system. Do not use electrical test equipment on these yellow wires or tamper with them in any way while working under the instrument panel.*

1   Remove the upper air intake splash shield (see Section 14).
2   Remove the headlight housing retaining bolts (see Section 14).
3   Remove the headlight bulb holders from the headlight housing (see Section 14).
4   Detach the headlight housing from the bracket and install the bracket on the new headlight housing.

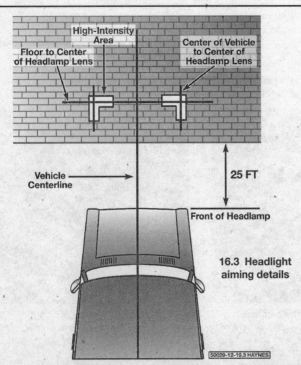

16.3 Headlight aiming details

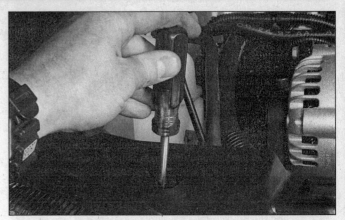

16.1b The headlight vertical and horizontal adjustment screws can be accessed through the trim piece

5   Installation is the reverse of removal.
6   Adjust the headlights (see Section 16).

16 Headlights - adjustment

Refer to illustrations 16.1a, 16.1b and 16.3
Caution: *The headlights must be aimed correctly. If adjusted incorrectly they could blind the driver of an oncoming vehicle and cause a serious accident or seriously reduce your ability to see the road. The headlights should be checked for proper aim every 12 months and any time a new headlight is installed or front end body work is performed. It should be emphasized that the following procedure is only an interim step which will provide temporary adjustment until the headlights can be adjusted by a properly equipped shop.*
1   Headlights have two spring loaded adjusting screws, one on the top controlling up-and-down movement and one on the side controlling left-and-right movement **(see illustrations)**.
2   There are several methods of adjusting the headlights. The simplest method requires masking tape, a blank wall and a level floor.
3   Position masking tape vertically on the wall in reference to the vehicle centerline and the centerlines of both headlights **(see illustration)**.
4   Position a horizontal tape line in reference to the centerline of all the headlights. **Note:** *It may be easier to position the tape on the wall with the vehicle parked only a few inches away.*
5   Adjustment should be made with the vehicle parked 25 feet away from the wall, sitting level, the gas tank half-full and no unusually heavy load in the vehicle.
6   Starting with the low beam adjustment, position the high intensity zone so it is two inches below the horizontal line and two inches to the right of the headlight vertical line. Adjustment is made by turning the top adjusting screw clockwise to raise the beam and counterclockwise to lower the beam. The adjusting screw on the side should be used in the same manner to move the beam left or right.
7   With the high beams on, the high intensity zone should be vertically centered with the exact center just below the horizontal line. **Note:** *It may not be possible to position the headlight aim exactly for both high and low beams. If a compromise must be made, keep in mind that the low beams are the most used and have the greatest effect on safety.*
8   Have the headlights adjusted by a dealer service department or service station at the earliest opportunity.

17 Bulb replacement

Front turn signal and side marker lights

1   On all models except Pontiac GT models, remove the headlight assembly (see Section 15). On Pontiac GT models, unscrew the side marker lens from the bumper fascia.

2   Twist the bulb socket a quarter turn counterclockwise, then remove the bulb assembly from the housing. **Note:** *On Chevrolet and some Pontiac models, the turn signal and side marker light sockets are screwed into separate housings. On other Pontiac models, they're in the same housing.*
3   Remove the bulb from the socket.
4   Installation is the reverse of removal.

Rear side marker lights and back-up lights (Pontiac models)

5   Remove the side marker or back-up light lens retaining screw(s) and pull the lens assembly out of the bumper fascia.
6   Detach the bulb socket from the lens assembly by turning it counterclockwise.
7   Remove the bulb from the socket.
8   Installation is the reverse of removal.

Rear brake, taillight and turn signal bulbs; Chevrolet back-up and side marker lights

Refer to illustrations 17.9, 17.10a, 17.10b and 17.10c

9   To replace a back-up light bulb (in the trunk lid), simply turn the bulb socket counterclockwise, remove the socket **(see illustration)** and remove the bulb from the socket.
10   To replace a brake, taillight, turn signal or side marker bulb (in the body), detach the trunk carpet liner **(see illustration)**, peel it back, turn the bulb socket counterclockwise, remove the socket **(see illustrations)** and remove the bulb from the socket.

Trunk light

Refer to illustrations 17.11a and 17.11b

11   Pinch the ends of the socket retainer together as shown **(see illus-**

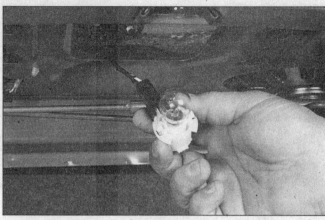

17.9 To replace a back-up light bulb on a Chevrolet model (located in the trunk lid), turn the bulb socket counterclockwise and detach it from the lens housing, pull it down and remove the bulb from the socket (on Pontiac models, the back-up lights are located in the rear bumper fascia and can be replaced by unscrewing the lens from the fascia)

tration) and pull down the retainer, socket and bulb **(see illustration)**.
12   Remove the bulb from the socket.
13   Installation is the reverse of removal.

License plate light

Refer to illustrations 17.14 and 17.15

14   Detach the two screws which attach the lens to the trunk lid **(see illustration)**.

17.10a To gain access to the rear brake, taillight, turn signal and side marker light bulbs, release this push-in retainer and peel back the trunk liner . . .

17.10b . . . then remove the brake, taillight, or turn signal bulb . . .

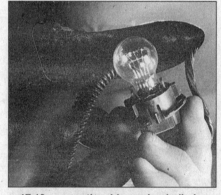

17.10c . . . or the side marker bulb, by turning the bulb socket counterclockwise and pulling it out of the lens assembly

17.11a To replace the trunk lid light bulb, squeeze the ends of the retainer together as shown . . .

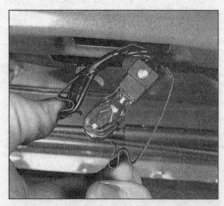

17.11b . . . then pull down the retainer and remove the bulb from the socket

17.14 To detach the license plate light bulb from the fascia, remove these two retaining screws (arrows)

12

17.15 Pull the license plate light assembly down, then detach the bulb and socket from the lens assembly

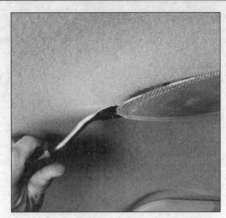

17.17 Carefully pry off the dome light lens

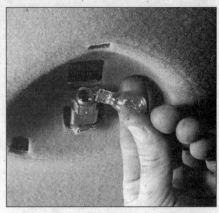

17.18 Remove the bulb from the dome light socket

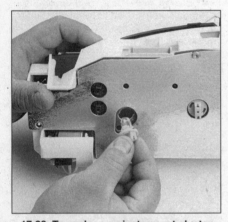

17.20 To replace an instrument cluster light bulb, turn it counterclockwise and pull it out

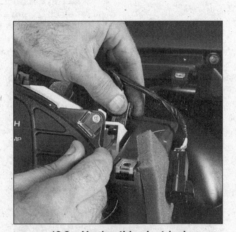

19.3a Unplug this electrical connector . . .

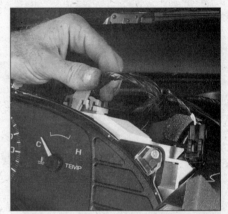

19.3b . . . and this one from the instrument cluster

15   Pull down the lens and socket, detach the socket from the lens **(see illustration)** and remove the bulb from the socket.
16   Installation is the reverse of removal.

Dome light

Refer to illustrations 17.17 and 17.18
17   Carefully pry off the lens **(see illustration)**.
18   Remove the bulb from the socket **(see illustration)**.
19   Installation is the reverse of removal.

Instrument cluster illumination

Refer to illustration 17.20
20   To gain access to the instrument cluster illumination lights, the instrument cluster will have to be removed (see Section 19). The bulbs can then be removed and replaced from the rear of the cluster **(see illustration)**.

18   Daytime Running Lights (DRL) - general information

The Daytime Running Lights (DRL) system used on all Canadian and later US models illuminates the headlights whenever the engine is running. The only exception is with the engine running and the parking brake engaged. Once the parking brake is released, the lights will remain on as long as the ignition switch is on, even if the parking brake is later applied.

The DRL system supplies reduced power to the headlights so they will be bright enough for daytime visibility while prolonging headlight life.

A malfunction with the daytime running lights can be caused by failure of the DRL relay, which is located in back of the front impact bar, or by a failure of the microprocessor unit in the instrument cluster assembly.

19   Instrument cluster - removal and installation

Refer to illustrations 19.3a, 19.3b, 19.4a and 19.4b
Warning: *The models covered by this manual are equipped with airbags. Always disable the airbag system before working in the vicinity of the impact sensors, steering column or instrument panel to avoid the possibility of accidental deployment of the airbag(s), which could cause personal injury (see Section 25). The yellow wires and connectors routed through the instrument panel are for this system. Do not use electrical test equipment on these yellow wires or tamper with them in any way while working under the instrument panel.*
Caution: *If the vehicle is equipped with a Delco Loc II or Theftlock audio system, make sure you have the correct activation code before disconnecting the battery.*
1   Detach the cable from the negative battery terminal and disable the airbag system (see Section 25).
2   Remove the instrument panel trim plate (Chevrolet models) or the instrument panel trim pad (Pontiac models) (see Chapter 11).
3   Unplug the electrical connectors from the instrument cluster **(see illustrations)**.
4   Remove the instrument cluster retaining screws **(see illustration)** and lift it from the dash **(see illustration)**.
5   Installation is the reverse of removal.

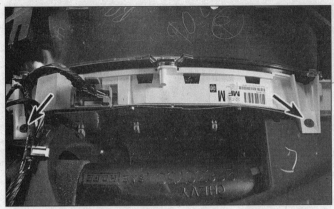

19.4a To detach the instrument cluster from the instrument panel, remove these two screws (arrows)

19.4b Make sure that everything is disconnected, then carefully lift the instrument cluster straight up and remove it from the instrument panel

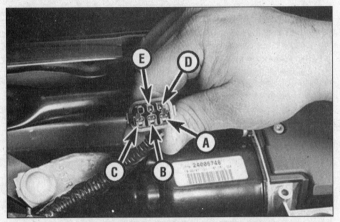

20.2 Use a voltmeter or test light to check for battery power at the wiper motor

A	Ground
B	Fuse output
C	Windshield wiper motor feed (High)
D	Windshield wiper switch signal (On)
E	Windshield wiper switch signal (Low/Pulse)

20 Wiper motor - check and replacement

Check

Refer to illustration 20.2

Note: *Refer to the wiring diagrams for wire colors and locations in the following checks. Keep in mind that power wires are generally larger in diameter and brighter colors, where ground wires are usually smaller in diameter and darker colors. When checking for voltage, probe a grounded 12-volt test light to each terminal at a connector until it lights; this verifies voltage (power) at the terminal.*

1   If the wipers work slowly, make sure the battery is in good condition and has a strong charge (see Chapter 1). If the battery is in good condition, remove the wiper motor (see below) and operate the wiper arms by hand. Check for binding linkage and pivots. Lubricate or repair the linkage or pivots as necessary. Reinstall the wiper motor. If the wipers still operate slowly, check for loose or corroded connections, especially the ground connection. If all connections look OK, replace the motor.

2   If the wipers fail to operate when activated, check the fuse. If the fuse is OK, connect a jumper wire between the wiper motor and ground, then retest. If the motor works now, repair the ground connection. If the motor still doesn't work, turn the wiper switch to the HI position and check for voltage at the motor **(see illustration)**. If there's voltage at the motor, remove the motor and check it off the vehicle with fused jumper wires from the battery. If the motor now works, check for binding linkage (see Step 1 above). If the motor still doesn't

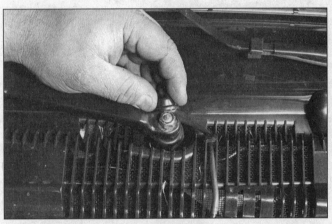

20.7 Use a small screwdriver to pry off the wiper arm nut cover, then remove the nut and pull the arm straight off its splined shaft

work, replace it. If there's no voltage at the motor, check for voltage at the wiper control module. If there's voltage at the wiper control module and no voltage at the at the wiper motor, check the switch for continuity (see Section 9). If the switch is OK, the wiper control module is probably bad.

3   If the interval (delay) function is inoperative, check the continuity of all the wiring between the switch and wiper control module. If the wiring is OK, check the resistance of the delay control knob of the multi-function switch (see Section 9). If the delay control knob is within the specified resistance, replace the wiper control module.

4   If the wipers stop at the position they're in when the switch is turned off (fail to park), check for voltage at the park feed wire of the wiper motor connector when the wiper switch is OFF but the ignition is ON. If no voltage is present, check for an open circuit between the wiper motor and the fuse panel.

5   If the wipers won't shut off unless the ignition is OFF, disconnect the wiring from the wiper control switch. If the wipers stop, replace the switch. If the wipers keep running, there's a defective limit switch in the motor; replace the motor.

6   If the wipers won't retract below the hood line, check for mechanical obstructions in the wiper linkage or on the vehicle's body which would prevent the wipers from parking. If there are no obstructions, check the wiring between the switch and motor for continuity. If the wiring is OK, replace the wiper motor.

Replacement

Refer to illustrations 20.7, 20.8a, 20.8b, 20.8c and 20.9

7   Pry off the cover for the wiper arm nuts, unscrew the nuts and remove both wiper arms **(see illustration)**.

12

20.8a To detach the cowl screen, disconnect the windshield washer lines . . .

20.8b . . . release the push-in retainers (left arrow points to one, others not visible in this photo), remove the single retaining screw (right arrow) . . .

20.8c . . . and remove the cowl screen

8   Disconnect the windshield washer line, remove the single retaining screw, release the five push-in retainers, and remove the cowl cover **(see illustrations)**.

9   Unplug the electrical connector from the windshield wiper motor **(see illustration)**.

10   Remove the three wiper motor assembly retaining bolts and remove the wiper motor assembly.

11   Detach the wiper transmission assembly from the wiper motor crank arm.

12   Detach the wiper motor crank arm from the wiper motor assembly.

13   Remove the three wiper motor retaining screws and detach the wiper motor from the bracket.

14   Installation is the reverse of removal.

21 Horn - check and replacement

Check

Refer to illustration 21.2

Note: *Check the CIG (cigarette lighter/horn) fuse before beginning electrical diagnosis.*

1   Remove the front bumper fascia (see Chapter 11).

2   Unplug the horn electrical connector **(see illustration)**.

3   To test the horn(s), connect battery voltage to the horn terminal with a pair of jumper wires. If the horn doesn't sound, replace it.

4   If the horn does sound, check for voltage at the terminal when the horn button is depressed. If there's voltage at the terminal, check for a bad ground at the horn.

5   If there's no voltage at the horn, check the relay (see Section 6). Note that most horn relays are either the four-terminal or externally grounded three-terminal type.

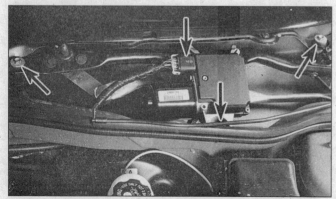

20.9 To detach the wiper motor assembly, unplug the electrical connector (arrow) and remove the three retaining screws (arrows)

6   If the relay is OK, check for voltage to the relay power and control circuits. If either of the circuits is not receiving voltage, inspect the wiring between the relay and the fuse panel.

7   If both relay circuits are receiving voltage, depress the horn button and check the circuit from the relay to the horn button for continuity to ground. If there's no continuity, check the circuit for an open. If there's no open circuit, replace the horn button.

8   If there's continuity to ground through the horn button, check for an open or short in the circuit from the relay to the horn.

Replacement

Refer to illustration 21.10

9   Remove the front bumper fascia (see Chapter 11).

10   Unplug the electrical connector **(see illustration 21.2)** and remove the bracket bolt **(see illustration)**.

11   Installation is the reverse of removal.

22 Electric side view mirrors - description and check

Note: *Check the INT LAMP fuse before beginning electrical diagnosis.*

1   Electric rear view mirrors use two motors to move the glass; one for up-and-down adjustments and one for left-to-right adjustments.

2   The mirror control switch, which is located in the left (driver's side) front door, can be rotated counterclockwise or clockwise to send voltage to the left or right mirror, respectively. With the ignition key turned to On (engine off), roll down the windows and operate the mirror control switch through all functions (left-right and up-down) for both the left and right side mirrors.

3   Listen carefully for the sound of the electric motors running in the mirrors.

4   If the motors can be heard but the mirror glass doesn't move, the drive mechanism inside the mirror is malfunctioning. Remove and disassemble the mirror to locate the problem (see Chapter 11).

5   If the mirrors don't operate and no sound comes from the mirrors, check the fuse (see Section 3).

6   If the fuse is OK, remove the door trim panel (see Chapter 11) and remove the mirror control switch assembly without disconnecting the wires attached to it. Turn the ignition key to On and check for voltage at the switch. There should be voltage at one terminal. If there's no voltage at the switch, check for an opening or short in the wiring between the fuse panel and the switch.

7   If there's voltage at the switch, unplug the electrical connector and check the switch for continuity in all its operating positions (refer to the wiring diagrams). If the switch does not operate properly, replace it.

8   Plug in the switch connector. Locate the wire going from the switch to ground. Leaving the switch connected, connect a jumper wire between this wire and ground. If the mirror works normally with this wire in place, repair the faulty ground connection.

9   If the mirror still doesn't work, remove the cover and check the wires at the mirror for voltage with a test light. Check with the ignition

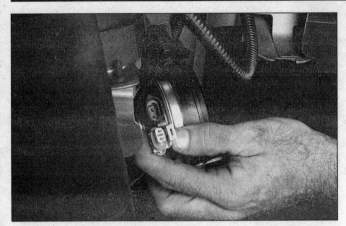

21.2 To test the horn, connect the horn to the battery with fused jumper wires; if the horn doesn't operate, replace it

21.10 To detach the horn from the bracket, remove this nut (arrow)

key turned to On and the mirror selector switch on the appropriate side. Operate the mirror switch in all its positions. There should be voltage at one of the switch-to-mirror wires in each switch position (except the neutral position).

10 If there isn't voltage at each switch position, check the wiring between the mirror and control switch for opens and shorts.

11 If there's voltage, remove the mirror and test it off the vehicle with jumper wires. Replace the mirror if it fails this test (see Chapter 11).

23 Cruise control system - description and check

1 The cruise control system maintains vehicle speed with an electronic servo motor located in the engine compartment, which is connected to the throttle linkage by a cable. The system consists of the electronic control module, brake switch, control switches, a relay, the vehicle speed sensor and associated wiring. Listed below are some general procedures that may be used to locate common cruise control problems.

2 Locate and check the CRUISE fuse (see Section 3).

3 Have an assistant operate the brake lights while you check their operation (voltage from the brake light switch deactivates the cruise control).

4 If the brake lights don't come on or don't shut off, correct the problem and retest the cruise control.

5 Inspect the cable linkage between the cruise control module and the throttle linkage. The cruise control module is located in the right rear corner of the engine compartment.

6 Visually inspect the wire harness and connector. Look for corrosion and damaged and broken wires.

7 The vehicle speed sensor is located on the transmission (see Chapter 7B). Raise the front of the vehicle and support it on jack stands. Unplug the electrical connector and touch one probe of a digital voltmeter to the orange wire of the connector and the other to a good ground. With the vehicle in Neutral and key On, measure the voltage while rotating one wheel with the other one blocked. If the voltage doesn't vary as the wheel rotates, the sensor is defective.

8 Test drive the vehicle to determine if the cruise control is now working. If it isn't, take it to a dealer service department or an automotive electrical specialist for further diagnosis and repair.

24 Power window system - description and check

1 The power window system operates electric motors, mounted in the doors, which lower and raise the windows. The system consists of the control switches, the motors, the regulators and the wiring harnesses.

2 The power windows can be lowered and raised from the master control switch by the driver or by remote switches located at the individual windows. Each window has a separate motor which is reversible. The position of the control switch determines the polarity

and therefore the direction of operation.

3 The circuit is protected by a fuse and a circuit breaker. Each motor is also equipped with an internal circuit breaker, this prevents one stuck window from disabling the whole system.

4 The power window system will only operate when the ignition key is turned to On. Some models have a window lockout switch at the master control switch which, when activated, disables the switches at the rear windows and, sometimes, the switch at the passenger's window also. Always check these items before troubleshooting a window problem.

5 These procedures are general in nature, so if you can't find the problem using them, take the vehicle to a dealer service department or other properly equipped repair facility.

6 If the power windows won't operate, always check the fuse and circuit breaker first.

7 If only the rear windows are inoperative, or if the windows only operate from the master control switch, check the rear window lockout switch for continuity in the unlocked position. Replace it if it doesn't have continuity.

8 Check the wiring between the switches and fuse panel for continuity. Repair the wiring, if necessary.

9 If only one window is inoperative from the master control switch, try the other control switch at the window. **Note:** *This doesn't apply to the drivers door window.*

10 If the same window works from one switch, but not the other, check the switch for continuity.

11 If the switch tests OK, check for a short or open in the circuit between the affected switch and the window motor.

12 If one window is inoperative from both switches, remove the trim panel from the affected door and check for voltage at the switch and at the motor while the switch is operated.

13 If voltage is reaching the motor, disconnect the glass from the regulator (see Chapter 11). Move the window up and down by hand while checking for binding and damage. Also check for binding and damage to the regulator. If the regulator is not damaged and the window moves up and down smoothly, replace the motor. If there's binding or damage, lubricate, repair or replace parts, as necessary.

14 If voltage isn't reaching the motor, check the wiring in the circuit for continuity between the switches and motors. You'll need to consult the wiring diagram for the vehicle. If the circuit is equipped with a relay, check that the relay is grounded properly and receiving voltage.

15 Test the windows after you are done to confirm proper repairs.

25 Airbag - general information

Warning: *The models covered by this manual are equipped with airbags. Airbag system components are located in the steering wheel, steering column, instrument panel and center console. The airbag(s) could accidentally deploy if any of the system components or wiring harnesses are disturbed, so be extremely careful when working in these*

12

25.2 The passenger's side airbag is located on top of the dash, right above the glovebox (instrument panel trim pad removed); notice that the big yellow connector, known as the Connector Position Assurance (CPA) connector, has been unplugged to ensure that the passenger's airbag isn't accidentally deployed while you're working in that area

25.3 The forward discriminating sensor is located to the right of the hood latch, right in front of the radiator (the other one is located on the dash, just ahead of the passenger's side airbag)

areas and don't disturb any airbag system components or wiring. You could be injured if an airbag accidentally deploys, or the airbag might not deploy correctly in a collision if any components or wiring in the system have been disturbed. The yellow wires and connectors routed through the instrument panel and center console are for this system. Do not use electrical test equipment on these yellow wires or tamper with them in any way while working in their vicinity.
Caution: *If the vehicle is equipped with a Delco Loc II or Theftlock audio system, make sure you have the correct activation code before disconnecting the battery.*

Description

Refer to illustration 25.2

1   The models covered by this manual are equipped with a Supplemental Inflatable Restraint (SIR) system, more commonly known as an airbag system. The SIR system is designed to protect the driver and passenger from serious injury in the event of a head-on or frontal collision.

2   The SIR system consists of two airbags - one in the steering wheel, one on top of the dashboard above the glovebox **(see illustration)** - and an impact sensor located below the hood latch. The arming sensor and the diagnostic/energy reserve module have been incorporated into one unit which is located under the instrument panel.

Sensors

Refer to illustration 25.3

3   There are two types of sensors. The two normally-open discriminating sensors, located next to the hood latch **(see illustration)** and in front of the passenger side airbag, are designed to close the airbag circuit when the change in velocity (i.e. the kind of sudden deceleration that can only occur as a result of an impact) exceeds a certain threshold. A third sensor, the dual-pole arming sensor, located inside the center portion of the dash, is designed to close at low-level changes in velocity (lower than the discriminating sensors). The airbag cannot deploy until one of the discriminating sensors, and the dual-pole sensor, have closed the inflator module circuit. Once the electrical circuit between the crash sensors and the diagnostic module is closed, it inflates the airbags.

Diagnostic/energy reserve module (DERM)

4   The diagnostic/energy reserve module, which is located in the center part of the dash, contains an on-board microprocessor which monitors the operation of the system. It performs a diagnostic check of the system every time the vehicle is started. If the system is operating properly, the AIRBAG warning light will blink on and off seven times. If there is a fault in the system, the light will remain on and the airbag

control module will store fault codes indicating the nature of the fault. If the AIRBAG warning light remains on after staring, or comes on while driving, the vehicle should be taken to your dealer immediately for service. The diagnostic/energy reserve module also contains a back-up power supply to deploy the airbags in the event battery power is lost during a collision. **Caution:** *Do NOT open the DERM case for any reason. Touching the connector pins or soldered components could cause damage from electrostatic discharge. A malfunctioning DERM must be replaced by authorized service personnel at a dealership.*

Operation

5   For the airbag(s) to deploy, an impact of sufficient force must occur within 30-degrees of the vehicle centerline. When this condition occurs, the circuit to the airbag inflator is closed and the airbag inflates. If the battery is destroyed by the impact, or is too low to power the inflators, a back-up power supply inside the diagnostic/energy reserve module supplies current to the airbags.

Self-diagnosis system

6   A self-diagnosis circuit in the module displays a light when the ignition switch is turned to the On position. If the system is operating normally, the light should go out after seven flashes. If the light doesn't come on, or doesn't go out after seven flashes, or if it comes on while you're driving the vehicle, there's a malfunction in the SIR system. Have it inspected and repaired as soon as possible. Do not attempt to troubleshoot or service the SIR system yourself. Even a small mistake could cause the SIR system to malfunction when you need it.

Servicing components near the SIR system

7   Nevertheless, there are times when you need to remove the steering wheel, radio or service other components on or near the instrument panel. At these times, you'll be working around components and wiring harnesses for the SIR system. SIR system wiring is easy to identify; they're all covered by a bright yellow conduit. Do not unplug the connectors for the SIR system wiring, except to disable the system. And do not use electrical test equipment on the SIR system wiring. **Always disable the SIR system before working near the SIR system components or related wiring.**

Disabling the SIR system

Refer to illustrations 25.10 and 25.11

8   Disconnect the cable from the negative battery terminal. **Caution:** *If the vehicle is equipped with a Delco Loc II or Theftlock audio system, make sure you have the correct activation code before disconnecting the battery.* Turn the steering wheel to the straight ahead position, place the ignition switch key in the Lock position and remove the key. Remove the airbag fuse from the fuse block (see Section 3).

9   Remove the steering column covers and the left sound insulator panel below the instrument panel (see Chapter 11).

25.10 The driver's side airbag Connector Position Assurance (CPA) connector is located at the base of the steering column; always unplug it before removing the steering wheel or working in the area of the steering wheel

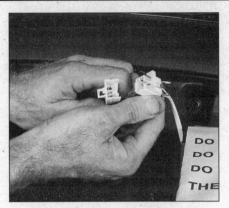

25.11 The passenger's side airbag Connector Position Assurance (CPA) connector is located to the left of the passenger's side airbag; always unplug it before removing the steering wheel or working in the area of the passenger's side airbag

25.15a To release the SIR coil from the steering shaft, remove this snap ring . . .

10   Unplug the yellow Connector Position Assurance (CPA) steering column harness connector **(see illustration)**.

11   Before working in or around the passenger's side of the instrument panel, unplug the yellow Connector Position Assurance (CPA) passenger airbag harness connector **(see illustration)**.

Enabling the SIR system

11   After you've disabled the airbag and performed the necessary service, plug in the steering column (driver's side) and passenger side CPA connectors. Reinstall the steering column lower trim panel, and the sound insulator panel.

12   Install the airbag fuse. Connect the negative battery terminal.

Centering the SIR coil

Refer to illustrations 25.15a, 25.15b, 25.16 and 25.18

13   Anytime some part of the steering system is disassembled for service or replacement, the steering column should be immobilized to ensure that the SIR coil doesn't become uncentered (moved). This can occur, for example, if the steering column is separated from the steering gear, or if the centering spring is pushed down, allowing the hub to rotate while the coil is removed from the steering column. If the coil becomes accidentally uncentered, re-center it as follows before reassembling the steering system:

14   Make sure that the wheels are pointed straight ahead.

15   Remove the coil assembly snap ring **(see illustration)** and

remove the coil assembly **(see illustration)**.

16   Holding the coil assembly with its bottom side facing up, depress the spring lock **(see illustration)** and rotate the hub in the direction of the arrow until it stops (the arrow is on the back of the coil assembly). The coil ribbon should be wound up snug against the center hub.

17   Rotate the coil hub in the opposite direction two and three-quarters turns, then release the spring lock. The coil is now centered.

18   Install the SIR coil and secure it with the snap ring. The tab will be at the top and the marks aligned when the coil is properly installed **(see illustration).**

26   Wiring diagrams - general information

Since it isn't possible to include all wiring diagrams for every year covered by this manual, the following diagrams are those that are typical and most commonly needed.

Prior to troubleshooting any circuit, check the fuse and circuit breakers (if equipped) to make sure they're in good condition. Make sure the battery is properly charged and check the cable connections (see Chapter 1).

When checking a circuit, make sure that all electrical connectors are clean, with no broken or loose terminals. When unplugging an electrical connector, do not pull on the wires. Pull only on the connector housings themselves.

25.15b . . . then remove the coil

25.16 To center the SIR coil, hold it with its underside facing up, depress the spring lock and rotate the hub in the direction of the arrow on the coil assembly until it stops

25.18 When properly installed, the airbag coil will be centered with the marks aligned (circle) and the tab fitted between the projections on the top of steering column (arrow)

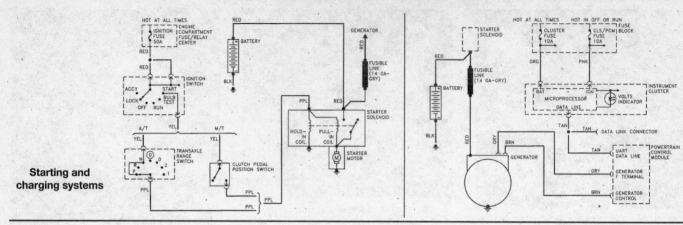

Starting and charging systems

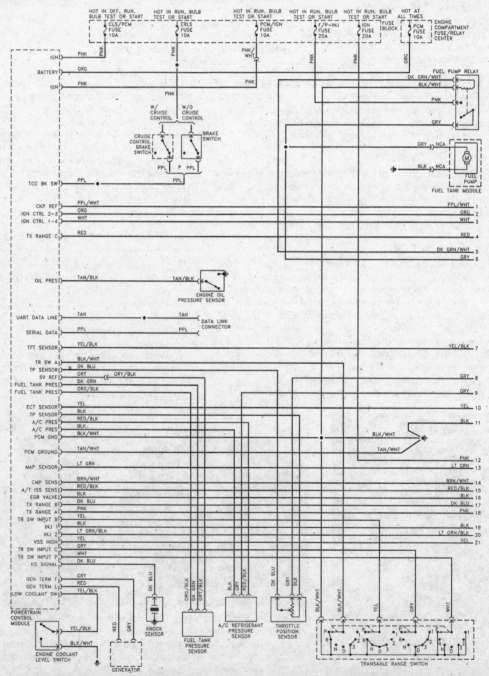

**Engine control system
(1 of 3)**

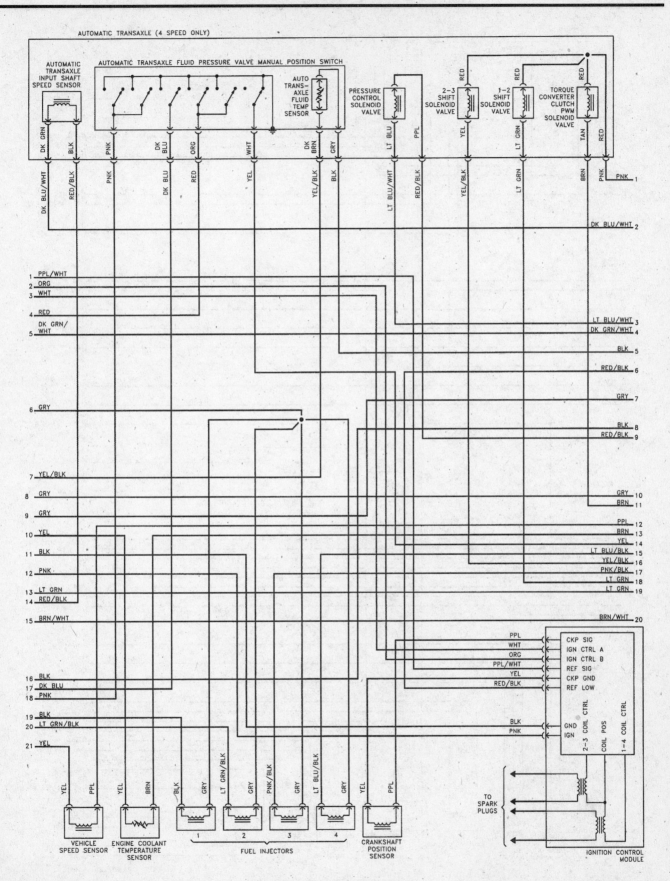

Engine control system (2 of 3)

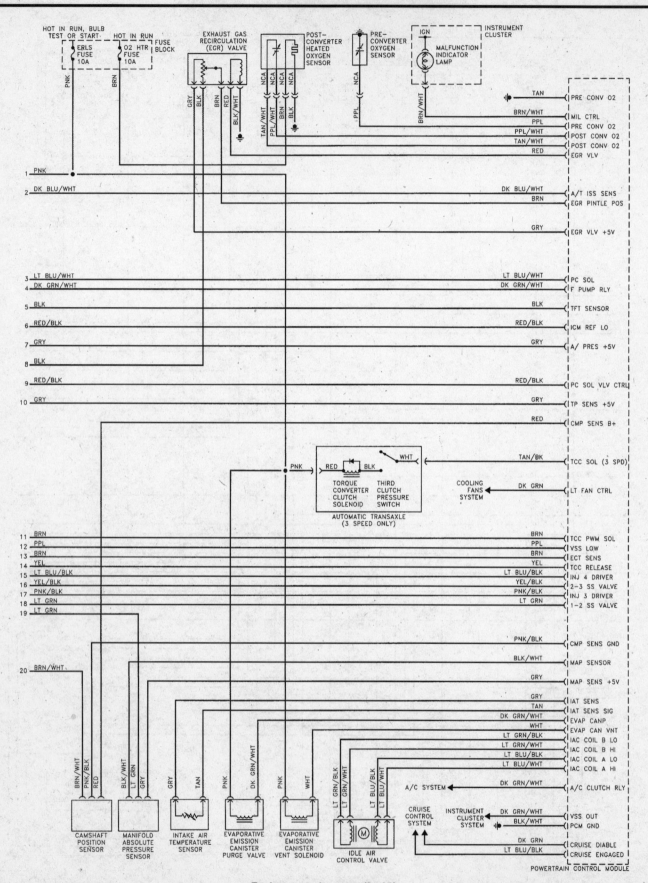

Engine control system (3 of 3)

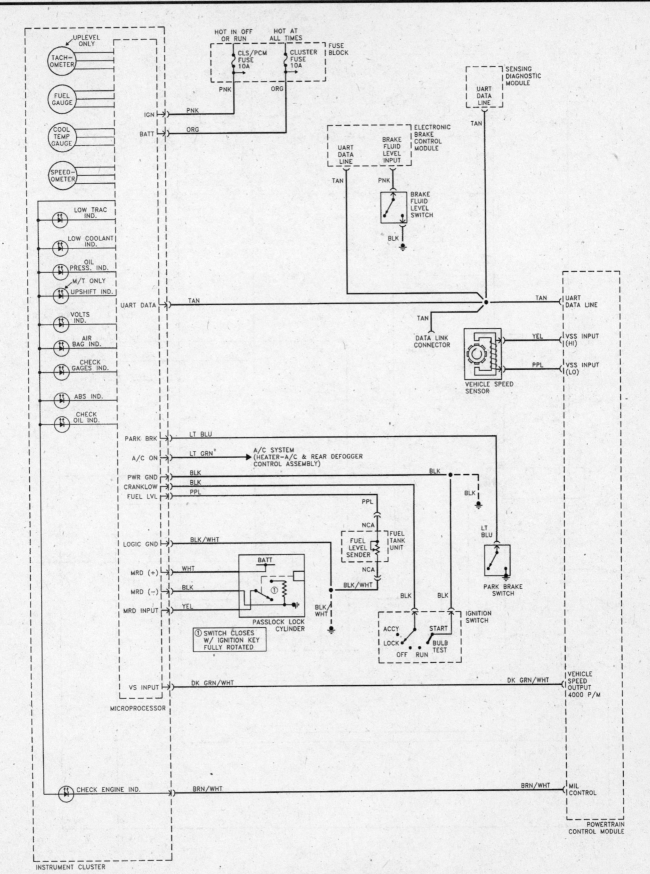

Instruments and engine warning system

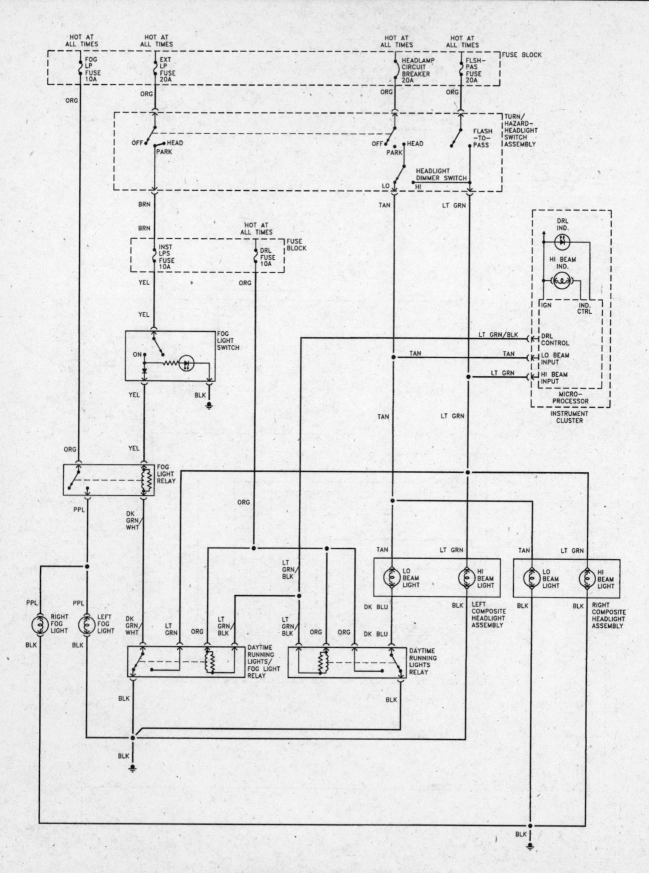

Headlight and foglight systems (1995 models)

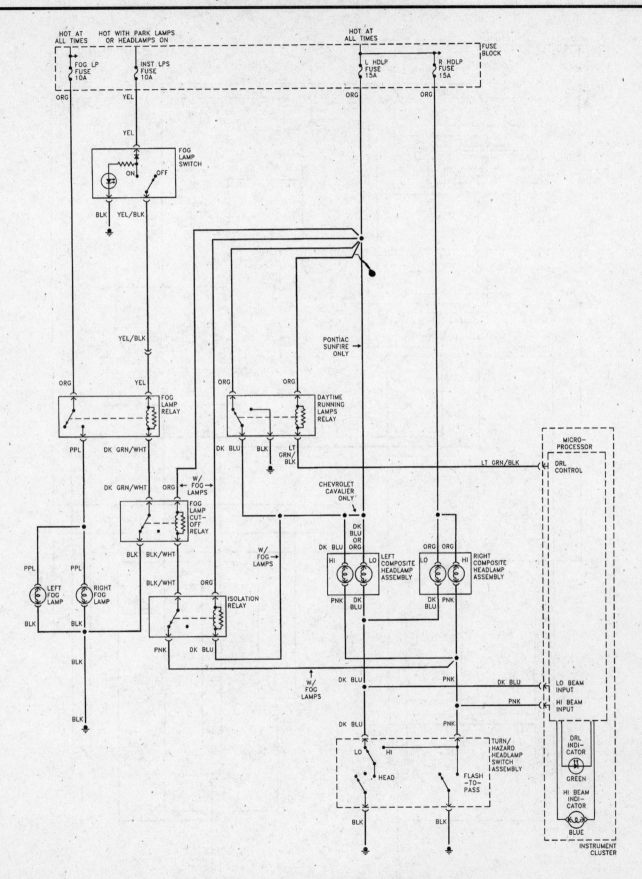

Headlight and foglight systems (1996 through 1998 models)

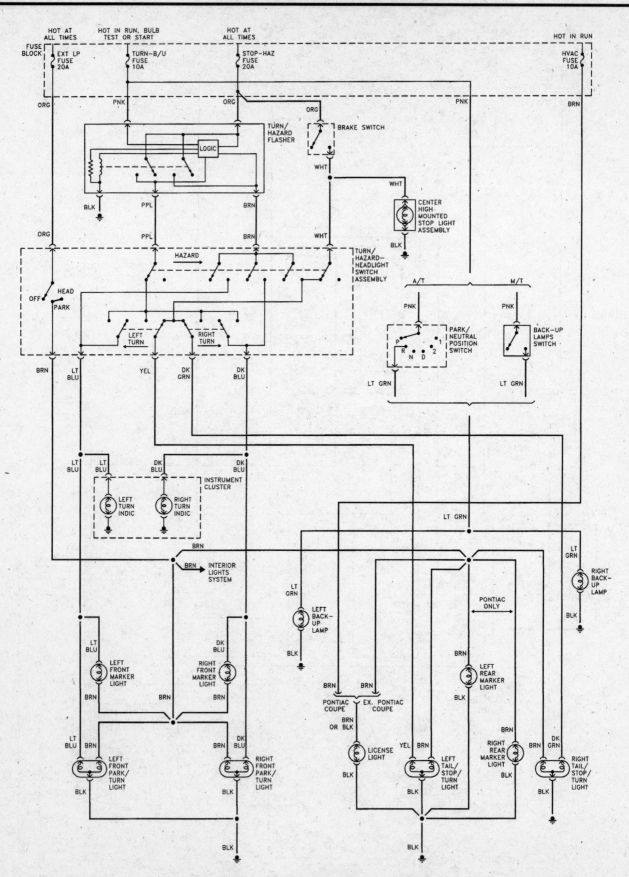

Exterior lighting systems, except headlights and foglights

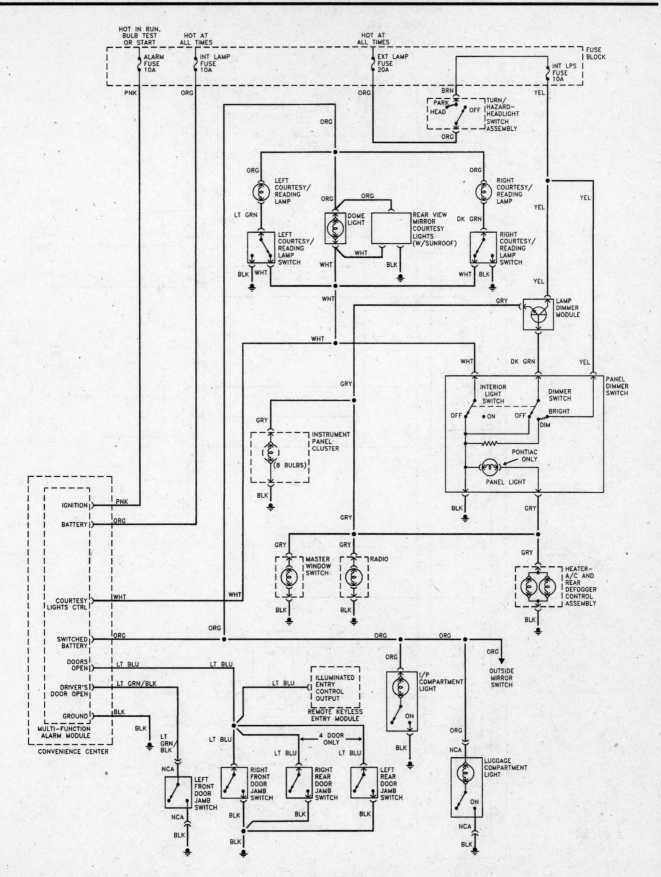

Interior lighting systems

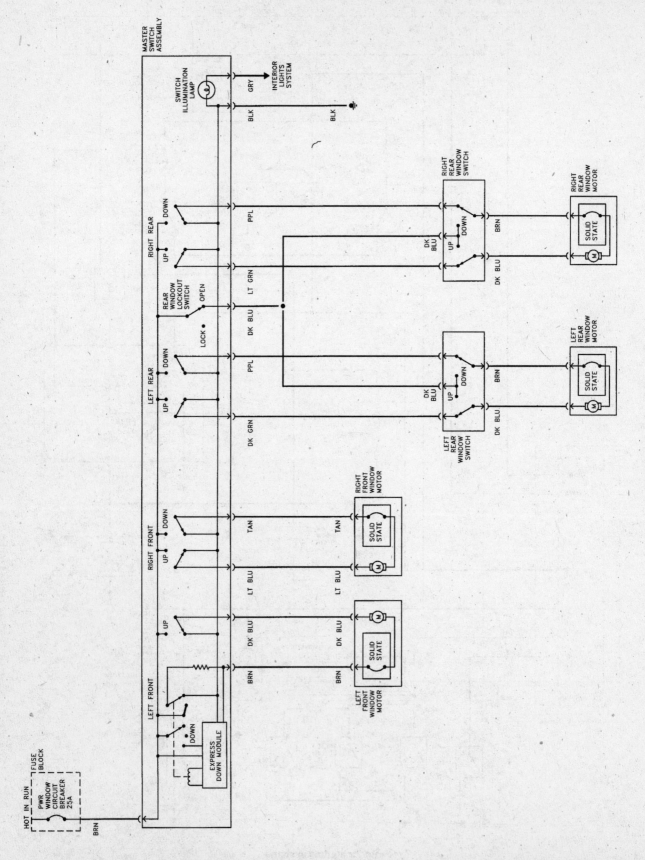
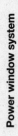

Power window system

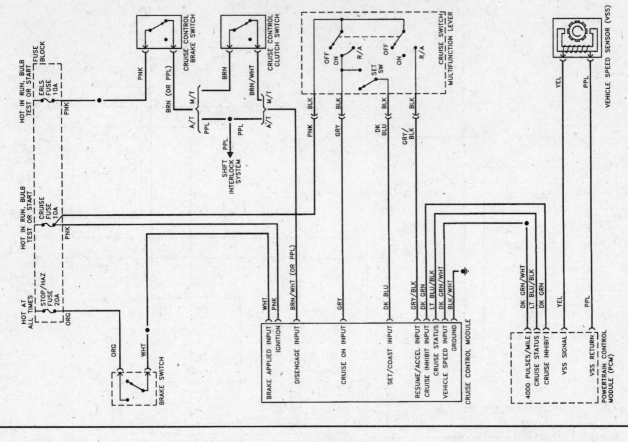

Cruise control system

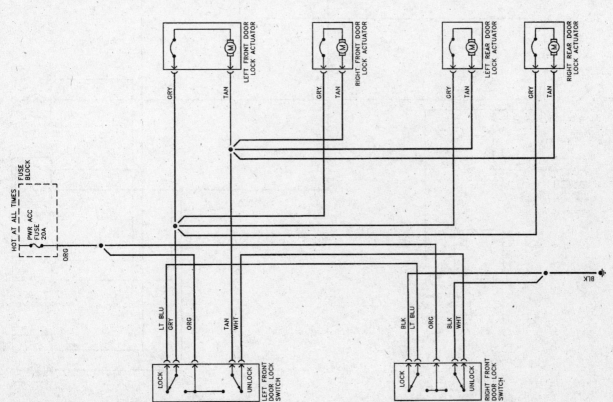

Power door lock system (without keyless entry)

12

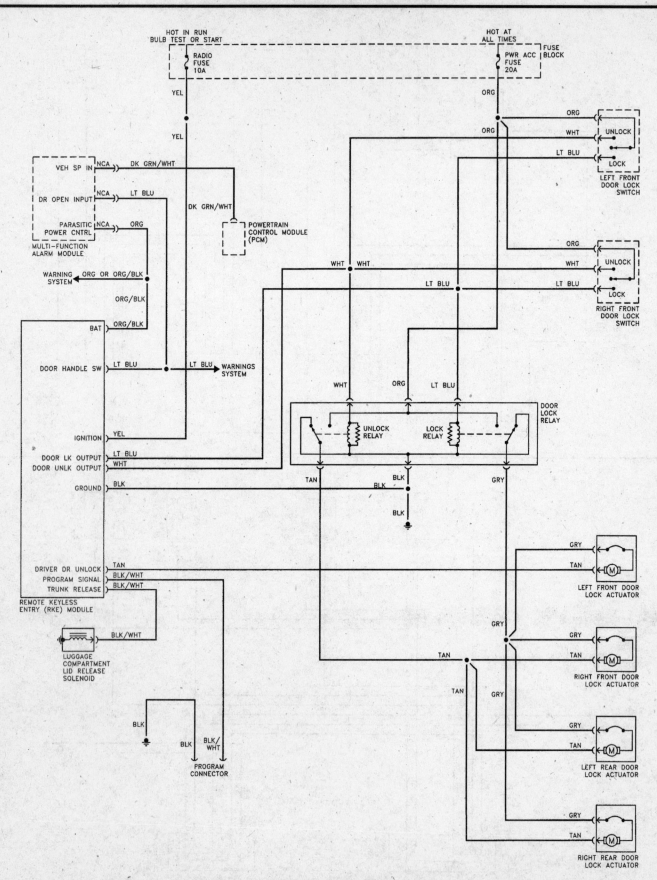

Power door lock system (with keyless entry)

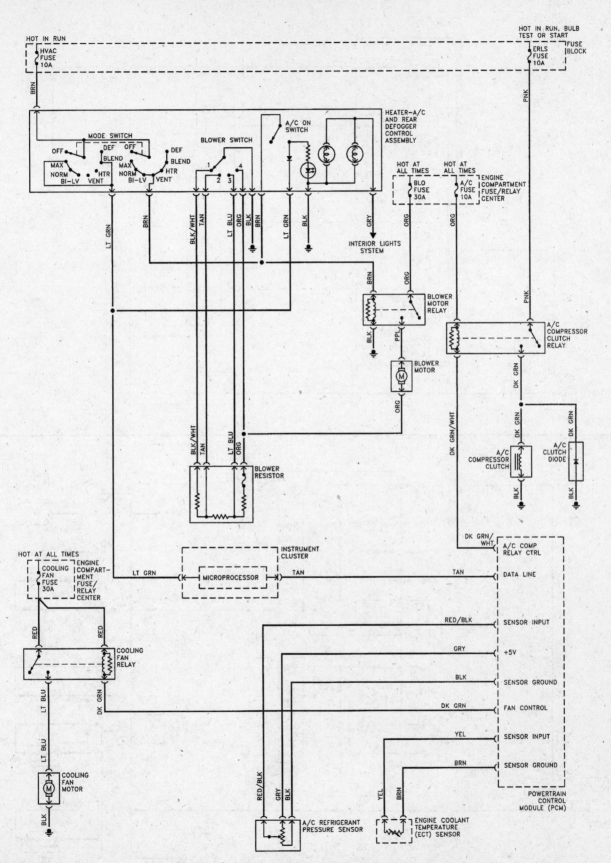

Heating, air conditioning and engine cooling systems

12

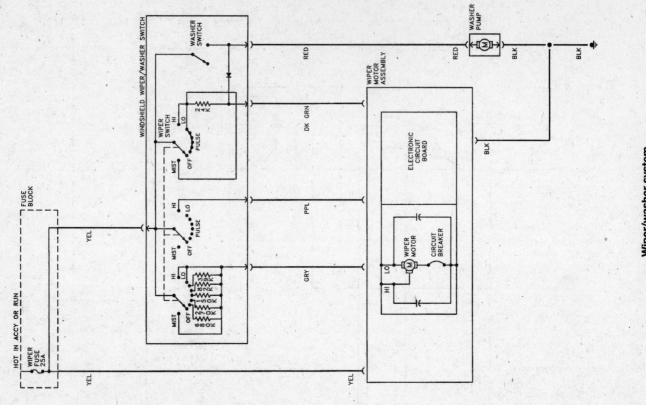

Wiper/washer system

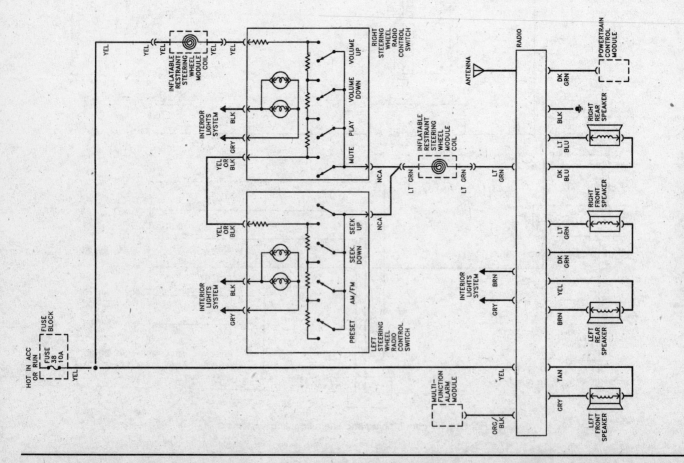

Radio, antenna and speakers

Index

Haynes Automotive Manuals

NOTE: New manuals are added to this list on a periodic basis. If you do not see a listing for your vehicle, consult your local Haynes dealer for the latest product information.

ACURA
*12020 **Integra** '86 thru '89 & **Legend** '86 thru '90

AMC
Jeep CJ - *see JEEP (50020)*
14020 **Mid-size models**, Concord, Hornet, Gremlin & Spirit '70 thru '83
14025 **(Renault) Alliance & Encore** '83 thru '87

AUDI
15020 **4000** all models '80 thru '87
15025 **5000** all models '77 thru '83
15026 **5000** all models '84 thru '88

AUSTIN-HEALEY
Sprite - *see MG Midget (66015)*

BMW
*18020 **3/5 Series** not including diesel or all-wheel drive models '82 thru '92
*18021 **3 Series** except 325iX models '92 thru '97
18025 **320i** all 4 cyl models '75 thru '83
18035 **528i & 530i** all models '75 thru '80
18050 **1500 thru 2002** except Turbo '59 thru '77

BUICK
Century (front wheel drive) - *see GM (829)*
*19020 **Buick, Oldsmobile & Pontiac Full-size** (Front wheel drive) all models '85 thru '98
Buick Electra, LeSabre and Park Avenue; Oldsmobile Delta 88 Royale, Ninety Eight and Regency; Pontiac Bonneville
19025 **Buick Oldsmobile & Pontiac Full-size** (Rear wheel drive)
Buick Estate '70 thru '90, Electra'70 thru '84, LeSabre '70 thru '85, Limited '74 thru '79
Oldsmobile Custom Cruiser '70 thru '90, Delta 88 '70 thru '85,Ninety-eight '70 thru '84
Pontiac Bonneville '70 thru '81, Catalina '70 thru '81, Grandville '70 thru '75, Parisienne '83 thru '86
19030 **Mid-size Regal & Century** all rear-drive models with V6, V8 and Turbo '74 thru '87
Regal - *see GENERAL MOTORS (38010)*
Riviera - *see GENERAL MOTORS (38030)*
Roadmaster - *see CHEVROLET (24046)*
Skyhawk - *see GENERAL MOTORS (38015)*
Skylark '80 thru '85 - *see GM (38020)*
Skylark '86 on - *see GM (38025)*
Somerset - *see GENERAL MOTORS (38025)*

CADILLAC
*21030 **Cadillac Rear Wheel Drive** all gasoline models '70 thru '93
Cimarron - *see GENERAL MOTORS (38015)*
Eldorado - *see GENERAL MOTORS (38030)*
Seville '80 thru '85 - *see GM (38030)*

CHEVROLET
*24010 **Astro & GMC Safari Mini-vans** '85 thru '93
24015 **Camaro V8** all models '70 thru '81
24016 **Camaro** all models '82 thru '92
Cavalier - *see GENERAL MOTORS (38015)*
Celebrity - *see GENERAL MOTORS (38005)*
24017 **Camaro & Firebird** '93 thru '99
24020 **Chevelle, Malibu & El Camino** '69 thru '87
24024 **Chevette & Pontiac T1000** '76 thru '87
Citation - *see GENERAL MOTORS (38020)*
*24032 **Corsica/Beretta** all models '87 thru '96
24040 **Corvette** all V8 models '68 thru '82
*24041 **Corvette** all models '84 thru '96
10305 **Chevrolet Engine Overhaul Manual**
24045 **Full-size Sedans** Caprice, Impala, Biscayne, Bel Air & Wagons '69 thru '90
24046 **Impala SS & Caprice and Buick Roadmaster** '91 thru '96
Lumina - *see GENERAL MOTORS (38010)*

24048 **Lumina & Monte Carlo** '95 thru '98
Lumina APV - *see GM (38035)*
24050 **Luv Pick-up** all 2WD & 4WD '72 thru '82
*24055 **Monte Carlo** all models '70 thru '88
Monte Carlo '95 thru '98 - *see LUMINA (24048)*
24059 **Nova** all V8 models '69 thru '79
*24060 **Nova and Geo Prizm** '85 thru '92
24064 **Pick-ups** '67 thru '87 - Chevrolet & GMC, all V8 & in-line 6 cyl, 2WD & 4WD '67 thru '87; Suburbans, Blazers & Jimmys '67 thru '91
*24065 **Pick-ups** '88 thru '98 - Chevrolet & GMC, all full-size pick-ups, '88 thru '98; Blazer & Jimmy '92 thru '94; Suburban '92 thru '98; Tahoe & Yukon '98
24070 **S-10 & S-15 Pick-ups** '82 thru '93, Blazer & Jimmy '83 thru '94,
*24071 **S-10 & S-15 Pick-ups** '94 thru '96 Blazer & Jimmy '95 thru '96
*24075 **Sprint & Geo Metro** '85 thru '94
*24080 **Vans - Chevrolet & GMC**, V8 & in-line 6 cylinder models '68 thru '96

CHRYSLER
25015 **Chrysler Cirrus, Dodge Stratus, Plymouth Breeze** '95 thru '98
25025 **Chrysler Concorde, New Yorker & LHS, Dodge Intrepid, Eagle Vision,** '93 thru '97
10310 **Chrysler Engine Overhaul Manual**
*25020 **Full-size Front-Wheel Drive** '88 thru '93
K-Cars - *see DODGE Aries (30008)*
Laser - *see DODGE Daytona (30030)*
*25030 **Chrysler & Plymouth Mid-size** front wheel drive '82 thru '95
Rear-wheel Drive - *see Dodge (30050)*

DATSUN
28005 **200SX** all models '80 thru '83
28007 **B-210** all models '73 thru '78
28009 **210** all models '79 thru '82
28012 **240Z, 260Z & 280Z** Coupe '70 thru '78
28014 **280ZX** Coupe & 2+2 '79 thru '83
300ZX - *see NISSAN (72010)*
28016 **310** all models '78 thru '82
28018 **510 & PL521 Pick-up** '68 thru '73
28020 **510** all models '78 thru '81
28022 **620 Series Pick-up** all models '73 thru '79
720 Series Pick-up - *see NISSAN (72030)*
28025 **810/Maxima** all gasoline models, '77 thru '84

DODGE
400 & 600 - *see CHRYSLER (25030)*
*30008 **Aries & Plymouth Reliant** '81 thru '89
30010 **Caravan & Plymouth Voyager Mini-Vans** all models '84 thru '95
*30011 **Caravan & Plymouth Voyager Mini-Vans** all models '96 thru '98
30012 **Challenger/Plymouth Saporro** '78 thru '83
30016 **Colt & Plymouth Champ** (front wheel drive) all models '78 thru '87
*30020 **Dakota Pick-ups** all models '87 thru '96
30025 **Dart, Demon, Plymouth Barracuda, Duster & Valiant** 6 cyl models '67 thru '76
*30030 **Daytona & Chrysler Laser** '84 thru '89
Intrepid - *see CHRYSLER (25025)*
*30034 **Neon** all models '95 thru '97
*30035 **Omni & Plymouth Horizon** '78 thru '90
*30040 **Pick-ups** all full-size models '74 thru '93
*30041 **Pick-ups** all full-size models '94 thru '96
*30045 **Ram 50/D50 Pick-ups & Raider and Plymouth Arrow Pick-ups** '79 thru '93
30050 **Dodge/Plymouth/Chrysler** rear wheel drive '71 thru '89
*30055 **Shadow & Plymouth Sundance** '87 thru '94
*30060 **Spirit & Plymouth Acclaim** '89 thru '95
*30065 **Vans - Dodge & Plymouth** '71 thru '96

EAGLE
Talon - *see Mitsubishi Eclipse (68030)*
Vision - *see CHRYSLER (25025)*

FIAT
34010 **124 Sport Coupe & Spider** '68 thru '78
34025 **X1/9** all models '74 thru '80

FORD
10355 **Ford Automatic Transmission Overhaul**
*36004 **Aerostar Mini-vans** all models '86 thru '96
*36006 **Contour & Mercury Mystique** '95 thru '98
36008 **Courier Pick-up** all models '72 thru '82
36012 **Crown Victoria & Mercury Grand Marquis** '88 thru '96
10320 **Ford Engine Overhaul Manual**
36016 **Escort/Mercury Lynx** all models '81 thru '90
*36020 **Escort/Mercury Tracer** '91 thru '96
*36024 **Explorer & Mazda Navajo** '91 thru '95
36028 **Fairmont & Mercury Zephyr** '78 thru '83
36030 **Festiva & Aspire** '88 thru '97
36032 **Fiesta** all models '77 thru '80
36036 **Ford & Mercury Full-size**, Ford LTD & Mercury Marquis ('75 thru '82); Ford Custom 500,Country Squire, Crown Victoria & Mercury Colony Park ('75 thru '87); Ford LTD Crown Victoria & Mercury Gran Marquis ('83 thru '87)
36040 **Granada & Mercury Monarch** '75 thru '80
36044 **Ford & Mercury Mid-size**, Ford Thunderbird & Mercury Cougar ('75 thru '82); Ford LTD & Mercury Marquis ('83 thru '86); Ford Torino,Gran Torino, Elite, Ranchero pick-up, LTD II, Mercury Montego, Comet, XR-7 & Lincoln Versailles ('75 thru '86)
36048 **Mustang V8** all models '64-1/2 thru '73
36049 **Mustang II** 4 cyl, V6 & V8 models '74 thru '78
36050 **Mustang & Mercury Capri** all models Mustang, '79 thru '93; Capri, '79 thru '86
*36051 **Mustang** all models '94 thru '97
36054 **Pick-ups & Bronco** '73 thru '79
36058 **Pick-ups & Bronco** '80 thru '96
36059 **Pick-ups, Expedition & Mercury Navigator** '97 thru '98
36062 **Pinto & Mercury Bobcat** '75 thru '80
36066 **Probe** all models '89 thru '92
36070 **Ranger/Bronco II** gasoline models '83 thru '92
*36071 **Ranger** '93 thru '97 & Mazda Pick-ups '94 thru '97
36074 **Taurus & Mercury Sable** '86 thru '95
*36075 **Taurus & Mercury Sable** '96 thru '98
*36078 **Tempo & Mercury Topaz** '84 thru '94
36082 **Thunderbird/Mercury Cougar** '83 thru '88
*36086 **Thunderbird/Mercury Cougar** '89 and '97
36090 **Vans** all V8 Econoline models '69 thru '91
*36094 **Vans** full size '92-'95
*36097 **Windstar Mini-van** '95-'98

GENERAL MOTORS
*10360 **GM Automatic Transmission Overhaul**
*38005 **Buick Century, Chevrolet Celebrity, Oldsmobile Cutlass Ciera & Pontiac 6000** all models '82 thru '96
*38010 **Buick Regal, Chevrolet Lumina, Oldsmobile Cutlass Supreme & Pontiac Grand Prix** front-wheel drive models '88 thru '95
*38015 **Buick Skyhawk, Cadillac Cimarron, Chevrolet Cavalier, Oldsmobile Firenza & Pontiac J-2000 & Sunbird** '82 thru '94
*38016 **Chevrolet Cavalier & Pontiac Sunfire** '95 thru '98
38020 **Buick Skylark, Chevrolet Citation, Olds Omega, Pontiac Phoenix** '80 thru '85
38025 **Buick Skylark & Somerset, Oldsmobile Achieva & Calais and Pontiac Grand Am** all models '85 thru '95
38030 **Cadillac Eldorado** '71 thru '85, **Seville** '80 thru '85, **Oldsmobile Toronado** '71 thru '85 & **Buick Riviera** '79 thru '85
*38035 **Chevrolet Lumina APV, Olds Silhouette & Pontiac Trans Sport** all models '90 thru '95
General Motors Full-size Rear-wheel Drive - *see BUICK (19025)*

(Continued on other side)

---

* Listings shown with an asterisk (*) indicate model coverage as of this printing. These titles will be periodically updated to include later model years - consult your Haynes dealer for more information.

Haynes North America, Inc., 861 Lawrence Drive, Newbury Park, CA 91320-1514 • (805) 498-6703

Haynes Automotive Manuals (continued)

NOTE: New manuals are added to this list on a periodic basis. If you do not see a listing for your vehicle, consult your local Haynes dealer for the latest product information.

GEO

Metro - *see CHEVROLET Sprint (24075)*
Prizm - *'85 thru '92 see CHEVY (24060), '93 thru '96 see TOYOTA Corolla (92036)*
*40030 **Storm** all models '90 thru '93
Tracker - *see SUZUKI Samurai (90010)*

GMC

Safari - *see CHEVROLET ASTRO (24010)*
Vans & Pick-ups - *see CHEVROLET*

HONDA

42010 **Accord CVCC** all models '76 thru '83
42011 **Accord** all models '84 thru '89
42012 **Accord** all models '90 thru '93
42013 **Accord** all models '94 thru '95
42020 **Civic 1200** all models '73 thru '79
42021 **Civic 1300 & 1500 CVCC** '80 thru '83
42022 **Civic 1500 CVCC** all models '75 thru '79
42023 **Civic** all models '84 thru '91
*42024 **Civic & del Sol** '92 thru '95
*42040 **Prelude CVCC** all models '79 thru '89

HYUNDAI

*43015 **Excel** all models '86 thru '94

ISUZU

Hombre - *see CHEVROLET S-10 (24071)*
*47017 **Rodeo** '91 thru '97; **Amigo** '89 thru '94; **Honda Passport** '95 thru '97
*47020 **Trooper & Pick-up**, all gasoline models Pick-up, '81 thru '93; Trooper, '84 thru '91

JAGUAR

*49010 **XJ6** all 6 cyl models '68 thru '86
*49011 **XJ6** all models '88 thru '94
*49015 **XJ12 & XJS** all 12 cyl models '72 thru '85

JEEP

*50010 **Cherokee, Comanche & Wagoneer Limited** all models '84 thru '96
50020 **CJ** all models '49 thru '86
*50025 **Grand Cherokee** all models '93 thru '98
50029 **Grand Wagoneer & Pick-up** '72 thru '91 Grand Wagoneer '84 thru '91, Cherokee & Wagoneer '72 thru '83, Pick-up '72 thru '88
*50030 **Wrangler** all models '87 thru '95

LINCOLN

Navigator - *see FORD Pick-up (36059)*
59010 **Rear Wheel Drive** all models '70 thru '96

MAZDA

61010 **GLC Hatchback (rear wheel drive)** '77 thru '83
61011 **GLC (front wheel drive)** '81 thru '85
*61015 **323 & Protegé** '90 thru '97
*61016 **MX-5 Miata** '90 thru '97
*61020 **MPV** all models '89 thru '94
Navajo - *see Ford Explorer (36024)*
61030 **Pick-ups** '72 thru '93
Pick-ups '94 thru '96 - see Ford Ranger (36071)
61035 **RX-7** all models '79 thru '85
*61036 **RX-7** all models '86 thru '91
61040 **626 (rear wheel drive)** all models '79 thru '82
*61041 **626/MX-6 (front wheel drive)** '83 thru '91

MERCEDES-BENZ

63012 **123 Series Diesel** '76 thru '85
*63015 **190 Series** four-cyl gas models, '84 thru '88
63020 **230/250/280** 6 cyl sohc models '68 thru '72
63025 **280 123 Series** gasoline models '77 thru '81
63030 **350 & 450** all models '71 thru '80

MERCURY

See FORD Listing.

MG

66010 **MGB** Roadster & GT Coupe '62 thru '80
66015 **MG Midget, Austin Healey Sprite** '58 thru '80

MITSUBISHI

*68020 **Cordia, Tredia, Galant, Precis & Mirage** '83 thru '93
*68030 **Eclipse, Eagle Talon & Ply. Laser** '90 thru '94
*68040 **Pick-up** '83 thru '96 & **Montero** '83 thru '93

NISSAN

72010 **300ZX** all models including Turbo '84 thru '89
*72015 **Altima** all models '93 thru '97
*72020 **Maxima** all models '85 thru '91
*72030 **Pick-ups** '80 thru '96 **Pathfinder** '87 thru '95
72040 **Pulsar** all models '83 thru '86
*72050 **Sentra** all models '82 thru '94
*72051 **Sentra & 200SX** all models '95 thru '98
*72060 **Stanza** all models '82 thru '90

OLDSMOBILE

*73015 **Cutlass** V6 & V8 gas models '74 thru '88
For other OLDSMOBILE titles, see BUICK, CHEVROLET or GENERAL MOTORS listing.

PLYMOUTH

For PLYMOUTH titles, see DODGE listing.

PONTIAC

79008 **Fiero** all models '84 thru '88
79018 **Firebird** V8 models except Turbo '70 thru '81
79019 **Firebird** all models '82 thru '92
For other PONTIAC titles, see BUICK, CHEVROLET or GENERAL MOTORS listing.

PORSCHE

*80020 **911** except Turbo & Carrera 4 '65 thru '89
80025 **914** all 4 cyl models '69 thru '76
80030 **924** all models including Turbo '76 thru '82
*80035 **944** all models including Turbo '83 thru '89

RENAULT

Alliance & Encore - *see AMC (14020)*

SAAB

*84010 **900** all models including Turbo '79 thru '88

SATURN

87010 **Saturn** all models '91 thru '96

SUBARU

89002 **1100, 1300, 1400 & 1600** '71 thru '79
*89003 **1600 & 1800** 2WD & 4WD '80 thru '94

SUZUKI

*90010 **Samurai/Sidekick & Geo Tracker** '86 thru '96

TOYOTA

92005 **Camry** all models '83 thru '91
92006 **Camry** all models '92 thru '96
92015 **Celica Rear Wheel Drive** '71 thru '85
*92020 **Celica Front Wheel Drive** '86 thru '93
92025 **Celica Supra** all models '79 thru '92
92030 **Corolla** all models '75 thru '79
92032 **Corolla** all rear wheel drive models '80 thru '87
92035 **Corolla** all front wheel drive models '84 thru '92
*92036 **Corolla & Geo Prizm** '93 thru '97
92040 **Corolla Tercel** all models '80 thru '82
92045 **Corona** all models '74 thru '82
92050 **Cressida** all models '78 thru '82
92055 **Land Cruiser** FJ40, 43, 45, 55 '68 thru '82
92056 **Land Cruiser** FJ60, 62, 80, FZJ80 '80 thru '96
*92065 **MR2** all models '85 thru '87
92070 **Pick-up** all models '69 thru '78
*92075 **Pick-up** all models '79 thru '95
*92076 **Tacoma** '95 thru '98, **4Runner** '96 thru '98, & **T100** '93 thru '98
*92080 **Previa** all models '91 thru '95
92085 **Tercel** all models '87 thru '94

TRIUMPH

94007 **Spitfire** all models '62 thru '81
94010 **TR7** all models '75 thru '81

VW

96008 **Beetle & Karmann Ghia** '54 thru '79
96012 **Dasher** all gasoline models '74 thru '81
*96016 **Rabbit, Jetta, Scirocco, & Pick-up** gas models '74 thru '91 & Convertible '80 thru '92
96017 **Golf & Jetta** all models '93 thru '97
96020 **Rabbit, Jetta & Pick-up** diesel '77 thru '84
96030 **Transporter 1600** all models '68 thru '79
96035 **Transporter 1700, 1800 & 2000** '72 thru '79
96040 **Type 3 1500 & 1600** all models '63 thru '73
96045 **Vanagon** all air-cooled models '80 thru '83

VOLVO

97010 **120, 130 Series & 1800 Sports** '61 thru '73
97015 **140 Series** all models '66 thru '74
*97020 **240 Series** all models '76 thru '93
97025 **260 Series** all models '75 thru '82
*97040 **740 & 760 Series** all models '82 thru '88

TECHBOOK MANUALS

10205 **Automotive Computer Codes**
10210 **Automotive Emissions Control Manual**
10215 **Fuel Injection Manual, 1978 thru 1985**
10220 **Fuel Injection Manual, 1986 thru 1996**
10225 **Holley Carburetor Manual**
10230 **Rochester Carburetor Manual**
10240 **Weber/Zenith/Stromberg/SU Carburetors**
10305 **Chevrolet Engine Overhaul Manual**
10310 **Chrysler Engine Overhaul Manual**
10320 **Ford Engine Overhaul Manual**
10330 **GM and Ford Diesel Engine Repair Manual**
10340 **Small Engine Repair Manual**
10345 **Suspension, Steering & Driveline Manual**
10355 **Ford Automatic Transmission Overhaul**
10360 **GM Automatic Transmission Overhaul**
10405 **Automotive Body Repair & Painting**
10410 **Automotive Brake Manual**
10415 **Automotive Detailing Manual**
10420 **Automotive Eelectrical Manual**
10425 **Automotive Heating & Air Conditioning**
10430 **Automotive Reference Manual & Dictionary**
10435 **Automotive Tools Manual**
10440 **Used Car Buying Guide**
10445 **Welding Manual**
10450 **ATV Basics**

SPANISH MANUALS

98903 **Reparación de Carrocería & Pintura**
98905 **Códigos Automotrices de la Computadora**
98910 **Frenos Automotriz**
98915 **Inyección de Combustible 1986 al 1994**
99040 **Chevrolet & GMC Camionetas** '67 al '87 Incluye Suburban, Blazer & Jimmy '67 al '91
99041 **Chevrolet & GMC Camionetas** '88 al '95 Incluye Suburban '92 al '95, Blazer & Jimmy '92 al '94, Tahoe y Yukon '95
99042 **Chevrolet & GMC Camionetas Cerradas** '68 al '95
99055 **Dodge Caravan & Plymouth Voyager** '84 al '95
99075 **Ford Camionetas y Bronco** '80 al '94
99077 **Ford Camionetas Cerradas** '69 al '91
99083 **Ford Modelos de Tamaño Grande** '75 al '87
99088 **Ford Modelos de Tamaño Mediano** '75 al '86
99091 **Ford Taurus & Mercury Sable** '86 al '95
99095 **GM Modelos de Tamaño Grande** '70 al '90
99100 **GM Modelos de Tamaño Mediano** '70 al '88
99110 **Nissan Camionetas** '80 al '96, **Pathfinder** '87 al '95
99118 **Nissan Sentra** '82 al '94
99125 **Toyota Camionetas y 4Runner** '79 al '95

Over 100 Haynes motorcycle manuals also available

5-98

Haynes North America, Inc., 861 Lawrence Drive, Newbury Park, CA 91320-1514 • (805) 498-6703

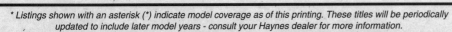